Theodor Tamir (Ed.)

Guided-Wave Optoelectronics

With Contributions by
R.C. Alferness   W.K. Burns   J.P. Donnelly
I.P. Kaminow   H. Kogelnik   F.J. Leonberger
A.F. Milton   T. Tamir   R.S. Tucker

Second Edition

With 236 Figures

Springer-Verlag
Berlin Heidelberg New York London
Paris Tokyo Hong Kong Barcelona

Theodor Tamir, Ph. D.

Polytechnic University, 333 Jay Street, Brooklyn, NY 11201, USA

Series Editors:

Dr. David H. Auston

Columbia University, Dept. of Electrical Engineering, New York, NY 10027, USA

Professor Dr. Walter Engl

Institut für Theoretische Elektrotechnik, Rhein.-Westf. Technische Hochschule,
Templergraben 55, D-5100 Aachen, Fed. Rep. of Germany

Professor Takuo Sugano

Department of Electronic Engineering, The Faculty of Engineering,
The University of Tokyo, 7-3-1, Hongo, Bunkyo-ku, Tokyo, 113, Japan

Managing Editor: Dr. Helmut K. V. Lotsch

Springer-Verlag, Tiergartenstrasse 17,
D-6900 Heidelberg, Fed. Rep. of Germany

ISBN 3-540-52780-X 2. Auflage Springer-Verlag Berlin Heidelberg New York
ISBN 0-387-52780-X 2nd Edition Springer-Verlag New York Berlin Heidelberg

ISBN 3-540-18795-2 1. Auflage Springer-Verlag Berlin Heidelberg New York
ISBN 0-387-18795-2 1st Edition Springer-Verlag New York Berlin Heidelberg

Library of Congress Cataloging-in-Publication Data. Guided-wave optoelectronics / Theodor Tamir (ed.) ;
with contributions by R. C. Alferness . . . [et al.]. – 2nd ed. p. cm. – (Springer series in electronics and photonics ; v. 26). Includes bibliographical references and index. ISBN 0-387-52780-X (New York : alk. paper)
1. Optoelectronic devices. 2. Integrated optics. 3. Optical wave guides. 4. Semiconductor lasers. I. Tamir,
Theodor, 1927– . II. Alferness, R. C. III. Series. TA1750.G85 1990 621.381'045–dc20 90.40987

Preface to the Second Edition

Because integrated optics and optoelectronics technology have been developing very rapidly during the past few years, significant advances have been made since the first edition of this book was published. Furthermore, interest in the book itself has been strong, leading to a demand for a new, updated version of the text. This has motivated us to issue the present revised paperback edition, whose lower price will make it more easily accessible to researchers in the area and to interested graduate students, in particular. The present edition is essentially similar to the original hardcover book, except that a new chapter (Chap. 7) has been added, which briefly reviews the recent advances in the area and provides new references. Typographical errors spotted in the original edition have also been corrected. Although great care has been exercised, some errors may still occur in the text and other improvements could be introduced in a possible future edition. The volume editor would therefore appreciate any comments from readers, who are urged to communicate their suggestions directly to him.

Brooklyn, New York   T. Tamir
April 1990

V

Preface to the First Edition

The first guided-wave components that employed signals in the form of light beams traveling along thin films were fabricated a little more than two decades ago. The parallel development of semiconductor lasers and the subsequent availability of low-loss optical fibers made possible the implementation of completely optical systems for communications, signal processing and other applications that had used only electronic circuitry in the past. Referred to as *integrated optics*, this technology has been reinforced by utilizing electronic components that act as controlling elements or perform other functions for which the optical counterparts are not as effective. The broader area thus generated was aptly named *optoelectronics* and it currently represents a fascinating, rapidly evolving and most promising technology. Specifically, the amalgamation of electronic and optics components into an integrated optoelectronics format is expected to provide a wide range of systems having miniaturized, high speed, broad band and reliable components for telecommunications, data processing, optical computing and other applications in the near and far future.

This book is intended to cover primarily the optical portion of the optoelectronics area by focusing on the theory and applications of components that use guided optical waves. Hence all aspects of integrated optics are discussed, but optoelectronic components having primarily electronic rather than optical functions have not been included. Each chapter has been written by experts who have actively participated in developing the specific areas addressed by them.

The book starts with an introductory chapter, which provides a general overview of integrated optics and optoelectronics. The next two chapters present the theory of optical waveguides, couplers and junctions, and describe the principal forms of these basic components. The fourth chapter discusses the fabrication and application of optical components based on lithium niobate substrates; these include devices that perform controlling functions, such as modulators, switches, filters, polarizers, etc. The last two chapters present the theory, fabrication and application of guided-wave components based on semiconductor (mainly gallium arsenide) substrates; in particular, Chap. 5 deals with semiconductor lasers that, together with all the other passive and active components discussed in Chap. 6, allow integration capabilities which are not achievable in lithium niobate substrates.

The book format described above enables the reader to use this volume as a coordinated treatise on the guided-wave optoelectronics area, or to refer to individual chapters for more specialized purposes. As a result, the text should be of interest to optical physicists, electrical engineers and professionals in related disciplines who wish to acquire a solid general background in this area, or to obtain the latest state-of-the-art details on specific topics. In addition, this volume can be used as a textbook or reference manual for graduate courses on integrated optics and optoelectronics.

The encouragement and expeditious handling of the manuscripts by Dr. H. Lotsch, Physics Editor of Springer-Verlag, have helped greatly in making this volume possible. His efforts and support are therefore greatly appreciated.

Brooklyn, January, 1988   *T. Tamir*

Contents

List of Contributors

Alferness, Rod C.
AT&T Bell Laboratories, Crawford Hill Laboratory, Box 400
Holmdel, NJ 07733, USA

Burns, William K.
Naval Research Laboratory, Code 6571, Washington, DC 20375, USA

Donnelly, Joseph P.
MIT Lincoln Laboratory, Box 73, Lexington, MA 02173, USA

Kaminow, Ivan P.
AT&T Bell Laboratories, Crawford Hill Laboratory, Box 400
Holmdel, NJ 07733, USA

Kogelnik, Herwig
AT&T Bell Laboratories, Crawford Hill Laboratory, Box 400
Holmdel, NJ 07733, USA

Leonberger, Frederick J.
United Technologies Research Center, Silver Lane,
East Hartford, CT 06108, USA

Milton, A. Fenner
General Electric Co., Electronics Lab., Electronics Park,
Syracuse, NY 13221, USA

Tamir, Theodor
Department of Electrical Engineering, Polytechnic University,
333 Jay Street, Brooklyn, NY 11202, USA

Tucker, Rodney S.
Department of Electrical and Electronic Engineering,
The University of Melbourne, Parkville, Victoria 3052, Australia

1. Introduction

T. Tamir

This book serves as a natural sequel to an earlier topics volume entitled
Integrated Optics, first published in 1975 and later reissued in updated edi-
tions [1.1]. Now, fifteen years later, the status of the technological area has
matured and its horizons have expanded considerably. To gain a proper per-
spective, Sect. 1.1 presents a brief overview on the evolution of integrated
optics and its current coverage. Section 1.2 then provides a descriptive outline of the other chapters in the book. In this context, we have listed only
references published in 1982 or later and refer the reader to the last (1982)
edition of [1.1] for earlier work.

1.1 Overview

The guiding of light beams along dielectric layers, first realized experimentally in the early sixties, has stimulated the growth of a new class of passive
and active components using light guided by films deposited on wafer-like
substrates. Because of their small dimensions and low-power requirements,
it was then projected that such optical components could replace electrical circuitry in integrated-electronics equipment. In addition, the optical
elements would provide the advantages of greater bandwidth and immunity
from electromagnetic interference. Towards the late sixties, guided-wave optical components became sufficiently effective to herald the beginning of a
highly promising and sophisticated technology for which the term *integrated
optics* was introduced [1.2]. The development of this area and its potential
applications have generated an increasing number of publications. In view
of their large number, we list here only a representative sampling of recent
review articles and textbooks [1.3–21].

Because the area of telecommunications is expected to require huge
bandwidths in the future, it has played a strong role in stimulating the
progress of integrated-optics technology [1.7, 14, 17]. Of course, this development was helped greatly by the realization of low-loss optical fibers, which
are crucial for implementing optical communications. Hence miniature optical transmitters and receivers were amongst the first to use integrated-
optics technology. However, the advantages provided by integrated optics
can benefit areas other than telecommunications. Thus, integrated-optics
components have been developed for spectrum analyzers, gyroscopes, digital

1

correlators, analog-to-digital converters, optical switches and multiplexers, signal processors and other applications [1.19, 21].

As outlined above, the basic idea behind integrated optics has been to serve as the optical counterpart of integrated electronics. This implies that integrated-optical systems should be able to manipulate or control signals in the form of light beams within miniaturized components fabricated on a single substrate. A few such truly integrated optical systems have indeed been achieved [1.3, 17, 19, 20], but many applications developed with integrated-optics intent use also non-optical elements and more than a single material. Such compromises were motivated either by the poor optical properties of certain substrates, or by advantages provided by electronic components in electro-optic systems.

To review the factors that affect the choice of materials and the use of non-optical elements, we recall that an integrated-optics system requires a combination of individual components. These generally include: sources to generate the optical power, optical waveguides to conduct that power, couplers to connect the various components, detectors to capture the resulting optical signals, and controlling elements (modulators, switches, etc.) to suitably modify those optical signals. The first three (source, guiding and coupling) components can be accommodated by using all-optical elements. However, detecting and controlling components can be achieved more easily by employing lower-frequency inputs involving photo-electric, electro-optic, acousto-optic, or other effects. As a result, integrated optics has been understood [1.1, 9, 10] to also include components using electrical, acoustical or other inputs for controlling purposes.

The principal materials used to implement the various components mentioned above are lithium niobate ($LiNbO_3$), which is a ferroelectric insulating crystal, and gallium arsenide (GaAs) and indium phosphide (InP), which are semiconductors. Glass was one of the first materials used in the earliest days of integrated optics [1.1, 2] and has also been used recently, but its application is limited to passive components [1.13]. By virtue of its low optical transmission loss and high electro-optic coefficient, lithium niobate is the premier material for many integrated-optics components, but it is unfortunately not suitable for fabricating lasers and detectors. By contrast, semiconductors have been successfully used for constructing all types of components, but their higher transmission loss and lower electro-optic coefficient makes them less effective for guiding and modulation purposes. As a result, integration has often been achieved by coupling semiconductor lasers and detectors onto a lithium niobate wafer on which the other components were fabricated. However, this procedure is properly regarded as *hybrid integration* because it involves two different materials.

On the other hand, semiconductors are used extensively in integrated-electronic circuits for a variety of signal processing applications. This has stimulated the development of monolithically integrated circuits which in-

2

clude both optical components (e.g., lasers and waveguides) and electronic elements (e.g., photo-detectors and field-effect transistors). The various elements of such a system can presumably be combined to achieve optimal operation in a single-material miniaturized chip [1.5, 6, 18]. Obviously, this technology goes beyond integrated optics and is currently referred to as *integrated optoelectronics*.

Because of similar inroads into related technologies, we find that integrated-optics techniques are utilized also in other areas. These are denoted by an apparently increasing number of names, such as *electro-optics, photonics, optical electronics, optronics, photo-electronics,* etc. As expected, a considerable overlap occurs in the technological coverage of these areas. However, they all express different facets of a rapidly developing effort that uses optical signals and processes in an increasingly wider range of applications.

1.2 Organization of the Book

In view of the developments described above, it was recognized that an updated version of [1.1] should include material that may be regarded as belonging outside integrated optics. However, a single volume could not possibly cover all the varieties of miniature optical components generated by integrated optics and electronics technology. The book emphasis was therefore placed on devices that involve a waveguiding function. Furthermore, the present state of the art dictates that the focus should properly be on individual components or processes, rather than on their integration. The book title *Guided-Wave Optoelectronics* was therefore chosen to represent an area with an individuality of its own from both theoretical and technological points of view.

The most basic component of guided-wave optoelectronic devices is, of course, the optical waveguide which supports the optical field. Chapter 2 therefore presents the theory of these dielectric guides and describes their most basic forms. That chapter is substantially an expanded version of the same author's chapter in [1.1], which was acclaimed as an eminent exposition of its subject matter. The new version starts with the same fundamental ray and electromagnetic theories of modes in planar dielectric guides, but it also includes a wealth of new material on multilayered and channel guides. The chapter concludes with a basic discussion of coupled-mode theory, which is needed for many of the topics discussed in later chapters.

At this stage, it is interesting to observe that the arrangement of Chaps. 3–6 is quite different from that in [1.1]. In particular, the third chapter of the previous monograph dealt with beam and waveguide couplers. By contrast, discussions of coupling configurations are now interspersed in most of the chapters. This occurs because coupling components and their associated problems appear in a wide variety of forms, which depend on the particular component (e.g., waveguide connector, modulator, etc.) under consideration. Moreover, the specific (GaAs, LiNbO$_3$, etc.) material technology affects the construction and performance of such couplers and therefore requires their discussion in several different contexts. While such an approach inevitably introduces some duplication, it minimizes the need for cross-referencing between too many chapters.

Chapter 3 treats waveguide junctions and transitions by first expanding on the coupled-mode theory whose basics were developed in Chap. 2. In particular, the theory is refined by contrasting the local normal modes of coupled waveguides with the (usually, less accurate but more familiar) coupled-mode representation. It then distinguishes between fast (abrupt) and slow (gradual) transitions and proceeds to discuss specific configurations, which include two-arm branches, horn transitions and branches having three arms.

Chapter 4 treats components and technology based on lithium niobate. It starts by describing the fabrication of LiNbO$_3$ waveguides by titanium in-diffusion and then addresses modulators, switches and couplers using the electro-optic effect. Polarization-controlled devices are discussed next and TE–TM converters are considered, in particular. Other devices are then described, including polarization-insensitive devices and optical switch arrays. The chapter concludes with a brief look at more complex applications, such as extended cavity lasers, fiber gyros and analog-to-digital converters.

Semiconductor devices serve as the subject of the last two chapters. The first of these (Chap. 5) deals with high-speed lasers, light-emitting diodes and amplifiers using index guiding to provide fundamental transverse mode behavior. The chapter begins with an outline of laser theory and considers structures for transverse-mode control. It then examines the question of longitudinal-mode control, after which it describes specific aspects, including linewidth, modulation response and high-frequency limitations. In conclusion, the chapter considers superluminescent diodes and laser amplifiers.

Because sources are covered in Chap. 5, most other semiconductor devices are treated in Chap. 6. This last chapter starts with semiconductor-waveguide theory and describes the pertinent material technology and fabrication. It then examines passive devices, including waveguides, couplers, bends and branches. Modulators are considered next from both theoretical and experimental points of view; their discussion includes phase modulation using the electro-optic effect, electro-absorption or nonlinearities. By concluding with a description of optoelectronic integrated circuits, this chapter provides a brief survey on the potentiality of integrating the many individual components in a variety of complex applications.

Finally, we note that an attempt was made to maintain consistency of notation throughout the various chapters. However, the available sym-

4

bols were not sufficient to cope with all the many varieties of geometries, three-dimensional modes, anisotropic properties, and other factors requiring a large quantity of letters, subscripts, Greek characters, etc. We had therefore to compromise between different notations that were dictated either by the complexity of the subject matter, or by differences in the scientific or technological background inherent to each chapter. To avoid possible confusion due to notation inconsistencies, the chapters were written as individual contributions that can be read independently. For this purpose, cross-references have been inserted primarily for relating concepts rather than mathematical details.

Acknowledgement. The editor wishes to acknowledge the help of the other contributors, whose constructive suggestions in preparing and organizing the material for this book have been invaluable.

References

1.1   T. Tamir: *Integrated Optics,* Topics Appl. Phys., Vol. 7 (Springer, Berlin, Heidelberg 1975; last edition: 1982)

1.2   S.E. Miller: Integrated optics: An introduction, Bell System Tech. J. **48**, 2059 (1969)

1.3   F.J. Leonberger: Applications of guided-wave interferometers, Laser Focus **18**, 125 (March 1982)

1.4   R.C. Alferness: Waveguide electrooptic modulators, IEEE Trans. **MTT-30**, 1121 (1982)

1.5   U. Koren, S. Margalit, T.R. Chen, K.L. Yu, A. Yariv, N. Bar-Chaim, K.Y. Lau, I. Ury: Recent developments in monolithic integration of InGaAsP/InP optoelectronic devices, IEEE J. **QE-18**, 1653 (1982)

1.6   N. Bar-Chaim, S. Margalit, A. Yariv, I. Ury: GaAs integrated optoelectronics, IEEE Trans. **ED-29**, 1372 (1982)

1.7   Y. Suematsu: Long-wavelength optical fiber communication, Proc. IEEE, **71**, 692 (1983)

1.8   H.A. Haus: *Waves and Fields in Optoelectronics* (Prentice-Hall, Englewood Cliffs, NJ 1984)

1.9   R.G. Hunsperger: *Integrated Optics: Theory and Technology,* Springer Ser. Opt. Sci., Vol. 33 (Springer, Berlin, Heidelberg 1985)

1.10 H.P. Nolting, R. Ulrich (eds.): *Integrated Optics,* Proc. 3rd European Conf., Springer Ser. Opt. Sci., Vol. 48 (Springer, Berlin, Heidelberg 1985)

1.11 P.K. Cheo: *Fiber Optics Devices and Systems* (Prentice-Hall, Englewood Cliffs, NJ 1985)

1.12 K.Y. Lau, A. Yariv: Ultra-high speed semiconductor lasers, IEEE J. **QE-21**, 121 (1985)

1.13 T. Findakly: Glass waveguides by ion exchange: a review, Opt. Engg. **24**, 244 (1985)

1.14 H. Kogelnik: High-speed lightwave transmission in optical fibers, Science **228**, 1043 (1985)

1.15 D.L. Lee: *Electromagnetic Principles of Integrated Optics* (Wiley, New York 1986)

1.16 R.C. Alferness: Optical guided-wave devices, Science **234**, 825 (1986)

1.17 E. Garmire: Fundamentals of waveguides; integrated optics technology, in *Electromagnetic Surface Excitations,* ed. by R.F. Wallis and G.I. Stegeman, Springer Ser. Wave Phen., Vol. 3 (Springer, Berlin, Heidelberg 1986) pp. 189 and 202

1.18 O. Wada, T. Sakurai, T. Nakagami: Recent progress in optoelectronic integrated circuits (OEIC's), IEEE J. **QE-22**, 805 (1986)
1.19 D.G. Hall: Integrated optics: The shape of things to come, Photonics Spectra, 87 (Aug. 1986)
1.20 B. Carlson: A foundry approach to integrated optics, Photonics Spectra,152 (March 1987)
1.21 L.D. Hutcheson (Ed.): Integrated optics: Evolution and prospects, Optics News **14**, 8 (Feb. 1988)

2. Theory of Optical Waveguides

H. Kogelnik

With 32 Figures

Optical waveguides, also known as "dielectric" waveguides, are the structures that are used to confine and guide the light in the guided-wave devices and circuits of integrated optics. This chapter is devoted to the theory of these waveguides. Other chapters of this book discuss their fabrication by such techniques as sputtering, diffusion, ion implantation or epitaxial growth. A well-known optical waveguide is, of course, the optical fiber which usually has a circular cross-section. In contrast, the guides of interest to integrated optics are usually planar structures such as planar films or strips. Our discussion will focus on these planar guides even though most of the fundamentals are applicable to all optical waveguide types.

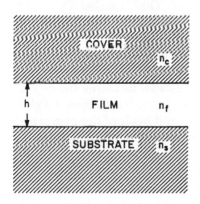

Fig. 2.1. Cross-section of a planar slab waveguide consisting of a thin film of thickness (or height) h and refractive index n_f, sandwiched between substrate and cover materials

The simplest dielectric guide is the planar slab guide shown in Fig. 2.1, where a planar film of refractive index n_f is sandwiched between a substrate and a cover material with lower refractive indices n_s and n_c ($n_f > n_s \geq n_c$). Often the cover material is air, in which case $n_c = 1$. As an illustration, we have listed in Table 2.1 the refractive indices of some dielectric waveguide materials used in integrated optics. Typical differences between the indices of the film and the substrate range from 10^{-3} to 10^{-1}, and a typical film thickness is $1\,\mu$m. The light is confined by total internal reflection at the film-substrate and film-cover interfaces.

Dielectric waveguides have already been the subject of several textbooks, [2.1–9], and we can refer the reader to these for a history on the subject as well as for a more complete list of references.

Table 2.1. Refractive index n of optical waveguide materials

Dielectric material	$\lambda\,[\mu m]$	n
Fused silica (SiO$_2$)	0.633	1.46
Typical microscope-slide glass	0.633	1.51
Sputtered Corning 7059 glass	0.633	1.62
LiTaO$_3$ (n_0)	0.80	2.15
(n_e)	0.80	2.16
LiNbO$_3$ (n_0)	0.80	2.28
(n_e)	0.80	2.19
GaAs	0.90	3.6
InP	1.51	3.17

In this text, we have several aims. We hope to give both an introduction to the subject as well as a collection of important results sufficiently detailed to be of use to the experimenter. We also aim to provide a compact theoretical framework of sufficient rigor and generality to be used as the basis for future work and to analyze virtually all waveguide types and devices of interest in integrated optics. In Sect. 2.1 we discuss the ray-optical picture of light propagation in slab waveguides. This is meant to provide both a first physical understanding as well as an introduction to the concepts and the terminology of dielectric waveguides in general. Some basic results of interest to the experimenter are also developed here. Section 2.2 is a collection of the general fundamentals of the electromagnetic theory of dielectric waveguides and their modes of propagation, including the orthogonality, symmetry, and variational properties of the modes. Section 2.3 gives a detailed listing of the formulas for the modes and fields of the planar slab waveguides, both for the guided TE and TM modes as well as for the radiation modes. Section 2.4 discusses graded index profiles giving the modal solutions for the parabolic, the "$1/\cosh^2$" and the exponential profiles. Brief treatments of graded profiles with an abrupt index step and of the WKB method are also included. Section 2.5 gives a discussion of channel waveguides and the application of the effective index method to these structures. The final Section 2.6 is devoted to the development of the general coupled-mode formalism for dielectric waveguides, including its application to the treatment of waveguide deformations and periodic waveguides.

We shall assume throughout this chapter that the guided light is coherent and monochromatic and that the waveguides consist of dielectric media that are lossless and isotropic. For a discussion of lossy and metal-clad optical waveguides, we refer the reader to the cited texts and to the papers by *Anderson* [2.10], *Reisinger* [2.11], *Kaminow* et al. [2.12] and the literature cited therein. Anisotropic dielectric waveguides have been treated by *Nelson* and *McKenna* [2.13], *Yamamoto* et al. [2.14], *Ramaswamy* [2.15, 16] and others.

2.1 Ray Optics of the Slab Waveguide

In this section, we propose to discuss and develop the ray-optical model of light propagation in a slab waveguide. Ray-optical techniques in connection with slab waveguides have been explored and used by *Tien* [2.17], *Maurer* and *Felsen* [2.18], *Lotsch* [2.19] and others. We have chosen the slab waveguide, sketched in Fig. 2.1, for two reasons: first, because it is relatively easy to understand and analyze, and second because it is of considerable practical interest in integrated optics. We shall use the ray-optical picture to introduce the basic concepts and terminology of dielectric-waveguide theory, including the nature of the modes of propagation, waveguide cutoff, the propagation constants and the effective guide thickness. In addition, we employ this picture to derive and provide a number of results of interest to the experimenter, such as plots of the propagation constant and of the effective width of slab waveguides. The ray-optical picture is a very simple picture with great intuitive appeal, but it is not as complete a description as that provided by electromagnetic theory, which we discuss later in Sects. 2.2 and 3. However, the results we present here are in perfect agreement with the latter.

Our picture of light guidance in a slab waveguide is one of light rays tracing a zig-zag path in the film, with total internal reflection of the light occurring at the film-substrate and film-cover interfaces. As reflection and refraction at these dielectric interfaces play an important role in the guiding process, let us briefly review the relevant laws and their consequences.

2.1.1 Refraction and Reflection

Consider an interface separating two lossless, isotropic, homogeneous dielectric media of refractive index n_1 and n_2 (Fig. 2.2), and a coherent light wave incident at an angle θ_1 between the wave normal and the normal to the interface. In general, the wave, having a complex amplitude A at the

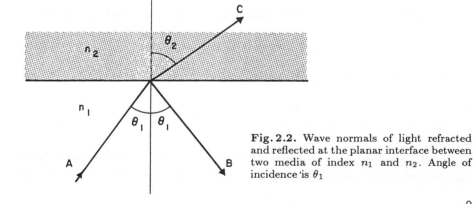

Fig. 2.2. Wave normals of light refracted and reflected at the planar interface between two media of index n_1 and n_2. Angle of incidence is θ_1

interface, is partially reflected and refracted as shown. The exit angle θ_2 of the refracted wave C is given by Snell's law

$$n_1 \sin \theta_1 = n_2 \sin \theta_2 \quad . \tag{2.1.1}$$

The reflected wave has a complex amplitude B at the interface linearly related to A by a complex reflection coefficient R

$$B = R \cdot A \quad . \tag{2.1.2}$$

The reflection coefficient depends on the angle of incidence and the polarization of the light, and is given by the Fresnel formulas. For TE polarization (i.e., electric fields perpendicular to the plane of incidence spanned by the wave normal and the normal to the interface) we have

$$R_{\text{TE}} = \frac{n_1 \cos \theta_1 - n_2 \cos \theta_2}{n_1 \cos \theta_1 + n_2 \cos \theta_2} = \frac{n_1 \cos \theta_1 - \sqrt{n_2^2 - n_1^2 \sin^2 \theta_1}}{n_1 \cos \theta_1 + \sqrt{n_2^2 - n_1^2 \sin^2 \theta_1}} \quad . \tag{2.1.3}$$

The corresponding formula for TM polarization (with the magnetic fields perpendicular to the plane of incidence) is

$$R_{\text{TM}} = \frac{n_2 \cos \theta_1 - n_1 \cos \theta_2}{n_2 \cos \theta_1 + n_1 \cos \theta_2} = \frac{n_2^2 \cos \theta_1 - n_1 \sqrt{n_2^2 - n_1^2 \sin^2 \theta_1}}{n_2^2 \cos \theta_1 + n_1 \sqrt{n_2^2 - n_1^2 \sin^2 \theta_1}} \quad . \tag{2.1.4}$$

The so-called critical angle θ_c is given by

$$\sin \theta_c = n_2/n_1 \quad . \tag{2.1.5}$$

As long as $\theta_1 < \theta_c$ we have only partial reflection and a real valued R. As soon as the critical angle is exceeded ($\theta_1 > \theta_c$), we have $|R| = 1$ and total reflection of the light occurs. R is now complex valued and a phase shift is imposed on the reflected light. We write

$$R = \exp(2\text{j}\phi) \quad , \tag{2.1.6}$$

and extract from the Fresnel formulas the following expressions for the phase shifts ϕ_{TE} and ϕ_{TM} corresponding to the two polarization states

$$\tan\phi_{\text{TE}} = \frac{\sqrt{n_1^2 \sin^2 \theta_1 - n_2^2}}{n_1 \cos \theta_1} \quad , \tag{2.1.7}$$

$$\tan \phi_{\text{TM}} = \frac{n_1^2}{n_2^2} \frac{\sqrt{n_1^2 \sin^2 \theta_1 - n_2^2}}{n_1 \cos \theta_1} \quad . \tag{2.1.8}$$

Figure 2.3 shows the dependence of ϕ_{TE} on the angle of incidence θ_1 for a selection of index ratios n_2/n_1 where the values 0.3, 0.5 and 0.7 correspond

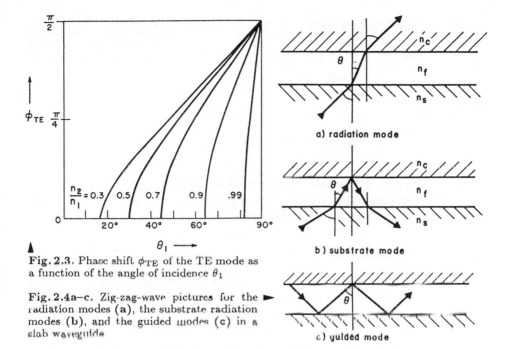

Fig. 2.3. Phase shift ϕ_{TE} of the TE mode as a function of the angle of incidence θ_1

Fig. 2.4a–c. Zig-zag-wave pictures for the radiation modes (a), the substrate radiation modes (b), and the guided modes (c) in a slab waveguide

approximately to interfaces between air and GaAs, LiNbO₃, and SiO₂, respectively. We note that the phase shift increases from 0 at the critical angle to $\pi/2$ at grazing incidence ($\theta_1 = 90°$). It increases with infinite slope at $\theta_1 = \theta_c$ and a slope of $(1 - n_2^2/n_1^2)^{-1/2}$ at $\theta_1 = 90°$. The behaviour of ϕ_{TM} is quite similar.

Consider, now, the ("asymmetric") slab waveguide structure shown in Fig. 2.4 with a film of index n_f and substrate and cover materials of index n_s and n_c. In general, we have $n_f > n_s > n_c$ and two critical angles of interest, θ_s for total reflection from the film-substrate interface and $\theta_c < \theta_s$ for total reflection from the film-cover interface. When we examine what happens as the angle of incidence θ is increased, we discover that there are three distinct cases which are sketched in Fig. 2.4. For small angles $\theta < \theta_s, \theta_c$ light incident from the substrate side is refracted according to Snell's law and escapes from the guide through the cover (a). There is essentially no confinement of light, and the electromagnetic mode corresponding to this picture is called a "radiation mode" which is discussed in more detail in Sect. 2.3. When θ is increased somewhat, such that $\theta_c < \theta < \theta_s$, we then find the situation depicted in (b). The light incident from the substrate is refracted at the film-substrate interface, totally reflected at the film-cover interface, refracted back into the substrate through which the light escapes from the structure. Again, there is no confinement and we talk of a "substrate radiation mode" (Sect. 2.3). Finally (c), when θ is large enough, we have $\theta_s, \theta_c < \theta$, i.e., total internal reflection at both interfaces. The light,

11

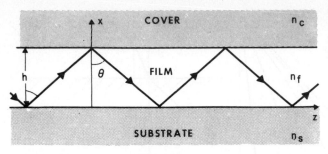

Fig. 2.5. Side-view of a slab waveguide showing wave normals of the zig-zag waves corresponding to a guided mode

once it is inside, is trapped and confined in the film and propagates in a zig-zag path. This case corresponds to a guided mode of propagation which we discuss in more detail in the following.

2.1.2 Guided Modes

In Fig. 2.5 we have sketched a side view of the slab waveguide and our choice of the coordinate system. We assume that the light in the guide propagates in the z direction, that confinement occurs transversely in the x direction, and that both the structure and the light are uniform in the y direction perpendicular to xz. Our physical picture of guided light propagation is, then, that of light traveling in zig-zag fashion through the film. More precisely, it is a picture of two superimposed uniform plane waves with wave normals which follow the zig-zag path indicated in the figure and which are totally reflected at the film boundaries. These waves are monochromatic and coherent with angular frequency ω, free-space wavelength λ, and they travel with a wave vector kn_f in the direction of the wave normal where the absolute value of k is

$$k = \frac{2\pi}{\lambda} = \frac{\omega}{c} \tag{2.1.9}$$

and c is the velocity of light in vacuum. The fields of these waves vary as

$$\exp[-jkn_f(\pm x \cos\theta + z \sin\theta)] \quad . \tag{2.1.10}$$

For a guided mode of the slab guide, the zig-zag picture predicts a propagation constant β (and the related phase velocity v_p)

$$\beta = \omega/v_p = kn_f \sin\theta \quad , \tag{2.1.11}$$

which is the z-component of the wave vector kn_f. However, not all angles θ are allowed; only a discrete set of angles (and sometimes none) lead to a self-consistent picture that corresponds to what we call the "guided modes". To examine this in more detail, let us look at a guide cross-section $z = $ const and add up the phase shifts that occur as we move up from the lower film boundary ($x = 0$) with one wave to the other boundary ($x = h$) and then back down again with the reflected wave to where we started from. For self-consistency, the sum of all these phase shifts must be a multiple of 2π. For a film of thickness h we have, specifically, a phase shift of $kn_f h \cos\theta$ for the first transverse passage through the film, a phase shift of $-2\phi_c$ on total reflection from the film-cover interface, another $kn_f h \cos\theta$ on the transverse passage down, and a phase shift of $-2\phi_s$ on total reflection from the film-substrate boundary. Thus, we have the self-consistency condition (also known as the "transverse resonance condition")

$$2kn_f h \cos\theta - 2\phi_s - 2\phi_c = 2\nu\pi \quad , \tag{2.1.12}$$

where ν is an integer $(0,1,2\dots)$ which identifies the mode number. As discussed before, the phase shifts ϕ_s and ϕ_c are functions of the angle θ, as described by (2.1.7 and 8) after the appropriate substitutions for n_1 and n_2. The above equation is essentially the dispersion equation of the guide yielding the propagation constant β as a function of frequency ω and film thickness h. From (2.1.5 and 11), we find for guided modes that β is bounded by the plane-wave propagation constants of substrate and film

$$kn_s < \beta < kn_f \quad . \tag{2.1.13}$$

It is often convenient to use an "effective guide index" defined by

$$N = \beta/k = n_f \sin\theta \quad , \tag{2.1.14}$$

which is bounded by

$$n_s < N < n_f \quad . \tag{2.1.15}$$

Figure 2.6 sketches a graphical solution of the dispersion equation (2.1.12) for the fundamental mode ($\nu = 0$) which gives us further information of the propagation characteristics of the guide. We have drawn here, as a function of the angle θ, both the phase shift on film traversal $kn_f h \cos\theta$ (dotted

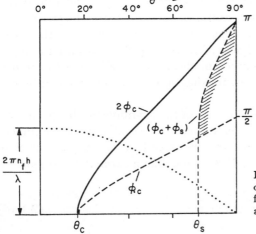

Fig. 2.6. Sketch of graphical solution of the dispersion equation for the fundamental modes of symmetric and asymmetric slab waveguides

13

curve) and the sum of the phase shifts ($\phi_s + \phi_c$) at the film boundaries. We show the latter for two cases, the symmetric guide where $\phi_c = \phi_s$ (solid curve), and the asymmetric guide (dashed curve). Consider the symmetric guide first, where the intersection between the solid and the dotted curve yields the zig-zag angle θ of the fundamental mode. We note that the zig-zags get steeper (θ smaller) as h/λ gets smaller, but there is always a solution even when the film thickness gets very small. This implies that there is no cutoff for the fundamental mode of a symmetric guide. Of course, as the guide gets thicker, it supports more and more guided modes. Considering the asymmetric guide, we look for an intersection between the dotted and the dashed curve. However, only the portion of the ($\phi_s + \phi_c$) curve emphasized by shading is above the critical angle θ_s of the film-substrate interface. For sufficiently thin films, we do not get an intersection of the curves above cut-off, which implies that an asymmetric guide cannot always support a guided mode, i.e., there is a cut-off condition even for the fundamental.

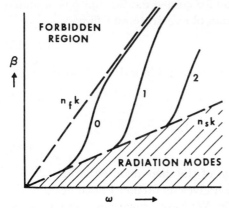

Fig. 2.7. Typical ω-β diagram of a dielectric waveguide

Figure 2.7 shows a sketch of an ω-β diagram that restates some of the above-discussed dispersion characteristics, which are typical for a dielectric waveguide. The first three guided modes ($\nu = 0, 1, 2$) are shown. At the cut-off frequency, the propagation constants assume the value of the lower bound $n_s k$, and as ω (or the thickness h) increases, β approaches its upper bound $n_f k$ and more and more guided modes exist. In addition to the discrete spectrum of the guided modes, the diagram also shows the continuous spectrum of the radiation modes.

To obtain a more precise ω-β diagram of the asymmetric slab guide, we have to evaluate (2.1.12) numerically. To make the results of such a numerical evaluation more broadly applicable, we introduce normalizations that combine several guide parameters. First, we define a normalized frequency and film thickness V by

$$V = kh\sqrt{n_f^2 - n_s^2} \quad , \tag{2.1.16}$$

and then a normalized guide index b related to the effective index N (and β) by

$$b = (N^2 - n_s^2)/(n_f^2 - n_s^2) \quad .$$ (2.1.17)

The index b is zero at cut-off and approaches unity far away from it. For small index differences $(n_f - n_s)$ we have the simple linear relation

$$N \approx n_s + b(n_f - n_s) \quad .$$ (2.1.18)

Finally, we introduce a measure for the asymmetry of the waveguide structure defined by

$$a = (n_s^2 - n_c^2)/(n_f^2 - n_s^2) \quad .$$ (2.1.19)

This measure applies to the TE modes and ranges in value from zero for perfect symmetry $(n_s = n_c)$ to infinity for strong asymmetry $(n_s \neq n_c$ and $n_s \approx n_f)$. As an illustration, Table 2.2 lists this asymmetry measure under a_E together with the refractive indices of three waveguide structures of practical interest.

Table 2.2. Asymmetry measures for the TE modes (a_E) and the TM modes (a_M) of slab waveguides

Waveguide	n_s	n_f	n_c	a_E	a_M
GaAlAs, double heterostructure	3.55	3.6	3.55	0	0
Sputtered glass	1.515	1.62	1	3.9	27.1
Ti-diffused LiNbO₃	2.214	2.234	1	43.9	1093
Outdiffused LiNbO₃	2.214	2.215	1	881	21206

For the TE modes, we use (2.1.7) together with the above normalizations to write the dispersion relation (2.1.12) in the form

$$V\sqrt{1-b} = \nu\pi + \tan^{-1}\sqrt{b/(1-b)} + \tan^{-1}\sqrt{(b+a)/(1-b)} \quad .$$ (2.1.20)

A numerical evaluation of (2.1.20) yields the normalized ω-β diagram shown in Fig. 2.8 which is taken from *Kogelnik* and *Ramaswamy* [2.20] and where the guide index b is plotted as a function of the normalized frequency V for four different values of the asymmetry measure and for the mode orders $\nu = 0, 1$ and 2.

Setting $b = \nu = 0$ in the dispersion relation (2.1.20), we determine the cutoff frequency V_0 of the fundamental mode as

$$V_0 = \tan^{-1}\sqrt{a} \quad .$$ (2.1.21)

This can also be written in the form

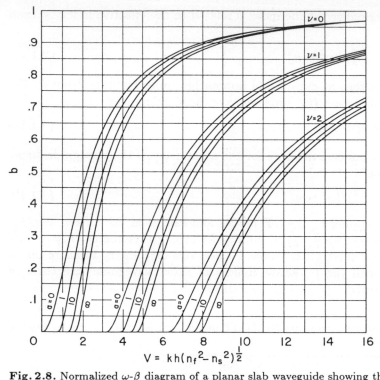

Fig. 2.8. Normalized ω-β diagram of a planar slab waveguide showing the guide index b as a function of the normalized thickness V for various degrees of asymmetry [2.20]

$$(h/\lambda)_0 = \frac{1}{2\pi}(n_{\mathrm{f}}^2 - n_{\mathrm{s}}^2)^{-1/2}\tan^{-1}\sqrt{a} \quad . \tag{2.1.22}$$

The cutoff frequency V_ν of the νth order mode is

$$V_\nu = V_0 + \nu\pi \quad , \tag{2.1.23}$$

from which we obtain an approximate formula for the number of guided modes allowed in the waveguide, which is

$$\nu = \frac{2h}{\lambda}\sqrt{n_{\mathrm{f}}^2 - n_{\mathrm{s}}^2} \quad . \tag{2.1.24}$$

For the TM mode, we get cut-off conditions of the same form as for the TE mode and ω-β diagrams that are very similar. In fact, when the index differences $(n_{\mathrm{f}} - n_{\mathrm{s}})$ are small, we can apply the diagram of Fig. 2.8 to the TM modes. However, these statements are only correct if we define the asymmetry measure in a somewhat different manner [2.20], namely, by

$$a = \frac{n_{\mathrm{f}}^4}{n_{\mathrm{c}}^4}\frac{n_{\mathrm{s}}^2 - n_{\mathrm{c}}^2}{n_{\mathrm{f}}^2 - n_{\mathrm{s}}^2} \quad . \tag{2.1.25}$$

Ilustrative values for this are given in Table 2.2 under a_{M}.

2.1.3 The Goos-Hänchen Shift

So far, we have described the light in the waveguide in terms of plane waves and their wave normals and phases. In this subsection and in the next, we consider also the energy of the light and its flow through the guide. To prepare for this, we have to be more precise about what we mean by a light ray. A light ray is defined here as the direction of the Poynting vector or the energy flow of light. Consistent with this is the view of a ray as the axis of a narrow beam of light or wave packet. The relation between wave normal and ray is essentially the spatial analog of the relation between the phase velocity and the group velocity. For the simple case of a plane wave in a homogeneous, isotropic medium the directions of the wave normal and the ray are the same, but in an anisotropic medium the ray generally points in a direction different from that of the wave normal.

The Goos-Hänchen shift that occurs on total reflection from a dielectric interface is another case where the ray behaves differently than the wave normal. Here the reflected ray (B) is shifted laterally relative to the incident ray or wave packet (A), as indicated in Fig. 2.9. This lateral ray shift has turned out to be an important element in the understanding of the flow of energy in dielectric waveguides in terms of the ray picture.

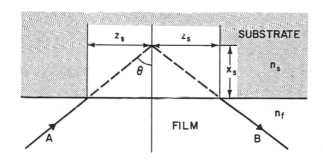

Fig. 2.9. Ray picture of total reflection at the interface between two dielectric media showing a lateral shift of the reflected ray (Goos-Hänchen shift)

To determine the lateral ray shift, shown as $2z_s$ in Fig. 2.9, consider a simple wave packet consisting of two plane waves incident at two slightly different angles. If the z-components of the corresponding wave vectors are $\beta \pm \Delta\beta$, we can write for the complex amplitude $A(z)$ of the incident wave packet at the interface $x = 0$

$$
\begin{aligned}
A &= [\exp(j\Delta\beta z) + \exp(-j\Delta\beta z)]\exp(-j\beta z) \\
&= 2\cos(\Delta\beta z)\exp(-j\beta z) \quad .
\end{aligned}
\tag{2.1.26}
$$

Before applying the reflection laws (2.1.2 and 6) to each individual plane wave, we have to remember that the phase shift ϕ occurring on total reflection is a function of θ (and β). For small $\Delta\phi$ and $\Delta\beta$, we can use an expansion of the form

$$\phi(\beta + \Delta\beta) = \phi(\beta) + \frac{d\phi}{d\beta}\Delta\beta \quad . \tag{2.1.27}$$

With this, we obtain for the amplitude $B(z)$ of the reflected wave packet at $x = 0$

$$
\begin{aligned}
B &= \{\exp[j(\Delta\beta z - 2\Delta\phi)] + \exp[-j(\Delta\beta z - 2\Delta\phi)]\}\exp[-j(\beta z - 2\phi)] \\
&= \cos[\Delta\beta(z - 2z_s)]\exp[-j(\beta z - 2\phi)] \quad , \tag{2.1.28}
\end{aligned}
$$

where

$$z_s = \frac{d\phi}{d\beta} \quad . \tag{2.1.29}$$

This gives us the lateral shift of the wave packet, i.e., of the ray, in compact and simple form [2.21, 22]. Using (2.1.11, 7 and 8), we obtain for the TE modes

$$k z_s = (N^2 - n_s^2)^{-1/2}\tan\theta \quad , \tag{2.1.30}$$

and for the TM modes

$$k z_s = (N^2 - n_s^2)^{-1/2}\tan\theta \Big/ \left(\frac{N^2}{n_s^2} + \frac{N^2}{n_f^2} - 1\right) \quad . \tag{2.1.31}$$

As sketched in Fig. 2.9, this lateral ray shift would indicate that the light penetrates to a depth x_s into the substrate before it is reflected, where

$$x_s = \frac{z_s}{\tan\theta} \quad . \tag{2.1.32}$$

If we compare this result with the electromagnetic field solutions to be given in Sect. 2.3, we find that these predict evanescent fields in the substrate whose decay constants are closely related to this penetration depth x_s of the ray.

2.1.4 Effective Guide Thickness

To obtain a zig-zag ray model of light propagation in the waveguide that is consistent with the flow of energy, we have to incorporate the Goos-Hänchen shifts at the film-substrate and film-cover interfaces, as first suggested by *Burke* [2.23]. Figure 2.10 shows a sketch of this ray model with lateral shifts $2z_s$ and $2z_c$, and ray penetration depths x_s and x_c. As a consequence of the ray penetration, the guide appears to possess an effective thickness

$$h_{\text{eff}} = h + x_s + x_c \tag{2.1.33}$$

which is larger than h. This is also indicated in the figure. We shall see in later sections that this effective thickness also turns up as a characteristic parameter in the electromagnetic theory of slab guides whenever questions of energy flow or energy exchange arise. The guided light spreads somewhat into the substrate and cover, and is essentially confined to a thickness h_{eff}.

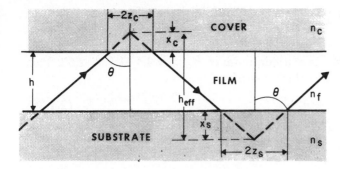

Fig. 2.10. Ray picture of zig-zag light propagation in a slab waveguide. Goos-Hänchen shifts are incorporated in the model, and the effective guide thickness h_{eff} is indicated

To illustrate the degree of light confinement provided by an asymmetric slab waveguide, we evaluate (2.1.33) numerically and plot the normalized effective thickness

$$H = k h_{\text{eff}} \sqrt{n_{\text{f}}^2 - n_{\text{s}}^2} \qquad (2.1.34)$$

as a function of the normalized frequency V. For TE modes we have

$$H = V + 1/\sqrt{b} + 1/\sqrt{b+a} \quad . \qquad (2.1.35)$$

The corresponding plots are shown in Fig. 2.11 for four values of the asym-

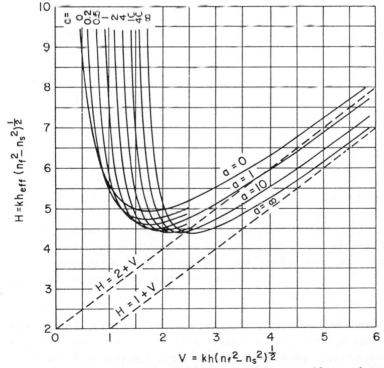

Fig. 2.11. Normalized effective thickness of a slab waveguide as a function of the normalized film thickness V for various degrees of asymmetry (after [2.20])

19

metry measure. Similar plots can be obtained for the TM modes [2.20]. In Fig. 2.11, we note the occurrence of minimum values of $H(V)$, for which we obtain maximum confinement of the light. For highly asymmetric guides ($a = \infty$), for example, we have a minimum of $H_{min} = 4.4$ at $V = 2.55$. This implies a minimum effective thickness of

$$(h_{eff}/\lambda)_{min} = 0.7(n_f^2 - n_s^2)^{-1/2} \quad . \tag{2.1.36}$$

For a typical thin-film glass waveguide, we have $n_s = 1.5$, $n_f = 1.6$ and $(h_{eff}/\lambda)_{min} \approx 1.3$.

2.2 Fundamentals of the Electromagnetic Theory of Dielectric Waveguides

In this section we propose to collect the, by now well developed, fundamentals of the electromagnetic theory of dielectric waveguides, and to discuss some of the general properties of the waveguide modes. This includes a listing of the relevant forms of Maxwell's equations and of the wave equations, a deduction of modal field properties following from symmetry considerations, a proof of the orthogonality of the modes, and a discussion of mode expansion and normalization, of the variation theorem, of power transport and stored energy, and of variational principles applicable to dielectric waveguides. We will keep this discussion as general as possible to allow for the variety of planar and strip waveguides considered for integrated optics applications.

2.1.1 Maxwell's Equations

Maxwell's equations for source-free, time-dependent fields are

$$\nabla \times \tilde{E} = -\partial \tilde{B}/\partial t \tag{2.2.1}$$

$$\nabla \times \tilde{H} = \partial \tilde{D}/\partial t \quad , \tag{2.2.2}$$

where t is time, $\nabla = (\partial/\partial x, \partial/\partial y, \partial/\partial z)$ is the del operator, and $\tilde{E}(t)$, $\tilde{H}(t)$, $\tilde{D}(t)$ and $\tilde{B}(t)$ are the time-dependent vectors of the dielectric and magnetic field, the electric displacement and the magnetic induction, respectively. We assume fields with a periodic time dependence which we write in the form

$$\tilde{E}(t) = E \exp(\mathrm{j}\omega t) + E^* \exp(-\mathrm{j}\omega t) \quad , \quad \text{etc.} \tag{2.2.3}$$

where E is a complex amplitude, ω the angular frequency, and the asterisk indicates a complex conjugate. Assuming a lossless medium with a scalar dielectric constant $\varepsilon(\omega)$ and a scalar magnetic permeability μ, we have the constitutive relations

$$D = \varepsilon E \quad , \tag{2.2.4}$$

$$B = \mu H \quad . \tag{2.2.5}$$

With this we get Maxwell's equations for the complex amplitudes of the form

$$\nabla \times E = -j\omega\mu H \quad , \tag{2.2.6}$$

$$\nabla \times H = j\omega\varepsilon E \quad . \tag{2.2.7}$$

These equations are subject to boundary conditions at surfaces where abrupt changes of the material constants occur. Figure 2.12 shows a sketch of such a boundary between two media distinguished by the indices 1 and 2, with the unit vector e_n chosen perpendicular to the surface. In the absence of surface charges and surface currents, we have the conditions

$$e_n \cdot (B_1 - B_2) = 0 \quad , \quad e_n \cdot (D_1 - D_2) = 0$$
$$e_n \times (E_1 - E_2) = 0 \quad , \quad e_n \times (H_1 - H_2) = 0 \tag{2.2.8}$$

for the fields E_1, E_2, etc., at the boundary. In dielectric waveguides we usually have a constant permeability $\mu = \mu_0$, which implies equality of the magnetic field vectors $H_1 - H_2$ at the boundary.

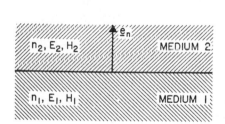

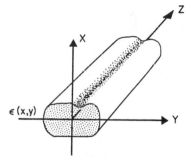

Fig. 2.13. Sketch of a dielectric waveguide and the choice of the coordinate system. The guide axis is chosen to coincide with the z-axis

Fig. 2.12. Boundary between two media of indices n_1 and n_2. The vector e_n is the normal to the surface

In Fig. 2.13 we have sketched a typical waveguide geometry, and indicated the choice of our coordinate system with the guide axis pointing in the z-direction. Relative to this geometry, we distinguish between the longitudinal field components E_z and H_z, and the transverse field components E_t and H_t,

$$E = E_t + E_z \quad , \quad H = H_t + H_z \quad . \tag{2.2.9}$$

To separate these components, we write Maxwell's equations in the form

21

$$\nabla_t \times \mathbf{E}_t = -j\omega\mu\mathbf{H}_z \quad ; \quad \nabla_t \times \mathbf{H}_t = j\omega\varepsilon\mathbf{E}_t \quad , \tag{2.2.10}$$

$$\nabla_t \times \mathbf{E}_z + \mathbf{e}_z \times \partial\mathbf{E}_t/\partial z = -j\omega\mu\mathbf{H}_t \quad , \tag{2.2.11}$$

$$\nabla_t \times \mathbf{H}_z + \mathbf{e}_z \times \partial\mathbf{H}_t/\partial z = j\omega\varepsilon\mathbf{E}_t \quad , \tag{2.2.12}$$

where

$$\nabla_t = (\partial/\partial x, \partial/\partial y, 0) \tag{2.2.13}$$

is the transverse del operator and $\mathbf{e}_z$ is a unit vector pointing in the z-direction. In dealing with waveguide problems, one usually focuses on the transverse components. Once these are known, one can determine the z-components via (2.2.10).

2.2.2 Modes of the Waveguide

As indicated in Fig. 2.13, a dielectric waveguide is characterized by a dielectric constant

$$\varepsilon = \varepsilon_0 n^2(x, y) \tag{2.2.14}$$

which is independent of the z coordinate. The function $n(x, y)$ is known as the refractive-index profile and generally assumes its highest values near the axis. A mode of the waveguide is defined as a field solution of the form

$$\mathbf{E}(x, y, z) = \mathbf{E}_\nu(x, y) \exp(-j\beta_\nu z) \quad ,$$
$$\mathbf{H}(x, y, z) = \mathbf{H}_\nu(x, y) \exp(-j\beta_\nu z) \quad , \tag{2.2.15}$$

where ν is a mode label (indicating the mode number, for example), and β_ν is the propagation constant of the mode. For guides providing confinement in two dimensions, such as strip guides or fibers, we need, of course, two labels, but only one is shown here, for simplicity. Inserting the modal fields of (2.2.15) into Maxwell's equations (2.2.10–12), we obtain

$$\nabla_t \mathbf{E}_{t\nu} = -j\omega\mu\mathbf{H}_{z\nu} \quad , \quad \nabla_t \times \mathbf{H}_{t\nu} = j\omega\varepsilon\mathbf{E}_{z\nu} \quad , \tag{2.2.16}$$

$$\nabla_t \times \mathbf{E}_{z\nu} - j\beta_\nu\mathbf{e}_z \times \mathbf{E}_{t\nu} = -j\omega\mu\mathbf{H}_{t\nu} \quad , \tag{2.2.17}$$

$$\nabla_t \times \mathbf{H}_{z\nu} - j\beta_\nu\mathbf{e}_z \times \mathbf{H}_{t\nu} = j\omega\varepsilon\mathbf{E}_{t\nu} \quad . \tag{2.2.18}$$

The general nature of the solutions to these equations was discussed in detail by *McKenna* [2.24] and in textbooks, [2.1–9]. One encounters a situation which is analogous to that encountered in quantum mechanics where one seeks solutions to Schrödinger's equation for various potential distributions and finds two types of solutions, one corresponding to bound states and the other to unbound states. For dielectric waveguides, we find guided modes (bound states) where the energy is confined near the axis, and radiation modes (unbound states) with their energy spread out through the medium surrounding the guide. The guided modes are associated with a discrete spectrum of propagation constants β_ν, while the radiation modes belong to a continuum. One also finds evanescent modes with imaginary propagation constants $\beta_\nu = -j\alpha_\nu$ and which decay as $\exp(-\alpha_\nu z)$. Solutions for specific waveguide examples are given in Sects. 2.3 and 4.

2.2.3 The Wave Equations for Planar Guides

In planar guides, the light is confined to one dimension only, which we choose to be in the x-direction. The refractive index $n(x)$ of a planar guide and the corresponding modal fields are functions of only this coordinate. One can, then, simplify the modal differential equations (2.2.16–18) by setting $\partial/\partial y = 0$. A planar guide supports transverse electric (TE) modes with zero longitudinal electric field ($E_z = 0$) and transverse magnetic modes (TM) with zero longitudinal magnetic field ($H_z = 0$). In the following, we derive the wave equations governing the two mode types, omitting the mode label ν to simplify the notation.

For TE modes we set $H_y = 0$ and get $E_z = 0$ from the second of (2.2.16). Equation (2.2.17) yields $E_x = 0$ and

$$\beta E_y = -\omega\mu H_x \quad . \tag{2.2.19}$$

Next we obtain

$$\partial E_y/\partial x = -j\omega\mu H_z \tag{2.2.20}$$

from the first of (2.2.16), and

$$\partial H_z/\partial x + j\beta H_x = -j\omega\varepsilon E_y \tag{2.2.21}$$

from (2.2.18). Combining the last three equations, we arrive at a wave equation for E_y

$$\partial^2 E_y/\partial x^2 = (\beta^2 - n^2 k^2)E_y \quad , \qquad \text{where} \tag{2.2.22}$$
$$k = \omega/c = \omega\sqrt{\varepsilon_0\mu_0} \tag{2.2.23}$$

is the propagation constant of free space.

For TM modes, we start by setting $E_y = 0$ and obtain in a similar fashion $H_z = 0$, $H_x = 0$,

$$\beta H_y = \omega\varepsilon E_x \quad , \tag{2.2.24}$$

$$\partial H_y/\partial x = j\omega\varepsilon E_z \quad , \tag{2.2.25}$$

$$\partial E_y/\partial x + j\beta E_x = j\omega\mu H_y \quad , \tag{2.2.26}$$

and a wave equation for H_y of the form

$$n^2 \frac{\partial}{\partial x}\left(\frac{1}{n^2}\frac{\partial H_y}{\partial x}\right) = (\beta^2 - n^2 k^2)H_y \quad . \tag{2.2.27}$$

2.2.4 Mode Properties Following from Symmetry

Several phase relationships between the modal field components follow directly from the symmetry of both the waveguide and Maxwell's equations. We shall use both time-reversal and z-reversal to construct new solutions $\tilde{E}_2(r,t)$, $\tilde{H}_2(r,t)$ from known solutions $\tilde{E}_1(r,t)$ $\tilde{H}_1(r,t)$.

As is well known, one can reverse the sign of t in Maxwell's equations to construct a new solution of the form

$$\tilde{E}_2(r,t) = \tilde{E}_1(r,-t) \quad ; \quad \tilde{H}_2(r,t) = -\tilde{H}_1(r,-t) \quad . \tag{2.2.28}$$

In terms of the complex amplitudes, this can be written as

$$E_2(r) = E_1^*(r) \quad ; \quad H_2(r) = -H_1^*(r) \quad . \tag{2.2.29}$$

Next, we consider z-reversal and assume that

$$\varepsilon(x,y,-z) = \varepsilon(x,y,z) \quad , \tag{2.2.30}$$

which is true for all isotropic dielectric waveguides. By reversing the sign of z in Maxwell's equations (2.2.10–12), we can construct new solutions of the form

$$E_{t2}(z) = E_{t1}(-z) \quad , \qquad E_{z2}(z) = -E_{z1}(-z) \quad ; \tag{2.2.31}$$

$$H_{t2}(z) = -H_{t1}(-z) \quad , \qquad H_{z2}(z) = H_{z1}(-z) \quad , \tag{2.2.32}$$

where we have omitted to indicate the (x,y) dependence in order to emphasize the z-reversal operation.

In applying these reversal operations to the modal fields of the form of (2.2.15) we have to distinguish between the propagating modes with real valued β, and the evanescent modes with imaginary β. In the first case, a forward traveling mode will vary as $\exp(-j\beta z)$ and either time reversal or z-reversal yield a backward traveling mode varying as $\exp(j\beta z)$. As the new solution must be unique, the application of (2.2.29), as well as (2.2.31, 32) must yield the same result. This requires that

$$E_{t\nu} = E_{t\nu}^* \quad , \qquad H_{t\nu} = H_{t\nu}^* \quad , \tag{2.2.33}$$

$$E_{z\nu}^* = -E_{z\nu} \quad , \qquad H_{z\nu}^* = -H_{z\nu} \quad . \tag{2.2.34}$$

We have constructed modal E and H fields with real valued transverse components and imaginary z-components. The general implications for a propagating mode are that the tangential components of its E and H fields are in phase, that their z-components are also in phase, and that the tangential and z-components are $90°$ out of phase.

The fields of a forward evanescent mode vary as $\exp(-\alpha z)$. Here the time reversal operation produces again a forward wave. Uniqueness requires that these two waves be identical; and because of (2.2.29), this requires that

$$E_\nu = E_\nu^* \quad , \quad H_\nu = -H_\nu^* \quad . \tag{2.2.35}$$

We have constructed an evanescent mode with a real valued electric field and an imaginary magnetic field. In general, E_ν and H_ν of an evanescent mode are $90°$ out of phase.

2.2.5 Orthogonality of the Modes

All modes of a dielectric waveguide are orthogonal to each other [2.3, 25]. This important property holds for both the guided and the radiation modes; it is the basis for much of the waveguide theory, including the theories of waveguide excitation, of waveguide discontinuities and waveguide perturbations. We will sketch here a derivation of the orthogonality relations which makes apparent their connection with power conservation and reciprocity.

We start with the complex Maxwell's equation (2.2.6, 7) for a lossless, scalar medium and consider two different solutions labeled 1 and 2. We form the dot products of H_2^* with (2.2.6) and of E_1 with the complex conjugate of (2.2.7), subtract the results and obtain

$$\nabla \cdot (E_1 \times H_2^*) = j\omega(\varepsilon E_1 \cdot E_2^* - \mu H_1 \cdot H_2^*) \quad , \tag{2.2.36}$$

where we have used the vector identity

$$\nabla \cdot (a \times b) = b \cdot (\nabla \times a) - a \cdot (\nabla \times b) \quad . \tag{2.2.37}$$

Note that (2.2.36) becomes the complex Poynting theorem if we make the labels 1 and 2 equal. Now we exchange the labels 1 and 2 in (2.2.36), take the complex conjugate and add the result to (2.2.36) to obtain

$$\nabla \cdot (E_1 \times H_2^* + E_2^* \times H_1) = 0 \quad . \tag{2.2.38}$$

This theorem is closely related to the Lorentz reciprocity theorem; in fact, we obtain a form of the latter if we replace field 2 in (2.2.38) by its corresponding time reversed field using (2.2.29) with the resulting

$$\nabla \cdot (E_1 \times H_2 - E_2 \times H_1) = 0 \quad . \tag{2.2.39}$$

We proceed by applying the theorem of (2.2.38) to waveguide modes and identify the fields 1 and 2 with two forward modes

$$E_1 = E_\nu(x,y)\exp(-j\beta_\nu z) \quad , \quad E_2 = E_\mu(x,y)\exp(-j\beta_\mu z) \quad , \tag{2.2.40}$$

which yields

$$\nabla_t \cdot (\boldsymbol{E}_\nu \times \boldsymbol{H}_\mu^* + \boldsymbol{E}_\mu^* \times \boldsymbol{H}_\nu)_t - j(\beta_\nu - \beta_\mu)$$
$$\times (\boldsymbol{E}_{t\nu} \times \boldsymbol{H}_{t\mu}^* + \boldsymbol{E}_{t\mu}^* \times \boldsymbol{H}_{t\nu})_z = 0 \quad . \tag{2.2.41}$$

Here we have, again, separated the transverse (t) and the longitudinal (z) components and used the transverse del operator ∇_t. The next step is to integrate (2.2.41) over a cross-section $z = $ const of the waveguide. Applying the divergence theorem to the first term we get

$$\int\!\!\!\int_{-\infty}^{+\infty} dx\, dy\, \nabla_t \cdot \boldsymbol{g} = \oint_c ds\, \boldsymbol{g} \cdot \boldsymbol{e}_t \tag{2.2.42}$$

where

$$\boldsymbol{g} = (\boldsymbol{E}_\nu \times \boldsymbol{H}_\mu^* + \boldsymbol{E}_\mu^* \times \boldsymbol{H}_\nu)_t \quad , \tag{2.2.43}$$

and the line integral extends over an infinitely large curve enclosing the waveguide, with $\boldsymbol{e}_t$ being a unit vector perpendicular to that curve. It is easy to see that this line integral vanishes if at least one of the two modes is a guided mode with fields decaying exponentially towards infinity. The line integral also vanishes when both modes are radiation modes; the argument to show this is somewhat more complicated and involves the oscillatory nature of the radiation modes [2.3].

The terms remaining after integration are

$$\int\!\!\!\int_{-\infty}^{+\infty} dx\, dx (\boldsymbol{E}_{t\nu} \times \boldsymbol{H}_{t\mu}^* + \boldsymbol{E}_{t\mu}^* \times \boldsymbol{H}_{t\nu}) = 0 \quad , \quad \beta_\nu \neq \beta_\mu \tag{2.2.44}$$

where we have dropped the factor $(\beta_\mu - \beta_\nu)$ as we assume that $\beta_\nu \neq \beta_\mu$. The z-reversal symmetry allows a further simplification. In order to achieve this, we apply (2.2.44) to a backward traveling mode (labeled $-\nu$) instead of the corresponding forward mode (labeled ν). According to (2.2.31, 32), the fields of the backward-traveling mode are given by

$$\boldsymbol{E}_{t,-\nu}(x, y) = \boldsymbol{E}_{t,\nu}(x, y) \tag{2.2.45}$$

$$\boldsymbol{H}_{t,-\nu}(x, y) = -\boldsymbol{H}_{t,\nu}(x, y) \quad . \tag{2.2.46}$$

With this (2.2.44) becomes

$$\int\!\!\!\int_{-\infty}^{+\infty} dx\, dy (\boldsymbol{E}_{t\nu} \times \boldsymbol{H}_{t\mu}^* - \boldsymbol{E}_{t\mu}^* \times \boldsymbol{H}_{t\nu}) = 0 \quad , \quad \beta_\mu \neq \beta_\nu \quad . \tag{2.2.47}$$

Adding the two results yields the simple orthogonality relation

$$\int\!\!\!\int_{-\infty}^{+\infty} dx\, dy\, \boldsymbol{E}_{t\nu} \times \boldsymbol{H}_{t\mu}^* = 0 \quad , \quad \beta_\mu \neq \beta_\nu \quad . \tag{2.2.48}$$

2.2.6 Mode Expansion and Normalization

The orthogonality of the modes allows us to express an arbitrary given field distribution as a superposition of waveguide modes. Doing this, we shall only deal with the transverse field components; the z-components follow from Maxwell's equations, i.e., (2.2.16). To simplify the notation for this subsection, we shall leave out the label t and designate the transverse fields of the forward modes as $E_{\nu\mu}(x,y)$ and $H_{\nu\mu}(x,y)$.

Assume, first, that only forward propagating modes are present in a given field with the transverse components $E_t(x,y)$ and $H_t(x,y)$ in a guide cross-section $z = $ const. This can be represented as a superposition of waveguide modes of the form

$$E_t(x,y) = \sum_\nu \sum_\mu a_{\nu\mu} E_{\nu\mu}(x,y)$$

$$+ \int_0^\infty \int_0^\infty d\nu\, d\mu\, a(\nu,\mu) E(\nu,\mu;x,y) \quad , \tag{2.2.49}$$

$$H_t(x,y) = \sum_\nu \sum_\mu a_{\nu\mu} H_{\nu\mu}(x,y)$$

$$+ \int_0^\infty \int_0^\infty d\nu\, d\mu\, a(\nu,\mu) H(\nu,\mu;x,y) \quad . \tag{2.2.50}$$

The summation here extends over the discrete and finite set of guided modes and the integration extends over the continuous spectrum of radiation modes. The discrete spectra in the above expressions are similar to those encountered in hollow metal waveguides and the continuous spectrum is similar to the angular spectrum of plane waves of a field in free space. It is thus natural to call the continuous labels ν and μ "spatial frequencies" and to assign to them the dimension [cm^{-1}]. In order to preserve the same dimensionality for the field distributions of both the guided and of the radiation modes, we have to keep the discrete coefficients $a_{\nu\mu}$ dimensionless and assign the dimension [cm^2] to the coefficients $a(\nu,\mu)$ of the continuum. It is advantageous to use only positive spatial frequencies; their lower limit, written as 0 in the above equations, actually depends on the particular choice of the labels ν and μ.

In (2.2.49,50), we have not explicitly indicated the necessary summation over modes of different polarization (e.g., TE and TM modes) and over degenerate modes with the same spatial frequency (e.g., the even and odd radiation modes to be discussed in Sect. 2.3).

It is convenient to normalize the modal fields by means of the cross power $\overline{P}(\nu,\overline{\nu},\mu,\overline{\mu})$ such that

27

$$\overline{\boldsymbol{P}} = 2 \iint\limits_{-\infty}^{+\infty} dx\, dy\, \boldsymbol{E}_{\nu\mu} \times \boldsymbol{H}^*_{\overline{\nu}\,\overline{\mu}} = \delta_{\nu\overline{\nu}}\delta_{\mu\overline{\mu}} \tag{2.2.51}$$

for discrete modes, where $\delta_{\nu\overline{\nu}}$ is the Kronecker delta, and

$$\overline{\boldsymbol{P}} = 2 \iint\limits_{-\infty}^{+\infty} dx\, dy\, \boldsymbol{E}(\nu,\mu) \times \boldsymbol{H}^*(\overline{\nu},\overline{\mu}) = \delta(\nu - \overline{\nu})\delta(\mu - \overline{\mu}) \tag{2.2.52}$$

for continuous modes, where $\delta(\nu - \overline{\nu})$ is the delta function. These normalizations express, of course, also the orthogonality demanded by (2.2.48).

When the modes are not normalized, the cross-power $\overline{\boldsymbol{P}}$ of the continuous modes is given by

$$\overline{P}(\nu,\overline{\nu},\mu,\overline{\mu}) = \boldsymbol{p}_{\nu\mu}\delta(\nu - \overline{\nu})\delta(\mu - \overline{\mu}) \quad . \tag{2.2.53}$$

We mention this specifically to make a point about the dimension of the factor $p_{\nu\mu}$. $\overline{\boldsymbol{P}}$ is measured in W; because ν and μ are measured in cm^{-1}, the dimension of the delta functions $\delta(\nu)$ and $\delta(\mu)$ is cm. The factor $p_{\nu\mu}$ is, therefore, measured in W/cm^2. A normalization as in (2.2.52) means that we have set $p_{\nu\mu}$ equal to $1\,\text{W/cm}^2$. For simplicity, we have not indicated this dimension in (2.2.52), but it should always be remembered when this relation is used below.

The orthonormality relations of (2.2.51 and 52) allow us to determine the coefficients of the mode expansion in a simple manner. We have

$$a_{\nu\mu} = 2 \iint\limits_{-\infty}^{+\infty} dx\, dy\, \boldsymbol{E}_{\text{t}} \times \boldsymbol{H}^*_{\nu\mu} = 2 \iint\limits_{-\infty}^{+\infty} dx\, dy\, \boldsymbol{E}^*_{\nu\mu} \times \boldsymbol{H}_{\text{t}} \quad , \tag{2.2.54}$$

$$a(\nu,\mu) = 2 \iint\limits_{-\infty}^{+\infty} dx\, dy\, \boldsymbol{E}_{\text{t}} \times \boldsymbol{H}^*(\nu,\mu)$$

$$= 2 \iint\limits_{-\infty}^{+\infty} dx\, dy\, \boldsymbol{E}^*(\nu,\mu) \times \boldsymbol{H}_{\text{t}} \quad . \tag{2.2.55}$$

The orthonormality relations also permit us to express the power carried by the total field in terms of the expansion coefficients. We obtain

$$P = \iint\limits_{-\infty}^{+\infty} dx\, dy (\boldsymbol{E}_{\text{t}} \times \boldsymbol{H}^*_{\text{t}} + \boldsymbol{E}^*_{\text{t}} \times \boldsymbol{H}_{\text{t}})_z$$

$$= \sum_{\nu}\sum_{\mu} a_{\nu\mu} a^*_{\nu\mu} + \iint\limits_{0}^{\infty} d\nu\, d\mu\, a(\nu,\mu)a^*(\nu,\mu) \quad , \tag{2.2.56}$$

where P is measured in W. We identify $a_{\nu\mu}a_{\nu\mu}^*$ as the power carried by a discrete mode, and $a(\nu,\mu)a^*(\nu,\mu)$ as a spectral power density, which implies that a power $a(\nu,\mu)a^*(\nu,\mu)\Delta\nu\Delta\mu$ is carried in the spatial frequency band $\Delta\nu\Delta\mu$ of the continuous mode spectrum.

In the above discussion, we have noted a formal analogy between the discrete and the continuous modes. In the following, we shall take advantage of this to simplify our notation by writing a_ν for $a_{\nu\mu}$ and $a(\nu,\mu)$, E_ν for $E_{\nu\mu}$ and $E(\nu,\mu)$, etc., and by writing

$$\sum \quad \text{for} \quad \sum_\nu\sum_\mu + \iint d\nu\,d\mu \quad . \tag{2.2.57}$$

Using this notation, we propose to discuss now the mode expansion of a field which contains forward- and backward-propagating modes (real β). Subsequently, we will consider evanescent modes. In (2.2.45, 46), we have indicated that a backward mode has the same transverse E field distribution as the corresponding forward mode and a transverse H distribution with reversed sign. Because of this, the mode expansion has the form

$$E_{\mathrm{t}} = \sum(a_\nu + b_\nu)E_\nu \quad , \tag{2.2.58}$$

$$H_{\mathrm{t}} = \sum(a_\nu - b_\nu)H_\nu \quad , \tag{2.2.59}$$

where a_ν are the coefficients of the forward waves and b_ν are the coefficients of the backward waves. With the help of the orthonormality relations we determine the coefficients as

$$a_\nu = \int\limits_{-\infty}^{+\infty}\!\!\!\int dx\,dy(E_{\mathrm{t}} \times H_\nu^* + E_\nu^* \times H_{\mathrm{t}}) \quad , \tag{2.2.60}$$

$$b_\nu = \int\limits_{-\infty}^{+\infty}\!\!\!\int dx\,dy(E_{\mathrm{t}} \times H_\nu^* - E_\nu^* \times H_{\mathrm{t}}) \quad . \tag{2.2.61}$$

The power carried by the total field becomes

$$P = \sum(a_\nu a_\nu^* - b_\nu b_\nu^*) \quad , \tag{2.2.62}$$

as expected.

Let us now consider evanescent modes, which have imaginary propagation constants. As none of these modes can carry power by itself, we obtain cross power products which are imaginary, and we have to change the orthonormality relation to

$$\overline{P} = 2\int\limits_{-\infty}^{+\infty}\!\!\!\int dx\,dy\,E(\nu,\mu) \times H^*(\overline{\nu},\overline{\mu}) = \pm \mathrm{j}\delta(\nu - \overline{\nu})\delta(\mu - \overline{\mu}) \quad , \tag{2.2.63}$$

where the occurrence of the $+$ or $-$ sign depends on the particular guide configuration and its mode solutions. Keeping the mode expansion in the form of (2.2.58, 59), we obtain for the coefficients of the evanescent modes

$$\pm \mathrm{j} a_\nu = \int\limits_{-\infty}^{+\infty}\!\!\! dx\, dy (\boldsymbol{E}_\mathrm{t} \times \boldsymbol{H}_\nu^* - \boldsymbol{E}_\nu^* \times \boldsymbol{H}_\mathrm{t}) \quad , \tag{2.2.64}$$

$$\pm \mathrm{j} b_\nu = \int\limits_{-\infty}^{+\infty}\!\!\! dx\, dy (\boldsymbol{E}_\mathrm{t} \times \boldsymbol{H}_\nu^* + \boldsymbol{E}_\nu^* \times \boldsymbol{H}_\mathrm{t}) \quad . \tag{2.2.65}$$

For the power carried by the field, we get the formula

$$P = \pm \mathrm{j} \sum (a_\nu^* b_\nu - a_\nu b_\nu^*) \quad , \tag{2.2.66}$$

which reflects the fact that an evanescent wave cannot carry power by itself, but a combination of forward and backward evanescent waves can lead to tunneling of power through a short region.

2.2.7 The Variation Theorem for Dielectric Waveguides

The variation theorem [2.26] connects the variations $\delta \boldsymbol{E}$ and $\delta \boldsymbol{H}$ of the electromagnetic field solutions to the perturbations $\delta(\omega\varepsilon)$ and $\delta(\omega\mu)$ of the frequency and of the constants of the medium which are the cause of these variations. The theorem follows directly from the complex Maxwell's equations (2.2.6, 7) and can be written in the general form

$$\begin{aligned} \nabla\cdot(\boldsymbol{E}^* \times \delta\boldsymbol{H} + \delta\boldsymbol{E} \times \boldsymbol{H}^*) \\ = -\mathrm{j}[\delta(\omega\varepsilon)\boldsymbol{E}\cdot\boldsymbol{E}^* + \delta(\omega\mu)\boldsymbol{H}\cdot\boldsymbol{H}^*] \quad . \end{aligned} \tag{2.2.67}$$

To apply this theorem to dielectric waveguides [2.27], we consider a waveguide mode with the fields

$$\boldsymbol{E} = \boldsymbol{E}_\nu \exp(-\mathrm{j}\beta_\nu z) \quad , \quad \boldsymbol{H} = \boldsymbol{H}_\nu \exp(-\mathrm{j}\beta_\nu z) \tag{2.2.68}$$

and express their variations in the form

$$\delta\boldsymbol{E} = (\delta\boldsymbol{E}_\nu - \mathrm{j}z\delta\beta_\nu\cdot\boldsymbol{E}_\nu)\exp(-\mathrm{j}\beta_\nu z) \quad , \tag{2.2.69}$$

and similarly for $\delta\boldsymbol{H}$. If we insert this in the above theorem, we obtain the relation

$$\nabla_\mathrm{t}\cdot\boldsymbol{g} - \mathrm{j}\delta\beta_\nu\cdot\boldsymbol{S}_z = -\mathrm{j}[\delta(\omega\varepsilon)\boldsymbol{E}_\nu\cdot\boldsymbol{E}_\nu^* + \delta(\omega\mu)\boldsymbol{H}_\nu\cdot\boldsymbol{H}_\nu^*] \quad , \tag{2.2.70}$$

where ∇_t is the transverse del operator,

$$\boldsymbol{S} = \boldsymbol{E}_\nu \times \boldsymbol{H}_\nu^* + \boldsymbol{E}_\nu^* \times \boldsymbol{H}_\nu \tag{2.2.71}$$

is the time-averaged Poynting vector, and

$$g = E_\nu^* \times \delta H_\nu + \delta E_\nu \times H_\nu^* - j\delta\beta_\nu \cdot zS \quad . \tag{2.2.72}$$

We proceed by integrating (2.2.70) over the cross-section of the guide. As in the derivation of the orthogonality relation in Sect. 2.2.5, we find that the integral over $\nabla_t \cdot g$ vanishes and obtain

$$\delta\beta_\nu \cdot P = \int\!\!\!\int\limits_{-\infty}^{+\infty} dx\, dy [\delta(\omega\varepsilon) E_\nu \cdot E_\nu^* + \delta(\omega\mu) H_\nu \cdot H_\nu^*] \quad , \tag{2.2.73}$$

where

$$P = \int\!\!\!\int\limits_{-\infty}^{+\infty} dx\, dy\, S_z \tag{2.2.74}$$

is the power carried by the mode. This is the variation theorem for dielectric waveguides. One of the applications of this theorem is the determination of the change $\Delta\beta$ of the propagation constant due to a perturbation $\Delta\varepsilon(x,y)$ of the dielectric constant of the waveguide. Here we have $\delta\omega = 0$ and $\delta\mu = 0$ and the theorem yields

$$\Delta\beta \cdot P = \omega \int\!\!\!\int\limits_{-\infty}^{+\infty} dx\, dy\, \Delta\varepsilon E_\nu \cdot E_\nu^* \quad . \tag{2.2.75}$$

2.2.8 Power Flow and Stored Energy in a Dielectric Waveguide

In this subsection we give a brief discussion of the power flow carried and the energy stored by the fields of a waveguide mode. A more detailed discussion which also explores connections to the zig-zag wave model has been given in [2.27].

We have already encountered in earlier sections the time-averaged Poynting vector associated with a mode, which is defined by

$$S(x,y) = E_\nu \times H_\nu^* + E_\nu^* \times H_\nu \quad , \tag{2.2.76}$$

and the time-averaged power transported by the mode

$$P = \int\!\!\!\int\limits_{-\infty}^{+\infty} dx\, dy\, S_z \quad . \tag{2.2.77}$$

The time-average density of the energy stored by the modal fields is given by

$$w(x,y) = \frac{d(\omega\varepsilon)}{d\omega} E_\nu \cdot E_\nu^* + \frac{d(\omega\mu)}{d\omega} H_\nu \cdot H_\nu^* \quad , \tag{2.2.78}$$

31

which is written here in a form that allows for dispersive medium constants $\varepsilon(\omega)$ and $\mu(\omega)$ [2.28]. The energy W stored per unit guide length is obtained by integrating w over the guide cross-section

$$W = \int\limits_{\infty}^{+\infty}\!\!\int dx\, dy\, w(x,y) \quad . \tag{2.2.79}$$

The group velocity v_g of the mode is the velocity at which signals carried by the mode propagate. It is determined from the dispersion relation $\omega(\beta)$ of the guide by

$$v_g = d\omega/d\beta \quad . \tag{2.2.80}$$

Applying the variation theorem (2.2.73) to cases where all variations are caused by perturbations in ω alone, we deduce the simple relation

$$P = (d\omega/d\beta)W = v_g W \quad , \tag{2.2.81}$$

which is known to hold for many other waveguide structures.

For the remainder of this subsection, we assume dispersion-free media and distinguish between the electric and magnetic energy (via superscripts ε and μ). We define

$$W_t^\varepsilon = \int\limits_{-\infty}^{+\infty}\!\!\int dx\, dy\, \varepsilon \boldsymbol{E}_t \cdot \boldsymbol{E}_t^* \quad , \quad W_z^\varepsilon = \int\limits_{-\infty}^{+\infty}\!\!\int dx\, dy\, \varepsilon \boldsymbol{E}_z \cdot \boldsymbol{E}_z^* \quad ; \tag{2.2.82}$$

$$W_t^\mu = \int\limits_{-\infty}^{+\infty}\!\!\int dx\, dy\, \mu \boldsymbol{H}_t \cdot \boldsymbol{H}_t^* \quad , \quad W_z^\mu = \int\limits_{-\infty}^{+\infty}\!\!\int dx\, dy\, \mu \boldsymbol{H}_z \cdot \boldsymbol{H}_z^* \quad . \tag{2.2.83}$$

Here we have also distinguished between the energy portions stored by the transverse (t) and longitudinal (z) field components, and have left out the mode label ν to simplify the notation. Forming dot products of the modal differential equations (2.2.16–18) with the appropriate field components and combining the results one obtains

$$\nabla_t \cdot \boldsymbol{E}_z \times \boldsymbol{H}_t^* - j\beta \boldsymbol{e}_z \cdot \boldsymbol{E}_t \times \boldsymbol{H}_t^* = j\omega\varepsilon \boldsymbol{E}_z \cdot \boldsymbol{E}_z^* - j\omega\mu \boldsymbol{H}_t \cdot \boldsymbol{H}_t^* \quad , \tag{2.2.84}$$

$$\nabla_t \cdot \boldsymbol{E}_t \times \boldsymbol{H}_z^* + j\beta \boldsymbol{e}_z \cdot \boldsymbol{E}_t \times \boldsymbol{H}_t^* = j\omega\varepsilon \boldsymbol{E}_t \cdot \boldsymbol{E}_t^* - j\omega\mu \boldsymbol{H}_z \cdot \boldsymbol{H}_z^* \quad . \tag{2.2.85}$$

When we integrate this over the guide cross-section, we find, as in Sect. 2.2.5, that the first terms vanish. This leads to the simple relations

$$\beta P = 2\omega(W_t^\mu - W_z^\varepsilon) \quad , \tag{2.2.86}$$

$$\beta P = 2\omega(W_t^\varepsilon - W_z^\mu) \quad . \tag{2.2.87}$$

Subtracting these two relations we find

$$W^\varepsilon = W_t^\varepsilon + W_z^\varepsilon = W_t^\mu + W_z^\mu = W^\mu \quad , \tag{2.2.88}$$

stating the equality of the stored electric energy W^ε and the stored magnetic energy W^μ. The same relation also follows from the complex Poynting theorem.

By adding (2.2.86 and 87) we obtain another interesting relation, namely

$$P = \frac{\omega}{\beta}(W_t - W_z) = v_{\rm p}(W_t - W_z) \quad , \qquad \text{where} \tag{2.2.89}$$

$$W_t = W_t^\varepsilon + W_t^\mu \quad , \qquad W_z = W_z^\varepsilon + W_z^\mu \quad . \tag{2.2.90}$$

This expression relates the phase velocity $v_{\rm p}$ of the mode to the power flow P and the quantity $(W_t - W_z)$, which can be identified as the electromagnetic momentum flow in the waveguide [2.27]. It should be contrasted to the relation (2.2.81) for the group velocity $v_{\rm g}$. Combining these two relations we get

$$\frac{v_{\rm p}}{v_{\rm g}} = \frac{W_t + W_z}{W_t - W_z} \tag{2.2.91}$$

which ties the difference between $v_{\rm p}$ and $v_{\rm g}$ to the presence of longitudinal field components.

2.2.9 Variational Properties of the Propagation Constant

There are many optical waveguide problems for which exact mode solutions are not available. The variational principle provides a powerful tool for the numerical solution of these problems. Variational methods have found extensive use elsewhere in physics and engineering. In quantum mechanics, for example, one uses a variational expression for the energy of the ground state which assumes a minimum value when the correct electron wave functions are inserted [2.29]. Another example is the use of variational techniques in the analysis of microwave waveguides and resonators [2.30–32], a discipline closely related to that of optical waveguides.

The basic root of the variational properties of waveguides is the principle of least action [2.33]. The essence of the principle is the formulation of a stationary expression for a key quantity, for example for the propagation constant β. The stationary expression takes the form of cross-sectional integrals involving the waveguide modes $\psi(x, y)$. This expression is distinguished by the property that it yields a result for β that is insensitive to inaccuracies in the trial function used for ψ. Mathematically this is expressed as

$$\delta\beta = 0 \qquad\qquad (2.2.92)$$

to first order in $\delta\psi$, assuming a trial function $\psi + \delta\psi$ is inserted in the stationary expression.

In addition, it can be proved in many cases, that the value for β obtained from the stationary expression assumes its *maximum value* when the correct mode function ψ is inserted. An approximation to ψ will, therefore, produce a lower bound for β. Larger β values correspond to better approximations.

The Scalar 1D Case

A simple example is the variational expression for the quantity β^2 of the TE mode of a planar slab waveguide, which has the form

$$\beta^2 = \int_{-\infty}^{+\infty} dx\, \psi(d^2/dx^2 + n^2 k^2)\psi \Big/ \int_{-\infty}^{+\infty} dx\, \psi^2 \quad . \qquad (2.2.93)$$

This expression follows from the scalar wave equation (2.2.22) where the mode function ψ corresponds to the $E_y(x)$ component of the modal field.

For this and the following variational expressions it is assumed that the mode functions are continuous and that they possess first derivatives. Surface integrals have to be added when this is not the case [2.30–32].

The Scalar 2D Case

The scalar two-dimensional wave equation

$$\nabla^2\psi + (n^2 k^2 - \beta^2)\psi = 0 \qquad\qquad (2.2.94)$$

is a frequently used approximation in the analysis of channel waveguides. The scalar mode function $\psi(x, y)$ corresponds to a transverse component of the electric field. The quantities ψ and $n(x, y)$ are now both functions of the two transverse dimensions x and y. The associated variational expression for β^2 is given by

$$\beta^2 = \int_{-\infty}^{+\infty} dx\, dy\, \psi(\nabla^2 + n^2 k^2)\psi \Big/ \int_{-\infty}^{+\infty} dx\, dy\, \psi^2 \quad . \qquad (2.2.95)$$

An early application of this stationary expression to optical waveguides has been reported by *Matsuhara* [2.34], who employed it for the analysis of rectangular channel waveguides. Applications to diffused channel guides have been discussed in [2.35 and 36], and an application to strip-loaded diffused guides has been described in [2.37].

The Vector Case

An exact treatment of most optical waveguide problems requires a formulation in terms of the exact vector fields. Stationary expressions are also available for this more complex case. A mixed field expression for the quantity β has the form [2.30, 31, 38]

$$\beta = \int\limits_{-\infty}^{+\infty} dx\, dy (\omega\varepsilon \boldsymbol{E}^* \cdot \boldsymbol{E} + \omega\mu_0 \boldsymbol{H}^* \cdot \boldsymbol{H} + \mathrm{j}[\boldsymbol{E}^* \cdot \nabla \times \boldsymbol{H} - \boldsymbol{H}^* Z . \nabla \times \boldsymbol{E}])$$

$$\bigg/ \int\limits_{-\infty}^{+\infty} dx\, dy (\boldsymbol{E} \times \boldsymbol{H}^* + \boldsymbol{E}^* \times \boldsymbol{H}) \quad . \tag{2.2.96}$$

In this expression, the propagation constant β is stationary with respect to independent variations of all six components of the modal fields $\boldsymbol{E}$ and $\boldsymbol{H}$. However, there appears to be no proof available at present that will assure that β is a maximum relative to all these variations.

This expression allows a simple derivation of the variation theorem (2.2.75) discussed earlier. For this purpose we take $\boldsymbol{E}$ and $\boldsymbol{H}$ as the modal fields of an unperturbed guide with the profile $\varepsilon(x, y)$. Then we consider a perturbation $\varepsilon + \Delta\varepsilon(x, y)$. We insert this in the stationary expression and use $\boldsymbol{E}$ and $\boldsymbol{H}$ as trial functions. The result is

$$\Delta\beta = \omega \int\limits_{-\infty}^{+\infty} dx\, dy\, \Delta\varepsilon\, \boldsymbol{E}^* \cdot \boldsymbol{E}/P \quad ,$$

which is the variation theorem.

In their analysis of rectangular channel guides, *Akiba* and *Haus* [2.39] have used a vectorial stationary expression for the quantity ω^2, which has the form

$$\omega^2 = \int\limits_{-\infty}^{+\infty} dx\, dy (\nabla \times \boldsymbol{F}^*) \cdot (\nabla \times \boldsymbol{F}) \bigg/ \int\limits_{-\infty}^{+\infty} dx\, dy\, \mu_0\varepsilon \boldsymbol{E}^* \cdot \boldsymbol{E} \quad , \tag{2.2.97}$$

where $\boldsymbol{F} = \boldsymbol{E}(x, y)\exp(-\mathrm{j}\beta z)$. Here β is considered a given quantity and ω a variable. In this expression, the number of independent field variables is reduced to the three components of $\boldsymbol{E}$, a considerable advantage for numerical computations.

A vectorial variational expression for β^2 involving only the two transverse components of $\boldsymbol{H}$ has been given in [2.32 and 40].

2.3 Modes of the Planar Slab Guide

In this section we list formulas for the fields of the modes of a planar slab waveguide. We consider the structure and coordinate system shown in Fig. 2.14 where a film of thickness h and uniform refractive index n_f is sandwiched between a substrate of uniform index n_s and a cover of uniform

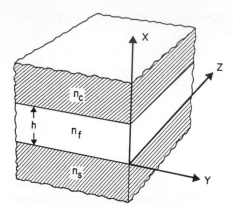

Fig. 2.14. Sketch of an "asymmetric" slab waveguide and the choice of the coordinate system. Note that the z-axis lies in the film-substrate interface

index n_c. This structure has also been called the "asymmetric" slab guide. The modal fields can be derived from the wave equations of Sect. 2.2.3, and the corresponding solutions have been discussed by *McKenna* [2.24], *Tien* [2.17], *Marcuse* [2.2], and others. We follow essentially *Marcuse*'s treatment, but cast the results in a simple form which employs the peak values of the fields in substrate, film and cover as well as the phase shifts at the film-substrate and film-cover boundaries that play an important role in the zig-zag wave picture. We have to distinguish between modes of TE (Sect. 2.3.1) and modes of TM polarization (Sect. 2.3.2). Another distinction is between guided modes and radiation modes, the latter divided into the categories of substrate radiation modes, substrate-cover (also called "air") radiation modes, and evanescent modes. Section 2.3.3 deals with multilayer waveguides.

In accordance with the wave equation we define various transverse decay (γ_i) and propagation constants (κ_i) by

$$\kappa_c^2 = n_c^2 k^2 - \beta^2 = -\gamma_c^2 \quad , \tag{2.3.1}$$

$$\kappa_f^2 = n_f^2 k^2 - \beta^2 \quad , \tag{2.3.2}$$

$$\kappa_s^2 = n_s^2 k^2 - \beta^2 = -\gamma_s^2 \quad , \tag{2.3.3}$$

where the subscripts s, f, and c refer to substrate, film, and cover respectively.

Table 2.3 shows the ranges of propagation constants β corresponding to the various mode types and categories, and also the associated ranges of the transverse propagation constant κ_s in the substrate. Because of its convenient range, the latter is chosen as the spatial frequency for the continuous mode spectrum.

As we are dealing with a planar problem, all field solutions can be made independent of the y-coordinate.

Table 2.3

Modes	β	κ_s
guided	$kn_f \rightarrow kn_s$	imaginary
substrate	$kn_s \rightarrow kn_c$	$0 \rightarrow k\sqrt{n_s^2 - n_c^2}$
substrate-cover	$kn_c \rightarrow 0$	$k\sqrt{n_s^2 - n_c^2} \rightarrow kn_s$
evanescent	imaginary	$kn_s \rightarrow \infty$

2.3.1 TE Modes

From Sect. 2.2.3, we have for the TE modes

$$H_y = E_x = E_z = 0 \quad , \tag{2.3.4}$$

$$H_x = -(\beta/\omega\mu)E_y \quad , \tag{2.3.5}$$

$$H_z = (j/\omega\mu)\partial E_y/\partial x \quad , \tag{2.3.6}$$

with the E_y component obeying the wave equation

$$\partial^2 E_y/\partial x^2 = (\beta^2 - n^2 k^2)E_y \quad . \tag{2.3.7}$$

The boundary conditions of (2.2.8) demand that E_y (and thereby automatically H_x) and $\partial E_y/\partial x$ (and thereby H_z) be continuous across the film boundaries at $x = 0$ and $x = h$.

For guided modes we have

$$\begin{aligned}
E_y &= E_c\exp[-\gamma_c(x-h)] \quad , \quad &\text{for} \quad h < x \quad , \quad &\text{(cover)} \quad , \\
E_y &= E_f\cos(\kappa_f x - \phi_s) \quad , \quad &\text{for} \quad 0 < x < h \quad , \quad &\text{(film)} \quad , \\
E_y &= E_s\exp(\gamma_s x) \quad , \quad &\text{for} \quad x < 0 \quad , \quad &\text{(substrate)} \quad .
\end{aligned} \tag{2.3.8}$$

Application of the boundary conditions yields the formulas for the phase shifts

$$\tan\phi_s = \gamma_s/\kappa_f \quad , \quad \tan\phi_c = \gamma_c/\kappa_f \quad , \tag{2.3.9}$$

and the dispersion relation

$$\kappa_f h - \phi_s - \phi_c = \nu\pi \quad , \tag{2.3.10}$$

where the mode label ν is an integer. This is in agreement with the dispersion relation obtained from the zig-zag wave picture (Sect. 2.1). We also get a relation between the peak fields E_s, E_f and E_c of the form

$$E_f^2(n_f^2 - N^2) = E_s^2(n_f^2 - n_s^2) = E_c^2(n_f^2 - n_c^2) \quad , \tag{2.3.11}$$

where $N = \beta/k$ is the effective refractive index.

In the above form, the modes are not normalized for power. We calculate the power P carried by a mode per unit guide width as follows

$$P = -2 \int\limits_{-\infty}^{+\infty} dx \, E_y H_x = \frac{2\beta}{\omega\mu} \int\limits_{-\infty}^{+\infty} dx \, E_y^2$$

$$= N\sqrt{\frac{\varepsilon_0}{\mu_0}} E_{\mathrm{f}}^2 \cdot h_{\mathrm{eff}} = E_{\mathrm{f}} H_{\mathrm{f}} \cdot h_{\mathrm{eff}} \quad , \tag{2.3.12}$$

where

$$h_{\mathrm{eff}} = h + \frac{1}{\gamma_{\mathrm{s}}} + \frac{1}{\gamma_{\mathrm{c}}} \tag{2.3.13}$$

is the effective thickness of the waveguide, as discussed in Sect. 2.1.4.

For the substrate radiation modes, the field distribution is

$$
\begin{aligned}
E_y &= E_{\mathrm{c}} \exp[-\gamma_{\mathrm{c}}(x - h)] && \text{for} \quad h < x \quad , \\
E_y &= E_{\mathrm{f}} \cos[\kappa_{\mathrm{f}}(x - h) + \phi_{\mathrm{c}}] && \text{for} \quad 0 < x < h \quad , \\
E_y &= E_{\mathrm{s}} \cos(\kappa_{\mathrm{s}} x + \phi) && \text{for} \quad x < 0 \quad .
\end{aligned}
\tag{2.3.14}
$$

The boundary conditions require that

$$\tan\phi_{\mathrm{c}} = \gamma_{\mathrm{c}}/\kappa_{\mathrm{f}} \quad , \tag{2.3.15}$$

$$\kappa_{\mathrm{s}}\tan\phi = \kappa_{\mathrm{f}}\tan(\phi_{\mathrm{c}} - \kappa_{\mathrm{f}}h) \quad , \tag{2.3.16}$$

and

$$E_{\mathrm{f}}^2(n_{\mathrm{f}}^2 - N^2) = E_{\mathrm{c}}^2(n_{\mathrm{f}}^2 - n_{\mathrm{c}}^2) \quad , \tag{2.3.17}$$

$$E_{\mathrm{s}}^2 = E_{\mathrm{f}}^2\left[1 + \frac{n_{\mathrm{f}}^2 - n_{\mathrm{s}}^2}{n_{\mathrm{s}}^2 - N^2}\sin^2(\phi_{\mathrm{c}} - \kappa_{\mathrm{f}}h)\right] \quad . \tag{2.3.18}$$

Here we have no dispersion relation leading to discrete values for β, and we choose κ_{s} as the independent continuous variable.

The above field of a substrate mode is exactly the same as that created by a plane wave incident from the substrate side, with $\kappa_{\mathrm{s}} = kn_{\mathrm{s}}\cos\theta_{\mathrm{s}}$ used as a measure for the angle of incidence θ_{s}. The incident wave is refracted and partially reflected at the film-substrate boundary and totally reflected at the film-cover boundary. The phase shift incurred at that total reflection is $2\phi_{\mathrm{c}}$, and the phase-shift for reflection from the film-cover combination is 2ϕ. Interference between the incident and reflected waves creates the sinusoidal standing-wave patterns in film and substrate.

For the cross power $\overline{P}(\kappa_{\mathrm{s}}, \overline{\kappa}_{\mathrm{s}})$, needed to normalize the substrate modes, we calculate

$$\overline{P} = -2 \int\limits_{-\infty}^{+\infty} dx \, E_y(\kappa_{\mathrm{s}}) H_x(\overline{\kappa}_{\mathrm{s}})$$

$$= \frac{\pi\beta}{\omega\mu} E_{\mathrm{s}}^2 \delta(\kappa_{\mathrm{s}} - \overline{\kappa}_{\mathrm{s}}) = \pi E_{\mathrm{s}} H_{\mathrm{s}} \delta(\kappa_{\mathrm{s}} - \overline{\kappa}_{\mathrm{s}}) \quad . \tag{2.3.19}$$

The substrate-cover radiation modes are degenerate; we obtain two independent field solutions for each given κ_s. Great care must be exercised to select two solutions which are orthogonal to each other, as required by the mode-expansion formalism. A convenient choice are modal fields which become even or odd functions of $(x - h/2)$ in the limit of a "symmetric" waveguide where $n_s = n_c$. For simplicity, we also call these modes "even" and "odd" in the asymmetric case. Their fields are

for odd modes   for even modes

$E_y = E_c \sin[\kappa_c(x-h)+\phi_c];$ $E_y = \overline{E}_c \cos[\kappa_c(x-h)+\overline{\phi}_c],$ for $h < x,$

$E_y = E_f \sin(\kappa_f x - \phi);$ $E_y = \overline{E}_f \cos(\kappa_f x - \phi),$ for $0 < x < h,$

$E_y = E_s \sin(\kappa_s x - \phi_0);$ $E_y = \overline{E}_s \cos(\kappa_s x - \overline{\phi}_s),$ for $x < 0,$

$$\text{(2.3.20)}$$

where the same phase shift ϕ is used for both the even and odd modes. Using the boundary conditions one derives the relations for the phase shifts for the odd modes

$$\kappa_s \cot \phi_s = \kappa_f \cot \phi \quad , \tag{2.3.21}$$

$$\kappa_c \cot \phi_c = \kappa_f \cot(\kappa_f h - \phi) \tag{2.3.22}$$

and for the even modes

$$\kappa_s \tan \overline{\phi}_s = \kappa_f \tan \phi \quad , \tag{2.3.23}$$

$$\kappa_c \tan \overline{\phi}_c = \kappa_f \tan(\kappa_f h - \phi) \quad . \tag{2.3.24}$$

The connections between the peak fields are

$$E_s^2 = E_f^2 \left(\sin^2 \phi + \frac{\kappa_f^2}{\kappa_s^2} \cos^2 \phi \right) \quad , \tag{2.3.25}$$

$$E_c^2 = E_f^2 \left[\sin^2(\kappa_f h - \phi) + \frac{\kappa_f^2}{\kappa_c^2} \cos^2(\kappa_f h - \phi) \right] \quad , \tag{2.3.26}$$

$$\overline{E}_s^2 = \overline{E}_f^2 \left(\cos^2 \phi + \frac{\kappa_f^2}{\kappa_s^2} \sin^2 \phi \right) \quad , \tag{2.3.27}$$

$$\overline{E}_c^2 = \overline{E}_f^2 \left[\cos^2(\kappa_f h - \phi) + \frac{\kappa_f^2}{\kappa_c^2} \sin^2(\kappa_f h - \phi) \right] \quad . \tag{2.3.28}$$

For the cross power between even or between odd modes we calculate

$$\overline{P} = \frac{\pi \beta}{\omega \mu} [E_s^2 \delta(\kappa_s - \overline{\kappa}_s) + E_c^2 \delta(\kappa_c - \overline{\kappa}_c)] \quad . \tag{2.3.29}$$

When needed, the second delta function in this expression can be rewritten as

$$\delta(\kappa_c - \overline{\kappa}_c) = (\kappa_c/\kappa_s)\delta(\kappa_s - \overline{\kappa}_s) \quad , \tag{2.3.30}$$

which follows from the fact that $\kappa_s^2 - \overline{\kappa}_s^2 = \kappa_c^2 - \overline{\kappa}_c^2$.

The cross-power between an even and an odd mode is

$$\overline{P} = \frac{\pi\beta}{\omega\mu}[E_s\overline{E}_s\delta(\kappa_s - \overline{\kappa}_s) + E_c\overline{E}_c\delta(\kappa_c - \overline{\kappa}_c)] \quad , \tag{2.3.31}$$

which is not necessarily zero when $\kappa_s = \overline{\kappa}_s$. As we postulate orthogonality between even and odd modes, we have to set

$$\kappa_s E_s\overline{E}_s + \kappa_c E_c\overline{E}_c = 0 \tag{2.3.32}$$

to make the cross-power vanish. After some manipulation, this condition can be rewritten as

$$\cos(\phi_s - \overline{\phi}_s) + \cos(\phi_c - \overline{\phi}_c) = 0 \quad , \tag{2.3.33}$$

which becomes a condition for the phase shift ϕ of the form

$$\tan 2\phi = \sin(2\kappa_f h) \Big/ \left[\cos 2\kappa_f h + \frac{\kappa_s}{\kappa_c}(1 - \kappa_f^2/\kappa_s^2)/(1 - \kappa_f^2/\kappa_c^2) \right] \tag{2.3.34}$$

For the symmetric guide we have $\kappa_s = \kappa_c$, and

$$\phi = \kappa_f h/2 \quad . \tag{2.3.35}$$

2.3.2 TM Modes

Section 2.2.3 gives the relations for the TM modes

$$E_y = H_x = H_z = 0 \quad , \tag{2.3.36}$$

$$E_x = (\beta/\omega\varepsilon)H_y \quad , \tag{2.3.37}$$

$$E_z = -(j/\omega\varepsilon)\partial H_y/\partial x \quad , \tag{2.3.38}$$

with the H_y component obeying the wave equation

$$\partial^2 H_y/\partial x^2 = (\beta^2 - n^2 k^2)H_y \quad . \tag{2.3.39}$$

The boundary conditions at $x = 0$ and $x = h$ demand the continuity of H_y (and thereby of εE_x) and of $n^{-2}\partial H_y/\partial x$ (and thereby of E_z).

The field solutions for the guided modes are

$$\begin{aligned}
H_y &= H_c \exp[-\gamma_c(x - h)] & \text{for} \quad h &< x \quad , \\
H_y &= H_f \cos(\kappa_f x - \phi_s) & \text{for} \quad 0 &< x < h \quad , \\
H_y &= H_s \exp(\gamma_s x) & \text{for} \quad x &< 0 \quad .
\end{aligned} \tag{2.3.40}$$

Using the boundary conditions, we obtain

$$\tan \phi_s = (n_f/n_s)^2 \gamma_s/\kappa_f \quad , \quad \tan \phi_c = (n_f/n_c)^2 \gamma_c/\kappa_f \quad , \qquad (2.3.41)$$

and also the now familiar dispersion relation

$$\kappa_f h - \phi_s - \phi_c = \nu \pi \quad , \qquad (2.3.42)$$

where ν is an integer. We also get a relation between the peak fields of the form

$$H_f^2(n_f^2 - N^2)/n_f^2 = H_s^2(n_f^2 - n_s^2)q_s/n_s^2 = H_c^2(n_f^2 - n_c^2)q_c/n_c^2 \quad , (2.3.43)$$

where the reduction factors q_s and q_c are defined as in Sect. 2.1 by

$$q_s = (N/n_f)^2 + (N/n_s)^2 - 1 \quad , \qquad (2.3.44)$$

$$q_c = (N/n_f)^2 + (N/n_c)^2 - 1 \quad . \qquad (2.3.45)$$

For the power per unit guide width carried by a mode we calculate

$$P = 2 \int_{-\infty}^{+\infty} dx \, E_x \, H_y = \frac{2\beta}{\omega \varepsilon_0} \int_{-\infty}^{+\infty} dx \, H_y^2/n^2$$

$$= N \sqrt{\frac{\mu_0}{c_0}} H_f^2 \cdot h_{\text{eff}}/n_f^2 = E_f H_f \cdot h_{\text{eff}} \qquad (2.3.46)$$

where the effective thickness for the TM modes is defined as

$$h_{\text{eff}} = h + \frac{1}{\gamma_s q_s} + \frac{1}{\gamma_c q_c} \quad . \qquad (2.3.47)$$

The fields of the TM substrate radiation modes are

$$\begin{aligned}
H_y &= H_c \exp[-\gamma_c(x - h)] \quad \text{for} \quad h < x \quad , \\
H_y &= H_f \cos[\kappa_f(x - h) + \phi_c] \quad \text{for} \quad 0 < x < h \quad , \\
H_y &= H_s \cos(\kappa_s x + \phi) \quad \text{for} \quad x < 0 \quad .
\end{aligned} \qquad (2.3.48)$$

The boundary conditions require that

$$\tan \phi_c = (n_f/n_c)^2 \gamma_c/\kappa_f \quad , \qquad (2.3.49)$$

$$(\kappa_s/n_s^2)\tan \phi = (\kappa_f/n_f^2)\tan(\phi_c - \kappa_f h) \quad , \qquad \text{and} \qquad (2.3.50)$$

$$H_f^2(n_f^2 - N^2)/n_f^2 = H_c^2(n_f^2 - n_c^2)q_c/n_c^2 \quad , \qquad (2.3.51)$$

$$H_s^2 = H_f^2\left[1 + (n_s^2/n_f^2 q_s)\frac{n_f^2 - n_s^2}{n_s^2 - N^2}\sin^2(\phi_c - \kappa_f h)\right] \quad . \qquad (2.3.52)$$

41

Comparing the above expression with the corresponding formulas for the TE modes, we note the appearance of the reduction factors q_s and q_c.

As in the TE-mode case, we use κ_s as the continuous-mode label. For the cross power $\overline{P}(\kappa_s, \overline{\kappa}_s)$ we calculate

$$\overline{P} = 2 \int_{-\infty}^{+\infty} dx\, E_x(\kappa_s) H_y(\overline{\kappa}_s)$$

$$= \frac{\pi\beta}{\omega\varepsilon_0 n_s^2} H_s^2 \delta(\kappa_s - \overline{\kappa}_s) = \pi E_s H_s \delta(\kappa_s - \overline{\kappa}_s) \quad . \qquad (2.3.53)$$

As in the TE case, the substrate-cover radiation modes are degenerate and we distinguish between "even" and "odd" modes. Their fields are

for odd modes | for even modes

$$H_y = H_c \sin[\kappa_c(x-h)+\phi_c]; \quad H_y = \overline{H}_c \cos[\kappa_c(x-h)+\overline{\phi}_c], \quad \text{for } h < x,$$

$$H_y = H_f \sin(\kappa_f x - \phi); \quad H_y = \overline{H}_f \cos(\kappa_f x - \phi), \quad \text{for } 0 < x < h,$$

$$H_y = H_s \sin(\kappa_s x - \phi_s); \quad H_y = \overline{H}_s \cos(\kappa_s x - \overline{\phi}_s), \quad \text{for } x < h,$$

$$(2.3.54)$$

where ϕ is the same for the even and odd modes. The relations between the phase shifts, as derived from the boundary conditions, are

$$(\kappa_s/n_s^2)\cot\phi_s = (\kappa_f/n_f^2)\cot\phi \quad , \qquad (2.3.55)$$

$$(\kappa_c/n_c^2)\cos\phi_c = (\kappa_f/n_f^2)\cot(\kappa_f h - \phi) \qquad (2.3.56)$$

for the odd modes, and

$$(\kappa_s/n_s^2)\tan\overline{\phi}_s = (\kappa_f/n_f^2)\tan\phi \quad , \qquad (2.3.57)$$

$$(\kappa_c/n_c^2)\tan\overline{\phi}_c = (\kappa_f/n_f^2)\tan(\kappa_f h - \phi) \qquad (2.3.58)$$

for the even modes. The peak fields are connected by

$$H_s^2 = H_f^2[\sin^2\phi + (n_s^2\kappa_f/n_f^2\kappa_s)^2\cos^2\phi] \quad , \qquad (2.3.59)$$

$$H_c^2 = H_f^2[\sin^2(\kappa_f h - \phi) + (n_c^2\kappa_f/n_f^2\kappa_c)^2\cos^2(\kappa_f h - \phi)] \quad , \qquad (2.3.60)$$

$$\overline{H}_s^2 = \overline{H}_f^2[\cos^2\phi + (n_s^2\kappa_f/n_f^2\kappa_s)^2\sin^2\phi] \quad , \qquad (2.3.61)$$

$$\overline{H}_c^2 = \overline{H}_f^2[\cos^2(\kappa_f h - \phi) + (n_c^2\kappa_f/n_f^2\kappa_c)^2\sin^2(\kappa_f h - \phi)] \quad . \qquad (2.3.62)$$

The cross-power between even or between odd modes of different κ_s is

$$\overline{P} = \frac{\pi\beta}{\omega\varepsilon_0}\left[\frac{1}{n_s^2}H_s^2\delta(\kappa_s - \overline{\kappa}_s) + \frac{1}{n_c^2}H_c^2\delta(\kappa_c - \overline{\kappa}_c)\right] \quad . \tag{2.3.63}$$

As in the TE case, we postulate orthogonality between the even and the odd TM radiation modes and, again, obtain the condition

$$\cos(\phi_s - \overline{\phi}_s) + \cos(\phi_c - \overline{\phi}_c) = 0 \quad . \tag{2.3.64}$$

This translates into a condition for ϕ which has the same form as (2.3.34) for the TE mode but with κ_f replaced by κ_f/n_f^2, κ_s by κ_s/n_s^2, and κ_c by κ_c/n_c^2.

2.3.3 Multilayer Slab Guides

The preceding subsections are dealing with the simplest slab waveguide structure which consists of only three layers, the substrate, film and cover layers. The subject of this subsection are multilayer slab guides consisting of more than three layers. Multilayer guides are employed for a variety of purposes. These include the use of buffer layers to separate electrodes from a guide, and the use of metal layers to serve as guided-wave polarization filters that separate the TE from the TM mode [2.41, 42]. Multilayers are also employed to tailor the waveguide dispersion, to obtain phase matching for guided-wave second-harmonic generation [2.43], and to achieve strong selective filtering of higher-order transverse modes [2.44]. Semiconductor laser technology uses the 5-layer guide of a separate-confinement heterostructure (SCH) laser to achieve separate confinement for the charge carriers and the photons, and the 4-layer guides of large optical cavity (LOC) lasers for designs allowing higher laser power [2.45, 46].

The analysis of multilayer guides [2.6, 11, 41, 45] proceeds along lines similar to the analysis of three-layer guides discussed earlier in this section. The key difference is that multilayers require the repeated application of boundary conditions at the layer interfaces. To give a systematic structure to this task, we adapt the matrix theory developed for the determination of the transmission and reflection properties of multilayer stacks [2.47, 48].

Figure 2.15 shows a sketch of the orientation of a multilayer guide, with the layers chosen perpendicular to the x-axis as before. Our outline of the

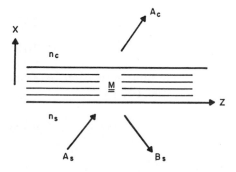

Fig. 2.15. Sketch of a multilayer stack waveguide with substrate index n_s and cover index n_c. The z-axis indicates the direction of mode propagation

analysis of this guide will concentrate on the TE modes. At the end of this section we will give a simple substitution rule that leads to results for the TM modes.

Multilayer Stack Theory

The theory of multilayer stacks [2.47, 48] starts with (2.19–21) and defines two field variables U and V by

$$U = E_y \quad , \quad V = \omega\mu H_z \quad , \qquad (2.3.65)$$

which describe the transverse variation of the optical field. These definitions are chosen because $U(x)$ and $V(x)$ are quantities that are continuous at the layer boundaries. From (2.19–21) we obtain the relations

$$U' = -jV \qquad (2.3.66)$$

$$V' = j(\beta^2 - n^2 k^2)U \quad , \qquad (2.3.67)$$

where the prime indicates differentiation with respect to x.

Both U and V obey the transverse wave equation

$$U'' = (\beta^2 - n^2 k^2)U \quad . \qquad (2.3.68)$$

U and V describe the transverse field distribution in a particular layer of constant refractive index n. The general solution of the wave equation in this layer is

$$U = A\exp(-j\kappa x) + B\exp(j\kappa x) \qquad (2.3.69)$$

$$V = \kappa[A\exp(-j\kappa x) - B\exp(j\kappa x)] \qquad (2.3.70)$$

where

$$\kappa^2 = n^2 k^2 - \beta^2 \qquad (2.3.71)$$

as before, see e.g. (2.3.1–3). The constants A and B can be replaced by the input values $U_0 = U(0)$ and $V_0 = V(0)$ at the input plane $x = 0$ of the layer. We have

$$A = \tfrac{1}{2}(U_0 + V_0/\kappa) \qquad (2.3.72)$$

$$B = \tfrac{1}{2}(U_0 - V_0/\kappa) \quad . \qquad (2.3.73)$$

A rearrangement of (2.3.69–73) leads to a simple matrix relation between the output quantities U, V and the input quantities U_0, V_0

$$\begin{pmatrix} U_0 \\ V_0 \end{pmatrix} = \mathbf{M}\begin{pmatrix} U \\ V \end{pmatrix} \quad , \qquad (2.3.74)$$

44

where the pairs (U_0, V_0) and (U, V) have been written as vectors, and $\mathbf{M}$ is the *characteristic matrix* of the layer. It has the form

$$\mathbf{M} = \begin{vmatrix} \cos(\kappa x) & (j/\kappa)\sin(\kappa x) \\ j\kappa\sin(\kappa x) & \cos(\kappa x) \end{vmatrix} \quad . \tag{2.3.75}$$

Note that $\det \mathbf{M} = 1$.

Next, consider a stack of n layers sandwiched between substrate and cover, as sketched in Fig. 2.15. We label the layers starting from the substrate. The layer thicknesses are h_i and the layer indices are n_i, where $i = 1$ to n. The *output* field variables for each layer are U_i and V_i. The characteristic matrices at the layers are

$$\mathbf{M}_i = \begin{vmatrix} \cos(\kappa_i h_i) & (j/\kappa_i)\sin(\kappa_i h_i) \\ j\kappa_i\sin(\kappa_i h_i) & \cos(\kappa_i h_i) \end{vmatrix} \quad , \tag{2.3.76}$$

where

$$\kappa_i^2 = n_i^2 k^2 - \beta^2 \quad . \tag{2.3.77}$$

The corresponding field variables are related by

$$\begin{pmatrix} U_{i-1} \\ V_{i-1} \end{pmatrix} = \mathbf{M}_i \begin{pmatrix} U_i \\ V_i \end{pmatrix} \quad . \tag{2.3.78}$$

Using matrix multiplication, we obtain a simple relation between the input variables U_0, V_0 at the substrate and the output variables U_n, V_n at the cover

$$\begin{pmatrix} U_0 \\ V_0 \end{pmatrix} = \mathbf{M} \begin{pmatrix} U_n \\ V_n \end{pmatrix} \quad , \tag{2.3.79}$$

where $\mathbf{M}$ is the *characteristic matrix* of the stack. It is given by the product of the individual layer matrices

$$\mathbf{M} \equiv \begin{vmatrix} m_{11} & m_{12} \\ m_{21} & m_{22} \end{vmatrix} = \mathbf{M}_1 \cdot \mathbf{M}_2 \cdot \mathbf{M}_3 \ \ldots \ \mathbf{M}_n \quad , \tag{2.3.80}$$

where m_{11}, m_{12} etc. are the matrix elements which we will use below.

Reflection and Transmission Coefficients

Once the characteristic matrix of a multilayer stack (such as an interference filter) is known, one can easily determine the reflection and transmission coefficients for light incident on the stack. We refer, again, to Fig. 2.15 and assume that the light is incident from the substrate side. This light is described by (2.3.69 and 70) with

$$\kappa_s^2 = n_s^2 k^2 - \beta^2 = -\gamma_s^2 \quad , \tag{2.3.81}$$

an incident amplitude A_s, and a reflected amplitude B_s.

The input variables are, therefore

$$U_0 = A_s + B_s \quad , \quad V_0 = \kappa_s(A_s - B_s) \quad . \tag{2.3.82}$$

On the cover side, we have light with a transmitted amplitude A_c, and the corresponding output variables

$$U_n = A_c \quad , \quad V_n = \kappa_c A_c \tag{2.3.83}$$

where

$$\kappa_c^2 = n_c^2 k^2 - \beta^2 = -\gamma_c^2 \quad . \tag{2.3.84}$$

Using the matrix relation (2.3.79), the amplitude transmission coefficient $t = A_c/A_s$ and the amplitude reflection coefficient $r = B_s/A_s$ can be expressed in terms of the elements of the characteristic stack matrix in the form

$$t = 2\kappa_s/(\kappa_s m_{11} + \kappa_c m_{22} + \kappa_s \kappa_c m_{12} + m_{21}) \quad , \tag{2.3.85}$$

$$r = (\kappa_s m_{11} - \kappa_c m_{22} + \kappa_s \kappa_c m_{12} - m_{21})t/2\kappa_s \quad . \tag{2.3.86}$$

This is a well known result, useful in the analysis of multilayer interference filters, antireflection coatings, and high-reflectivity mirrors.

Dispersion Relation of Multilayer Slab Waveguide

The transmission problem considered above is closely related to the problem of determining the dispersion relation for a mode guided by the multilayer structure. The key difference is that there is no incident light in the case of a mode where the optical field in the substrate and cover layers consists of evanescent waves. The corresponding fields are written in the form

$$U = A \exp(\gamma x) + B \exp(-\gamma x) \quad , \tag{2.3.87}$$

$$V = j\gamma[A \exp(\gamma x) - B \exp(-\gamma x)] \quad , \tag{2.3.88}$$

which obey the transverse wave equation (2.3.68). Instead of the transverse propagation constants κ_s and κ_c, we must now use transverse decay constants γ_s and γ_c, which are already defined in (2.3.81 and 84). We postulate that the fields should decay away from the multilayer, and obtain for the input and output field variables

$$U_0 = A_s \quad , \quad V_0 = j\gamma_s A_s \quad ; \quad V_n = B_c \quad , \quad V_n = -j\gamma_c B_c \quad . \tag{2.3.89}$$

Inserting this into the matrix relation (2.3.79) yields

$$A_s = (m_{11} - j\gamma_c m_{12})B_c \quad ,$$
$$j\gamma_s A_s = (m_{21} - j\gamma_c m_{22})B_c \quad . \tag{2.3.90}$$

Dividing these two relations leads to the desired dispersion relation for a multilayer slab guide

$$j(\gamma_s m_{11} + \gamma_c m_{22}) = m_{21} - \gamma_s \gamma_c m_{12} \quad .$$ (2.3.91)

It is expressed in terms of the decay constants γ_s and γ_c, and the elements of the characteristic matrix of the stack. The relation establishes the connection between the frequency $\omega = kc$ of the light and the propagation constant β of the mode guided by the multilayer. It is valid for an arbitrary number of layers and can also be used for the approximate analysis of waveguides with graded index profiles. As a useful illustration we shall discuss below the examples of the four-layer and the symmetric five-layer waveguides.

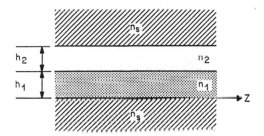

Fig. 2.16. Side view of a four-layer slab guide with equal substrate and cover index (n_s). The refractive indices of the two sandwiched films are n_1 and n_2, and the corresponding film heights are h_1 and h_2

Four-Layer Waveguides

Our first example is a four-layer slab guide such as that used in an LOC laser structure. A sketch of this guide is shown in Fig. 2.16 together with the key guide parameters. For simplicity, we choose a substrate and cover of equal index ($n_s = n_c$), which implies equal decay constants $\gamma_s = \gamma_c = \gamma$. For this case, the dispersion relation simplifies to

$$j\gamma(m_{11} + m_{22}) = m_{21} - \gamma^2 m_{12} \quad .$$ (2.3.92)

This relation implies the modal cut-off condition $m_{21} = 0$ which corresponds to the case of zero decay ($\gamma = 0$) of the fields in the substrate. The required elements of the characteristic matrix $\mathbf{M} = \mathbf{M}_1 \mathbf{M}_2$ of the two-layer stack are obtained from (2.3.76) by matrix multiplication with the result

$$m_{11} = \cos(\kappa_1 h_1)\cos(\kappa_2 h_2) - (\kappa_2/\kappa_1)\sin(\kappa_1 h_1)\sin(\kappa_2 h_2) \quad ,$$
$$m_{22} = \cos(\kappa_1 h_1)\cos(\kappa_2 h_2) - (\kappa_1/\kappa_2)\sin(\kappa_1 h_1)\sin(\kappa_2 h_2) \quad ,$$ (2.3.93)

$$m_{12} = (j/\kappa_1)\sin(\kappa_1 h_1)\cos(\kappa_2 h_2) + (j/\kappa_2)\cos(\kappa_1 h_1)\sin(\kappa_2 h_2) \quad ,$$
$$m_{21} = j\kappa_1\sin(\kappa_1 h_1)\cos(\kappa_2 h_2) + j\kappa_2\cos(\kappa_1 h_1)\sin(\kappa_2 h_2) \quad .$$

After inserting these elements into (2.3.92) we can rearrange the dispersion relation in the factored form

$$[\cos(\kappa_1 h_1) - (\kappa_1/\gamma)\sin(\kappa_1 h_1)][\cos(\kappa_2 h_2) + (\gamma/\kappa_2)\sin(\kappa_2 h_2)]$$
$$= -[\cos(\kappa_1 h_1) + (\gamma/\kappa_1)\sin(\kappa_1 h_1)]$$
$$\times [\cos(\kappa_2 h_2) - (\kappa_2/\gamma)\sin(\kappa_2 h_2)] \quad . \tag{2.3.94}$$

As a final step, we use familiar identities for the trigonometric functions to rewrite the dispersion relation of the four-layer slab guide in the form

$$\kappa_1 \tan[\kappa_1 h_1 - \tan^{-1}(\gamma/\kappa_1)] = -\kappa_2 \tan[\kappa_2 h_2 - \tan^{-1}(\gamma/\kappa_2)] \quad . \tag{2.3.95}$$

This agrees with the dispersion relation obtained by other methods [2.6, 41]. As a further check, we set $h_2 = 0$ and find the result to agree with the dispersion relation of the symmetric 3-layer slab guide as expected.

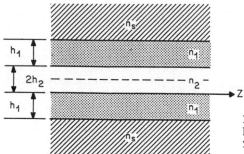

Fig. 2.17. Symmetric 5-layer guide. The heights of the sandwiched films are h_1, $2h_2$, and h_1, and their indices are n_1, n_2, and n_1 as shown

Symmetric Five-Layer Waveguides

Our second example is a symmetric 5-layer guide of the kind used in a separate confinement (SCH) laser. Its geometry is shown in Fig. 2.17. Substrate and cover indices are assumed to be equal ($n_s = n_c$), which implies equal decay constants $\gamma_s = \gamma_c = \gamma$. To exploit the symmetry of the guide, we pretend that the inner three layers are made up of two pairs of layers with thickness values h_1, h_2 and h_2, h_1, respectively. The dashed line in the figure indicates where the fictitious two-layer pairs meet. The corresponding layer matrices are the pairs $\mathbf{M}_1$, $\mathbf{M}_2$ and $\mathbf{M}_2$, $\mathbf{M}_1$. The characteristic matrix of the 3-layer stack is obtained by matrix multiplication

$$\mathbf{M} = \begin{vmatrix} M_{11} & M_{12} \\ M_{21} & M_{22} \end{vmatrix} = \mathbf{M}_1 \mathbf{M}_2 \mathbf{M}_2 \mathbf{M}_1 \quad , \tag{2.3.96}$$

where M_{11}, M_{12} etc. are the matrix elements.

The matrices of the layer pairs are

$$\mathbf{M}_1 \mathbf{M}_2 = \begin{vmatrix} m_{11} & m_{12} \\ m_{21} & m_{22} \end{vmatrix} \quad , \quad \mathbf{M}_2 \mathbf{M}_1 = \begin{vmatrix} m_{22} & m_{12} \\ m_{21} & m_{11} \end{vmatrix} \quad , \tag{2.3.97}$$

where we use the lower case to distinguish the matrix elements of the pairs

$(m_{11}, m_{12}$ etc.) from the matrix elements of the 3-layer stack. Expressions for the elements m_{11}, m_{12} etc. are given by (2.3.93). Inspecting these expressions, we find that the matrices of the two-layer pairs are related by a reversal of the elements m_{11} and m_{22}, as reflected in the equations above. Performing the matrix multiplication, we find the following relations

$$M_{11} = M_{22} = m_{11}m_{22} + m_{12}m_{21} \quad ,$$

$$M_{12} = 2m_{11}m_{12} \quad , \quad M_{21} = 2m_{21}m_{22} \quad . \tag{2.3.98}$$

This contains the relation $M_{11} = M_{22}$ which we expect for a symmetric stack.

With the latter simplification, the dispersion relation of the 5-layer guide becomes

$$2j\gamma M_{11} = M_{21} - \gamma^2 M_{12} \quad . \tag{2.3.99}$$

After inserting (2.3.98), the dispersion relation can be rewritten in the factored form

$$(m_{21} - j\gamma m_{11})(m_{22} - j\gamma m_{12}) = 0 \quad . \tag{2.3.100}$$

This means that we have the two dispersion relations

$$m_{21} = j\gamma m_{11} \quad , \quad \text{and} \tag{2.3.101}$$

$$m_{22} = j\gamma m_{12} \tag{2.3.102}$$

Because $\det (\mathbf{M}_1 \mathbf{M}_2) = 1$, the two relations cannot be simultaneously satisfied for the same ω and β values. They correspond to different sets of modes. Closer inspection shows that these sets are the even-order and odd-order modes.

Equation (2.3.101) corresponds to the even-order modes. If we insert the matrix elements from (2.3.93) and sort the terms with common functions of $(\kappa_2 h_2)$, we obtain

$$\kappa_2 \sin(\kappa_2 h_2) \cdot [\cos(\kappa_1 h_1) + (\gamma/\kappa_1) \sin(\kappa_1 h_1)]$$
$$= \kappa_1 \cos(\kappa_2 h_2)[(\gamma/\kappa_1) \cos(\kappa_1 h_1) - \sin(\kappa_1 h_1)] \quad . \tag{2.3.103}$$

Note that sorting for common functions of $(\kappa_1 h_1)$ yields the dispersion relation given in [2.45]. However, we proceed with (2.3.103), apply well known identities for trigonometric functions and arrive at the result

$$\kappa_2 \tan(\kappa_2 h_2) = -\kappa_1 \tan[\kappa_1 h_1 - \tan^{-1}(\gamma/\kappa_1)] \quad . \tag{2.3.104}$$

This agrees with the dispersion relation for even-order modes obtained by a different method [2.6].

For the odd-order modes we start with (2.3.102), proceed along similar lines as before, and obtain the dispersion relation

$$\kappa_2 \cot(\kappa_2 h_2) = \kappa_1 \tan[\kappa_1 h_1 - \tan^{-1}(\gamma/\kappa_1)] \quad . \tag{2.3.105}$$

TM Modes

The preceding discussion of multilayer guides is valid for the TE modes. We can modify the TE results to obtain results for the TM modes with the aid of a simple substitution.

For TM modes we have $E_y = H_z = H_x = 0$. The analysis starts with (2.2.24–26) and follows a parallel path to the TE analysis. One defines the field variables

$$U = H_y \quad , \quad V = \omega\varepsilon_0 E_z \quad , \tag{2.3.106}$$

and obtains the relations

$$U' = \mathrm{j}n^2 V \quad , \quad V' = \mathrm{j}(k^2 - \beta^2/n^2)U \quad . \tag{2.3.107}$$

Together, these lead to a transverse wave equation with the general solution

$$U = A\exp(-\mathrm{j}\kappa x) + B\exp(\mathrm{j}\kappa x) \quad ,$$
$$V = -(\kappa/n^2)[A\exp(-\mathrm{j}\kappa x) - B\exp(\mathrm{j}\kappa x)] \quad . \tag{2.3.108}$$

This already gives us the key for the required substitution. A comparison with equations (2.3.69, 70) shows that the general TM solution is the same as the general TE solution if we substitute

$$\mathrm{TE} \rightarrow \mathrm{TM} \quad , \quad \kappa \rightarrow -(\kappa/n^2) \quad , \tag{2.3.109}$$

and similarly for γ.

Note, that this substitution should *not* be applied to the phase terms (κx), $(\kappa_1 h_1)$ etc., as these are identical in the two general solutions.

The substitution can be applied to all results that follow from the general solution, i.e. almost all results of interest. This includes the formulas for the reflection and transmission coefficients, and, particularly, the dispersion relation. For TM modes, the latter assumes the form

$$-\mathrm{j}(m_{11}\gamma_\mathrm{s}/n_\mathrm{s}^2 + m_{22}\gamma_\mathrm{c}/n_\mathrm{c}^2) = m_{21} - (\gamma_\mathrm{s}\gamma_\mathrm{c}/n_\mathrm{s}^2 n_\mathrm{c}^2)m_{12} \tag{2.3.110}$$

as a result of the substitution. If we apply the substitution to the characteristic matrix of a single layer, as given in (2.3.76), we obtain the TM result

$$\mathbf{M}_i = \begin{vmatrix} \cos(\kappa_i h_i) & -\mathrm{j}(n_i^2/\kappa_i)\sin(\kappa_i h_i) \\ -\mathrm{j}(\kappa_i/n_i^2)\sin(\kappa_i h_i) & \cos(\kappa_i h_i) \end{vmatrix} \quad . \tag{2.3.111}$$

In similar fashion, we can adapt the other TE results obtained above to the case of TM waves.

2.4 Planar Guides with Graded-Index Profiles

Several fabrication processes, in particular diffusion and ion implantation, lead to dielectric waveguide layers with graded-index profiles where the refractive index $n(x)$ varies gradually over the cross-section of the guide. We list in this section the known mode solutions for 3 symmetric index profiles of different shapes, namely, the parabolic, the "$1/\cosh^2$", and the exponential profiles. We also discuss graded profiles with one sharp index discontinuity which correspond to the practical case of diffused guides with a sharp index step at the film-air boundary; and, finally, we briefly mention the application of the WKB method to profiles with a slow index gradation.

The discussion focuses on the TE modes governed by the wave equation (2.2.22) for the E_y component

$$d^2 F_y/dx^2 = (\beta^2 - n^2 k^2)E_y \quad .\tag{2.4.1}$$

The exact solutions available for these modes can be adapted to serve as approximations for the TM modes if the gradient of $n(x)$ is small enough.

The wave equation (2.4.1) has the same form as the Schrödinger equation of quantum mechanics, with $N^2 = \beta^2/k^2$ essentially corresponding to energy level and $n^2(x)$ to the potential energy well; hence, we can draw on the extensive literature of quantum mechanics both for specific solutions and for methods of analysis.

2.4.1 The Parabolic Profile (Harmonic Oscillator)

As sketched in Fig. 2.18, one can use the parabolic profile

$$n^2(x) = n_f^2(1 - x^2/x_0^2)\tag{2.4.2}$$

to approximate the actual index variation of a practical waveguide (shown dashed) near the guide axis ($x = 0$) where $n(0) = n_f$. For small x, we can write this as

$$n(x) \approx n_f(1 - \tfrac{1}{2}x^2/x_0^2) \quad .\tag{2.4.3}$$

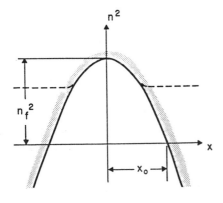

Fig. 2.18. Parabolic index profile (*solid curve*) and the practical guide profile (*dashed curve*) which is approximated by it

The index profile of (2.4.2) corresponds to the potential well of the harmonic oscillator (see, e.g., [2.29]), and the solutions of the wave equation are

$$E_y = \mathrm{H}_\nu(\sqrt{2}x/w)\exp(-x^2/w^2) \quad , \tag{2.4.4}$$

where the H_ν are the Hermite polynomials defined by

$$\mathrm{H}_\nu(x) = (-1)^\nu \exp(x^2)\frac{d^\nu}{dx^\nu}\exp(-x^2) \quad . \tag{2.4.5}$$

For the lowest orders we have

$$\begin{aligned}
\mathrm{H}_0(x) &= 1 \quad , \\
\mathrm{H}_1(x) &= 2x \quad , \\
\mathrm{H}_2(x) &= 4x^2 - 2 \quad , \\
\mathrm{H}_3(x) &= 8x^3 - 12x \quad .
\end{aligned} \tag{2.4.6}$$

The Hermite-Gaussian functions of (2.4.4) are also known as the standard description for the modes of laser beams and laser resonators [2.3, 49], where the parameter w is called the "beam radius". The latter is given by

$$w^2 = (\lambda x_0/\pi n_{\mathrm{f}}) \tag{2.4.7}$$

and indicates the degree of confinement of the fundamental mode. For the propagation constant β_ν and the effective index N_ν of a mode of order ν, one obtains

$$\beta_\nu^2 = n_{\mathrm{f}}^2 k^2 - (2\nu + 1)n_{\mathrm{f}}k/x_0 \quad , \tag{2.4.8}$$

$$N_\nu^2 = n_{\mathrm{f}}^2 - (\nu + \tfrac{1}{2})(n_{\mathrm{f}}\lambda/\pi x_0) \quad . \tag{2.4.9}$$

The parabolic profile corresponds to a closed well and predicts an infinite set of discrete modes. But as the order ν increases, the energy of a mode is spread out further from the guide axis, and eventually the distances x are so large that (2.4.2) can no longer be regarded as a good approximation for the actual guide profile.

2.4.2 The "1/cosh^2" Profile

The "1/cosh^2" profile is an open well described by

$$n^2(x) = n_{\mathrm{s}}^2 + 2n_{\mathrm{s}}\Delta n/\cosh^2(2x/h) \quad . \tag{2.4.10}$$

The corresponding potential is also employed in quantum mechanics [2.29] and is a special case of the *Pöschl-Teller* potential [2.50]. For small peak-index deviations Δn from the substrate index n_{s}, we have approximately

$$n(x) \approx n_{\mathrm{s}} + \Delta n/\cosh^2(2x/h) \quad , \tag{2.4.11}$$

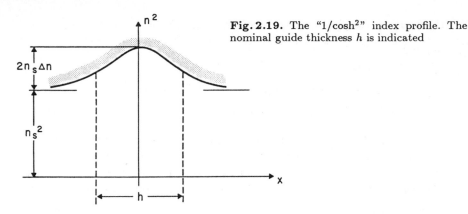

Fig. 2.19. The "1/cosh²" index profile. The nominal guide thickness h is indicated

indicating a guiding layer of thickness h, as sketched in Fig. 2.19. As in the case of the uniform slab guide discussed in Sect. 2.1, it is convenient to use a normalized thickness V, which we define as

$$V = kh\sqrt{2n_s\Delta n} \quad . \tag{2.4.12}$$

The solution of the wave equation (2.4.1) for the "1/cosh²" profile of (2.4.10) yields a very simple expression for the maximum mode number $s \ge \nu$ of guided (bound) modes supported by the waveguide, which is

$$s = \tfrac{1}{2}(\sqrt{1 + V^2} - 1) \quad . \tag{2.4.13}$$

For the propagation constants β_ν and the effective indices N_ν we obtain

$$\beta_\nu^2 = n_s^2 k^2 + 4(s - \nu)^2/h^2 \quad , \tag{2.4.14}$$

$$N_\nu^2 = n_s^2 + (s - \nu)^2(\lambda/\pi h)^2 \quad . \tag{2.4.15}$$

The modal field distribution is

$$E_y = u_\nu(2x/h)/\cosh^s(2x/h) \quad , \tag{2.4.16}$$

where the u_ν are hypergeometric functions [2.28]. For even mode number ν we have

$$\begin{aligned}
u_\nu = 1 &- \tfrac{1}{2}\nu(2s - \nu)\sinh^2(2x/h)/(1 \cdot 1!) \\
&+ \tfrac{1}{4}\nu(\nu - 2)(2s - \nu)(2s - \nu - 2)\sinh^4(2x/h)/(1 \cdot 3 \cdot 2!) + \ \ldots \quad ,
\end{aligned} \tag{2.4.17}$$

and for odd ν we get

$$\begin{aligned}
u_\nu = \sinh(2x/h)[1 &- \tfrac{1}{2}(\nu - 1)(2s - \nu - 1)\sinh^2(2x/h)/(3 \cdot 1!) \\
&+ \tfrac{1}{4}(\nu - 1)(\nu - 3)(2s - \nu - 1)(2s - \nu - 3) \\
&\times \sinh^4(2x/h)/(3 \cdot 5 \cdot 2!) + \ \ldots] \quad . \tag{2.4.18}
\end{aligned}$$

For the lower mode orders these functions simplify to

$$u_0 = 1 \quad ,$$
$$u_1 = \sinh(2x/h) \quad ,$$
$$u_2 = 1 - 2(s-1)\sinh^2(2x/h) \quad ,$$
$$u_3 = \sinh(2x/h)[1 - \tfrac{2}{3}(s-2)\sinh^2(2x/h)] \quad . \tag{2.4.19}$$

Nelson and *McKenna* [2.13] give a discussion of an asymmetric graded-index profile which is a generalization of the "$1/\cosh^2$" profile for cases where the indices of the substrate and the cover are unequal.

Another profile formally related to the "$1/\cosh^2$" potential has the form

$$n^2(x) = n_f^2[1 - \Delta\tan^2(x/x_0)] \quad . \tag{2.4.20}$$

This is also a special case of the *Pöschl-Teller* potential for which exact solutions are available, as discussed in detail by *Gordon* [2.51].

2.4.3 The Exponential Profile

The exponential profile is another case for which exact solutions are available [2.52]. We adapt these here to the symmetric profile sketched in Fig. 2.20 and described by

$$n^2(x) = n_s^2 + 2n_s\Delta n \exp(-2|x|/h) \quad , \tag{2.4.21}$$

which, for small Δn, can be approximated by

$$n(x) \approx n_s + \Delta n \exp(-2|x|/h) \quad . \tag{2.4.22}$$

Again we introduce a normalized layer thickness V by defining

$$V = kh\sqrt{2n_s\Delta n} \quad . \tag{2.4.23}$$

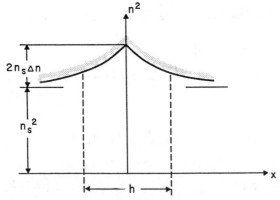

Fig. 2.20. The symmetric exponential profile and the corresponding nominal guide thickness h

54

The solutions for the modal fields of this profile can be expressed in terms of Bessel functions J_p of the first kind and of non-integral order p,

$$E_y = J_p(V \exp[-x/h]) \quad , \quad \text{for} \quad x > 0 \ ,$$
$$= J_p(V \exp[x/h]) \quad , \quad \text{for} \quad x < 0 \ . \tag{2.4.24}$$

The order p_ν is determined by matching the solutions at the boundary $x = 0$ for given V. For the even modes, we require that

$$J'_p(V) = 0 \quad , \tag{2.4.25}$$

and for the odd modes we have to demand

$$J_p(V) = 0 \quad . \tag{2.4.26}$$

Each of the two conditions yields approximately V/π discrete solutions for p_ν. In terms of p_ν, we can write for the propagation constant β_ν

$$\beta_\nu^2 = n_s^2 k^2 + p_\nu^2/h^2 \ , \tag{2.4.27}$$

and for the effective index N_ν

$$N_\nu^2 = n_s^2 + p_\nu^2/(kh)^2 \quad . \tag{2.4.28}$$

Figure 2.21 shows a plot of the quantity

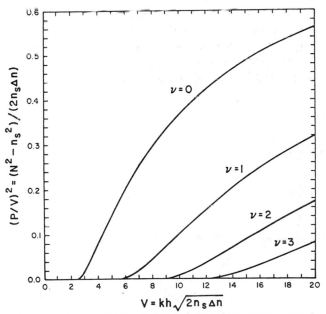

Fig. 2.21. Normalized ω-β diagram for planar guides with an exponential profile. Shown are the dispersion curves for the modes of odd order which correspond to the modes of guides with an asymmetric exponential profile of mode number $\nu = \nu_{\mathrm{asym}}$, as given by (2.4.30). (After [2.53])

$$(p_\nu/V)^2 = (N_\nu^2 - n_{\mathrm{s}}^2)/(2n_{\mathrm{s}}\Delta n) \approx (N_\nu - n_{\mathrm{s}})/\Delta n \qquad (2.4.29)$$

for the odd-order modes, which is taken from *Carruthers* et al. [2.53] and represents the results of a numerical solution of (2.4.26).

2.4.4 Index Profiles with Strong Asymmetry

In a class of practical waveguides, a layer of increased index is induced (e.g., by diffusion) in the substrate material. While the induced index change is relatively small, there is often a large step from the substrate index n_{s} to the cover index n_{c} which is usually that of air. The result is an index profile as that sketched in Fig. 2.22. We call such a profile "strongly asymmetric" because the corresponding asymmetry measure, defined in Sect. 2.1, assumes very large values. We shall discuss here a correspondence between strongly asymmetric profiles and symmetric profiles such as those discussed above, which can be used to obtain approximate mode solutions for the asymmetric profiles from solutions known for symmetric ones. This technique has been used by *Standley* and *Ramaswamy* [2.54] for the asymmetric parabolic profile, and essentially also by *Conwell* [2.52] and *Carruthers* et al. [2.53] for the asymmetric exponential profile. To obtain a first approximation, one assumes that the field in the cover region ($x < 0$) vanishes, as shown in the Fig. 2.22. The corresponding symmetric profile is indicated by the dashed line. The odd-order mode solutions of symmetric profiles have a zero field at $x = 0$ as sketched in the figure. For $x > 0$ we can therefore use these solutions for the asymmetric profile if we continue them with $E = 0$ for $x \leq 0$. The propagation constants corresponding to these modes can also be used unaltered for the asymmetric profile as a first approximation. As we only use the symmetric-profile modes of odd order, we renumber the modes

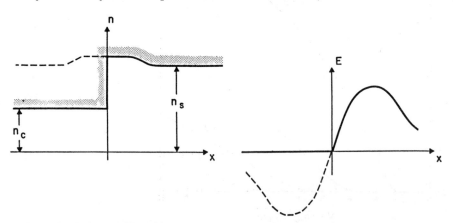

Fig. 2.22. Index profile with strong asymmetry (*solid curve*) and the associated symmetric profile (*dashed curve*). The corresponding approximate field distribution $E(x)$ is also shown

using the simple relation

$$2\nu_{\text{asym}} + 1 = \nu_{\text{sym}} \tag{2.4.30}$$

between the mode numbers ν_{asym} of the asymmetric profile and the mode numbers ν_{sym} of the corresponding symmetric profile.

Even though the actual field at the film-air interface is very small, it is not exactly zero. For some applications, such as the design of guided-wave filters with a surface corrugation, one needs to know the field values at the interface. *Haus* and *Schmidt* [2.55] have described how one can improve the above approximation and obtain estimates for these field values. They assumed evanescent fields in the cover region with the decay constant γ_c given as usual by

$$\gamma_c^2 = \beta^2 - n_c^2 k^2 \quad . \tag{2.4.31}$$

Matching the fields at the boundary $x = 0$ they obtain for the y-component of the electric field

$$E_y(0) = \frac{1}{\gamma_c} \left(\frac{dE_y}{dx} \right)_{x=0} \tag{2.4.32}$$

which relates the field at the surface $x = 0$ to the slope dE/dx. The latter is calculated from the approximate mode solutions discussed above.

Exact solutions for the asymmetric exponential profile were discussed by *Conwell* [2.52].

2.4.5 The WKB Method

The WKB method can be used to obtain approximate solutions of the wave equation (2.4.1) for the modes of profiles with slowly varying index $n(x)$. The method is thoroughly treated in the quantum mechanics literature (see, e.g., [2.29,56], and its application to dielectric waveguides has been discussed by *Gordon* [2.51], *Felsen* and *Marcuvitz* [2.57], and others). The method involves selection of a trial value for the propagation constant β (or the effective index N) of the waveguide and subsequent determination of the "turning points" x_1 and x_2 as indicated in Fig. 2.23 and described

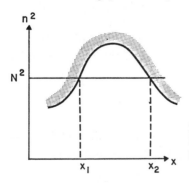

Fig. 2.23. Turning points x_1 and x_2 as determined by intersecting the line $n^2 = N^2$ with the index profile $n^2(x)$

mathematically by

$$n(x_i) = N = \beta/k \quad , \quad i = 1,2 \quad .$$
(2.4.33)

The values of β which obey the condition

$$\int_{x_1}^{x_2} dx \sqrt{n^2 k^2 - \beta^2} = \pi(\nu + \tfrac{1}{2})$$
(2.4.34)

with integer ν, are the propagation constants β_ν predicted by the WKB method. In terms of the effective index, this condition can be written as

$$\int_{x_1}^{x_2} dx \sqrt{n^2 - N^2} = \frac{\lambda}{2}(\nu + \tfrac{1}{2}) \quad .$$
(2.4.35)

For the modal fields, the WKB method predicts oscillatory field distributions in the region where $n(x) > N$ (i.e., $x_1 < x < x_2$) and exponentially decaying fields where $n(x) < N$ (i.e., $x > x_2$ and $x < x_1$) .

For the special case of the parabolic profile, the WKB predictions are known to agree exactly with the closed form solutions (2.4.8) available for this profile.

2.5 Channel Waveguides

The planar optical guides discussed in the preceding two sections provide no confinement of the light within the film plane, i.e. the y-z plane. There, confinement takes place in the x-dimension only. Optical channel guides can provide this additional confinement in the y dimension. Channel guides are used in many active and passive devices of integrated optics, including lasers, modulators, switches and directional couplers. The additional confinement can help to bring about desirable device characteristics such as savings in drive power and drive voltage. In addition, it is required for the design of single-mode structures that are compatible with single-mode fiber guides.

The following subsections will give an overview of channel guide geometries, discuss the vector wave equation suitable for the exact analysis of channel guides, and discuss methods of approximation including the separation of variables, the method of field shadows, and the effective index method.

2.5.1 Channel Guide Geometries

Figure 2.24 shows sketches of the x-y cross sections of six different types of channel guides. For simplicity, the figure shows abrupt transitions of the re-

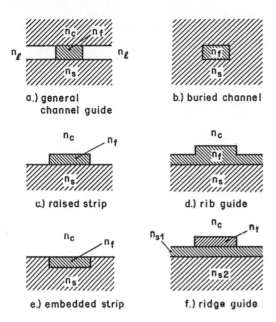

Fig. 2.24a–f. Cross-sections of six channel guide structures

a.) general channel guide

b.) buried channel

c.) raised strip

d.) rib guide

e.) embedded strip

f.) ridge guide

fractive index. However, it should be clear that fabrication techniques such as diffusion may produce guide cross sections with graded-index profiles. In all six examples the light is essentially confined to the film material with index n_f. We use the same notation as for slab waveguides, denoting the substrate index by n_s, and the cover or cap index by n_c. This is illustrated for a general guide geometry in Fig. 2.24a. The index of the lateral layer is n_l, which is different from the substrate and cap indices in many cases. Example (b) shows a buried guide with a uniform cladding. Example (c) illustrates the cross section of a raised strip guide. This can be fabricated by starting with a planar slab guide, masking the strip, and removing the surrounding film by sputtering or etching techniques. The rib guide (d) can be made with a similar technique but with incomplete removal of the surrounding film. Embedded strip guides (e) can be made by masked diffusion or ion implantation. A ridge guide (f) is fabricated by depositing (or etching) a strip of index n_f onto a planar waveguide. It has been suggested that both the rib guide and the ridge guide can be designed for relaxed resolution and edge roughness requirements. These two guides use a "propagating surround", i.e., lateral guide layers that allow at least one guided mode when they are operated as a planar slab waveguide.

2.5.2 The Vector Wave Equation

In a channel guide the refractive index $n = n(x, y)$ is a function of both transverse coordinates. As a result, the analysis of the waveguide modes becomes a more complex task than that for slab guides. A starting point for this analysis are the Maxwell's equations (2.2.6, 7) for the complex amplitudes, which we repeat here, for convenience,

$$\nabla \times \boldsymbol{E} = -j\omega\mu_0 \boldsymbol{H} \quad , \quad \nabla \times \boldsymbol{H} = j\omega\varepsilon \boldsymbol{E} \quad . \tag{2.5.1}$$

In the following we present a derivation which will yield the vector wave equation for the transverse fields of a channel guide [2.4, 6, 7]. As a first step, we take the curl of Maxwell's equations, which gives

$$\nabla \times \nabla \times \boldsymbol{E} = -j\omega\mu_0 \nabla \times \boldsymbol{H} = \omega^2 \varepsilon\mu_0 \boldsymbol{E} \quad , \tag{2.5.2}$$

$$\nabla \times \nabla \times \boldsymbol{H} = j\omega\nabla \times (\varepsilon\boldsymbol{E}) = \omega^2 \varepsilon\mu_0 \boldsymbol{H} + \nabla \ln\varepsilon \times (\nabla \times \boldsymbol{H}) \quad . \tag{2.5.3}$$

Here we have used the vector identity

$$\nabla \times (a\boldsymbol{b}) = a\nabla \times \boldsymbol{b} + (\nabla a) \times \boldsymbol{b} \quad , \tag{2.5.4}$$

as well as the relation

$$\nabla \ln\varepsilon = \nabla\varepsilon/\varepsilon \tag{2.5.5}$$

for the gradient of $\varepsilon(x, y)$. The divergence of (2.5.1) gives

$$\nabla \cdot \boldsymbol{H} = 0 \quad , \quad \nabla \cdot (\varepsilon\boldsymbol{E}) = 0 \quad . \tag{2.5.6}$$

The second of these divergence relations can be written in the more convenient form

$$\nabla \cdot \boldsymbol{E} = -\boldsymbol{E} \cdot \nabla \ln\varepsilon \quad . \tag{2.5.7}$$

We can use these relations, together with the vector identitiy

$$\nabla \times \nabla \times \boldsymbol{a} = -\nabla^2 \boldsymbol{a} + \nabla(\nabla \cdot \boldsymbol{a}) \quad , \tag{2.5.8}$$

to obtain wave equations for the fields of the guide which are of the form

$$\nabla^2 \boldsymbol{E} + \nabla(\boldsymbol{E} \cdot \nabla \ln\varepsilon) + \omega^2 \varepsilon\mu_0 \boldsymbol{E} = 0 \quad , \tag{2.5.9}$$

$$\nabla^2 \boldsymbol{H} + (\nabla \ln\varepsilon) \times (\nabla \times \boldsymbol{H}) + \omega^2 \varepsilon\mu_0 \boldsymbol{H} = 0 \quad . \tag{2.5.10}$$

We note that the transverse index gradients described by $\nabla\ln\varepsilon$ adds terms to the usual wave equation which tend to couple the components of the vector fields $\boldsymbol{E}$ and $\boldsymbol{H}$. Closer inspection shows that the longitudinal field components are decoupled from the transverse components. This becomes clear when we separate the fields into transverse ($\boldsymbol{E}_t$, $\boldsymbol{H}_t$) and longitudinal ($\boldsymbol{E}_z$, $\boldsymbol{H}_z$) components as in Sect. 2.2, and write the modal fields in the form

$$\boldsymbol{E} = (\boldsymbol{E}_t + \boldsymbol{E}_z)\exp(-j\beta z) \quad , \quad \boldsymbol{H} = (\boldsymbol{H}_t + \boldsymbol{H}_z)\exp(-j\beta z) \quad , \tag{2.5.11}$$

omitting the modal subscript ν for reasons of simplicity. With this we obtain the vector wave equations for the transverse fields

$$\nabla^2 \boldsymbol{E}_\mathrm{t} + \nabla(\boldsymbol{E}_\mathrm{t} \cdot \nabla \ln \varepsilon) + (\omega^2 \varepsilon \mu_0 - \beta^2)\boldsymbol{E}_\mathrm{t} = 0 \quad , \tag{2.5.12}$$

$$\nabla^2 \boldsymbol{H}_\mathrm{t} + (\nabla \ln \varepsilon) \times (\nabla \times \boldsymbol{H}_\mathrm{t}) + (\omega^2 \varepsilon \mu_0 - \beta^2)\boldsymbol{H}_\mathrm{t} = 0 \quad . \tag{2.5.13}$$

We note here that the quantities ε, $\boldsymbol{E}_\mathrm{t}$ and $\boldsymbol{H}_\mathrm{t}$ are independent of z. Therefore, we have $\nabla_\mathrm{t} \ln \varepsilon \equiv \nabla \ln \varepsilon$, $\nabla_\mathrm{t} \boldsymbol{E}_\mathrm{t} \equiv \nabla \boldsymbol{E}_\mathrm{t}$, and $\nabla_\mathrm{t} \times \boldsymbol{H}_\mathrm{t} \equiv \nabla \times \boldsymbol{H}_\mathrm{t}$, etc., where ∇_t is the transverse grad operator used earlier. We notice that the gradient term in the vector wave equations will generally couple the x- and y-components of the fields. However, no longitudinal field components appear in these equations.

In principle, we need to solve only one of the two vector wave equations for $\boldsymbol{E}_\mathrm{t}$ or $\boldsymbol{H}_\mathrm{t}$. The corresponding z-components follow from the divergence relations (2.5.6 or 7), which can be expressed as

$$j\beta E_z = \nabla \cdot \boldsymbol{E}_\mathrm{t} + \boldsymbol{E}_\mathrm{t} \cdot \nabla \ln \varepsilon \quad , \tag{2.5.14}$$

$$j\beta H_z = \nabla \cdot \boldsymbol{H}_\mathrm{t} \quad . \tag{2.5.15}$$

Once all components of $\boldsymbol{E}$ are determined, the $\boldsymbol{H}$ field follows from Maxwell's equations, and vice versa if $\boldsymbol{H}$ is found first.

2.5.3 Numerical Analysis

A variety of methods have been reported which are suitable for the numerical analysis of channel waveguides based directly on Maxwell's equation for the guide [2.6, 5]. For the case of buried rectangular guides, *Goell* [2.58] has employed cylindrical space harmonics to analyze guides with aspect (width to height) ratios between 1 and 2. *Schlosser* and *Unger* [2.59] have described a numerical method which is suited to rectangular guides with large aspect ratios. Additional numerical results for rectangular guides have been reported in [2.60, 61], and the use of variational techniques for the analysis of these guides has been presented in [2.34, 39]. For a recent detailed review of numerical methods the reader is referred to [2.81].

Figure 2.25 shows dispersion curves obtained by *Goell* for buried guides of index n_f surrounded by a cladding of index n_s. For small index differences $(n_\mathrm{f} - n_\mathrm{s})$, the results can be plotted in broadly applicable normalized form as shown.

Here, we have used the same normalizations as those used for planar slab waveguides in Sect. 2.1, where

$$b = (N^2 - n_\mathrm{s}^2)/(n_\mathrm{f}^2 - n_\mathrm{s}^2) \tag{2.5.16}$$

is the normalized guide index related to the effective index N, and

$$V = kh(n_\mathrm{f}^2 - n_\mathrm{s}^2)^{1/2} \tag{2.5.17}$$

is the normalized guide thickness (or height). The curves are labeled with the aspect ratio, i.e., the ratio between the guide width w and the guide

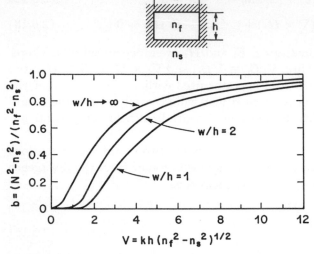

Fig. 2.25. Dispersion curves of a buried channel guide of height h and width w. The normalized guide index b is shown as a function of the normalized frequency (or guide thickness) V for the w/h ratios of 1, 2, and ∞. (After [2.58])

height h. The curve with the label $w/h = \infty$ corresponds to a planar slab waveguide, which is given for purposes of comparison. Note that, for given V, a reduction in the guide width corresponds to a decrease in β, i.e., an increase in the phase velocity ω/β of the guide. This is just as expected when one considers that for smaller widths one gets a larger portion of the modal field in the faster (lower index) cladding region.

2.5.4 Separation of Variables

In attempts to simplify the analysis one often attempts to separate the transverse variables x and y. In the case of the exact vector wave equations (2.5.12, 13) this attempt turns out to be a difficult if not impossible task. The scalar wave equation

$$\nabla^2 E_t + (n^2 k^2 - \beta^2) E_t = 0 \tag{2.5.18}$$

is often used for channel guides as a good approximation to the vector wave equation. In this case the separation

$$E_t(x, y) = X(x) Y(y) \tag{2.5.19}$$

succeeds if the refractive index squares, i.e., the permittivities, can be written in the additive form

$$n^2(x, y) = n_0^2 + n_x^2(x) + n_y^2(y) \quad . \tag{2.5.20}$$

Here it is required that X and n_x are functions of x only, and Y and n_y are functions of y only.

Under these conditions, the two-dimensional wave equations can be separated into two one-dimensional parts

$$\frac{d^2 X}{dx^2} + (k^2 n_x^2 - \beta_x^2)X = 0 \quad , \tag{2.5.21}$$

$$\frac{d^2 Y}{dy^2} + (k^2 n_y^2 - \beta_y^2)Y = 0 \quad . \tag{2.5.22}$$

These equations are now in a form which allow us to make use of known planar slab guide solutions, such as those given in Sect. 2.4. Once the solutions X, β_x, Y and β_y are determined, the propagation constant β of the channel guide is given by

$$\beta^2 = k^2 n_0^2 + \beta_x^2 + \beta_y^2 \quad . \tag{2.5.23}$$

The Parabolic Channel

A good example for the use of the separation of variables is the parabolic channel guide. This guide is characterized by an index profile of the form

$$n^2(x,y) = n_{\mathrm{f}}^2 \left(1 - \frac{x^2}{x_0^2} - \frac{y^2}{y_0^2}\right) \quad , \tag{2.5.24}$$

where x_0 and y_0 indicate the guide height and width, respectively. The modal properties of this guide follow directly from the properties of the parabolic slab guide discussed in Sect. 2.4.1. The fundamental mode is a Gaussian

$$E_{\mathrm{t}}(x,y) = E_0 \exp\left(-\frac{x^2}{w_x^2} - \frac{y^2}{w_y^2}\right) \quad . \tag{2.5.25}$$

Its beam radii w_x and w_y are determined by

$$w_x^2 = (\lambda x_0/\pi n_{\mathrm{f}}) \quad , \quad w_y^2 = (\lambda y_0/\pi n_{\mathrm{f}}) \quad . \tag{2.5.26}$$

As in (2.4.4), these parameters are also used for the Hermite-Gaussian functions which describe higher-order modes. The propagation constants of the modes of the guide are obtained from

$$\beta^2 = k^2 n_{\mathrm{f}}^2 - (2\nu + 1)n_{\mathrm{f}}k/x_0 - (2\mu + 1)n_{\mathrm{f}}k/y_0 \quad , \tag{2.5.27}$$

where ν and μ are the mode numbers in the x and y dimensions.

2.5.5 The Method of Field Shadows

Even when the scalar-wave approximation is made, very few practical index profiles are separable in the form (2.5.20). To allow the analysis of a

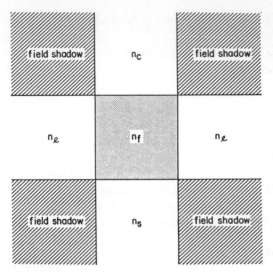

Fig. 2.26. Illustration of the method of field shadows showing the cross-section of a buried channel guide. The method ignores the fields in the shaded "shadow" areas

wider range of channel guides, *Marcatili* [2.62] has proposed a further approximation that permits that separation in many cases. The method is illustrated in Fig. 2.26 which shows the cross-section of a general channel guide. It calls on us to ignore the fields and the refractive indices in the shaded field shadow regions. This step will often result in separable index profiles. The method works well as long as the fields are well confined in the high index (n_f) region of the waveguide. Near cut-off the method is not applicable, because in this situation the fields penetrate into the shadow regions.

In the following, we illustrate the method for a rectangular buried channel guide. Numerical results for this guide are given in Sect. 2.5.3. Figure 2.27 shows the cross-sections of two planar slab gides, one with boundary planes perpendicular to the x-axis, the other with boundaries perpendicular to the y-axis. The n^2 values for the films and substrates are $n_f^2/2$ and $(n_s^2 - n_f^2/2)$ as shown. The composite guide shown is constructed by superimposing the permittivities (n^2 values) of the two slab guides. Because of the clever choice of the slab-guide indices, the superposition according to (2.5.20) will result in a channel guide of the index n_f and a substrate of index n_s as indicated. The index n_0 of the shadow region becomes

$$n_0^2 = 2n_s^2 - n_f^2 \quad . \tag{2.5.28}$$

The field-shadow method tells us to ignore the shadow region. Therefore, the index profile of the composite channel guide represents a good model for the buried channel guide and it allows the separation of variables. The modal field of the channel guide $E(x,y)$ is therefore given by the product

$$E(x,y) = X(x)Y(y)$$

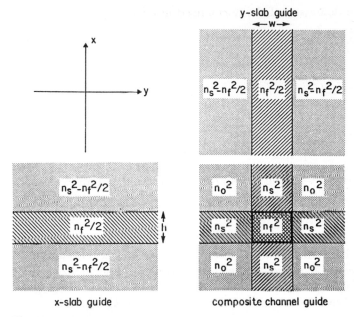

Fig. 2.27. Method of field shadows. The sketch shows the x-y cross-section of a composite guide made up by summing the permittivities (n^2) of an x-slab guide of height h and a y-slab guide of height w. The various n^2 values are indicated

of the fields $X(x)$ of the x-slab guide and $Y(y)$ of y-slab guide. The propagation constants β_x, β_y and effective indices N_x, N_y of the slab guides determine the corresponding quantities β and N of the channel guide via

$$\beta^2 = \beta_x^2 + \beta_y^2 \quad , \tag{2.5.29}$$

$$N^2 = N_x^2 + N_y^2 \quad . \tag{2.5.30}$$

In order to obtain quantitative answers, we refer to the slab guide results presented in Sect. 2.1.2, and particularly to the normalized dispersion curves of Fig. 2.8. Our goal is to cast the channel guide results into a similar normalized form to allow for easy scaling and broader applicability. The normalized guide-thickness values V_x and V_y of the slab guides contain the width w and height h of the channel guide. We have

$$V_x = kh\sqrt{n_f^2 - n_s^2} = V \quad , \tag{2.5.31}$$

$$V_y = kw\sqrt{n_f^2 - n_s^2} = \frac{w}{h}V \quad , \tag{2.5.32}$$

where V is the normalized height of the channel guide. Fig. 2.8 is used to determine the normalized guide indices $b_x(V_x)$ and $b_y(V_y)$ of the slab guides.

Now we invoke the definition of the normalized indices

$$b_x = \frac{N_x^2 - n_s^2 + n_f^2/2}{n_f^2 - n_s^2} \quad , \tag{2.5.33}$$

$$b_y = \frac{N_y^2 - n_s^2 + n_f^2/2}{n_f^2 - n_s^2} \quad , \tag{2.5.34}$$

$$b = \frac{N^2 - n_s^2}{n_f^2 - n_s^2} = \frac{N_x^2 + N_y^2 - n_s^2}{n_f^2 - n_s^2} \quad , \tag{2.5.35}$$

to derive the surprisingly simple relation for the normalized index b of the channel guide

$$b = b_x + b_y - 1 \quad . \tag{2.5.36}$$

This means that the normalized dispersion chart for slab guides shown in Fig. 2.8 can be used in a simple manner to determine the normalized guide index of a buried channel guide.

For the special case of a square channel where $b_x = b_y$ we find

$$b = 2b_x - 1 \quad . \tag{2.5.37}$$

We can use this relation to relabel the vertical axis of the slab-guide dispersion chart to obtain the dispersion chart for the square channel guide.

Improvements of the Method

The method of field shadows relies on two approximations: (1) the scalar approximation, and (2) the ignoring of the refractive index in the shadow regions.

Akiba and *Haus* [2.39] have obtained improved results by using a vector variational principle, where the above solutions are used as trial functions for the buried channel guide. *Kumar* et al. [2.63] have suggested that a perturbation calculation be used to correct for the error due to the permittivity differences in the shadow regions between the buried channel guide and the composite guide used above. The error in Δn^2 is

$$\Delta n^2 = n_s^2 - n_0^2 = n_f^2 - n_s^2 \quad . \tag{2.5.38}$$

The variation theorem (2.2.75) can be used to obtain a correction $\Delta\beta$ for the propagation constant due to this error. We get

$$\Delta\beta = \omega \int_{\text{shadows}} dx \, dy \, \varepsilon_0 \Delta n^2 |E|^2 / P \quad .$$

2.5.6 The Vector Perturbation Theorem

Channel-guide results obtained via the scalar-wave approximation can be improved with the vector perturbation theorem. This is quite useful in cases where small differences between two parameters play an important role. An example is the determination of the birefringence of a channel guide.

In order to derive this theorem, we assume that the scalar wave results $E_0(x, y)$ and β_0 are known. They obey the scalar wave equation (2.5.18)

$$\nabla^2 E_0 + (n^2 k^2 - \beta_0^2) E_0 = 0 \quad . \tag{2.5.39}$$

The exact solutions $E(x, y)$ and β must obey the vector wave equation (2.5.12)

$$\nabla^2 E + \nabla(E \cdot \nabla \ln \varepsilon) + (n^2 k^2 - \beta^2) E = 0 \quad . \tag{2.5.40}$$

If E_0 is a good approximation, we can write

$$E = E_0 + E_1 \quad , \tag{2.5.41}$$

and treat the quantities E_1, $\nabla \ln \varepsilon$, and $\Delta \beta^2$ as small perturbations, where

$$\Delta \beta^2 = \beta^2 - \beta_0^2 \quad . \tag{2.5.42}$$

Subtracting (2.5.39) from (2.5.40) and dropping perturbations of second order yields

$$\nabla^2 E_1 + \nabla(E_0 \cdot \nabla \ln \varepsilon) + \Delta \beta^2 E_0 + (n^2 k^2 - \beta^2) E_1 = 0 \tag{2.5.43}$$

We form the dot product of this with E_0 and the dot product of (2.5.39) with E_1 and subtract the results. We proceed by integrating over the guide cross-section and employing Green's theorem to obtain

$$\Delta \beta^2 = \int dx\, dy\, E_0 \cdot \nabla(E_0 \cdot \nabla \ln \varepsilon) / \int dx\, dy\, E_0^2 \quad . \tag{2.5.44}$$

This is the vector perturbation theorem. It gives us a correction to the propagation constant obtained from the scalar wave equation.

Channel Guide Birefringence

In the following we will illustrate the use of the above theorem for the determination of the birefringence of a channel guide. Waveguide birefringence indicates a difference in the phase velocities or propagation constants of two modes of different polarization. The scalar-wave approximation does not distinguish the polarization states of the field, and cannot, therefore, predict waveguide birefringence due to guide geometry. The scalar wave equation yields a modal field $\psi(x, y)$ and a propagation constant β. We use this to write down the following approximations for the vector components of the TE-like and TM-like modes:

$$\begin{aligned}
\text{TE}: \quad & E_x = 0 \quad , \quad E_y = \psi \quad , \\
\text{TM}: \quad & E_x = \psi \quad , \quad E_y = 0 \quad .
\end{aligned} \tag{2.5.45}$$

The associated propagation constants are given by

$$\beta_{TE}^2 = \beta_0^2 + \Delta\beta_{TE}^2 \quad ,$$
$$\beta_{TM}^2 = \beta_0^2 + \Delta\beta_{TM}^2 \quad , \tag{2.5.46}$$

where estimates for the correction terms $\Delta\beta_{TE}$ and $\Delta\beta_{TM}$ are calculated by means of the vector perturbation theorem. The birefringence of the guide can be expressed in the form

$$\beta_{TE}^2 - \beta_{TM}^2 = \Delta\beta_{TE}^2 - \Delta\beta_{TM}^2 \quad . \tag{2.5.47}$$

Inserting the components of (2.5.45) and integrating by parts allows us to simplify the theorem (2.5.44) to

$$\Delta\beta_{TE}^2 = \frac{1}{2} \int dx\, dy\, \psi^2 \frac{\partial^2}{\partial y^2} \ln\varepsilon \Big/ \int dx\, dy\, \psi^2 \quad ,$$

$$\Delta\beta_{TM}^2 = \frac{1}{2} \int dx\, dy\, \psi^2 \frac{\partial^2}{\partial x^2} \ln\varepsilon \Big/ \int dx\, dy\, \psi^2 \quad . \tag{2.5.48}$$

This leads to the following formula for the geometrical birefringence of the channel guide

$$\beta_{TE}^2 - \beta_{TM}^2 = \frac{1}{2} \int dx\, dy\, \psi^2 \left(\frac{\partial^2}{\partial x^2} - \frac{\partial^2}{\partial y^2} \right) \ln\varepsilon \Big/ \int dx\, dy\, \psi^2 \quad . \tag{2.5.49}$$

As an illustration for the use of this theorem we consider the parabolic channel guide discussed in Sect. 2.5.4. The permittivity profile of this guide is (2.5.24)

$$\varepsilon = \varepsilon_0 n_f^2 \left(1 - \frac{x^2}{x_0^2} - \frac{y^2}{y_0^2} \right) \quad .$$

From this we find approximations for the required derivatives of $\ln\varepsilon$ which are

$$\frac{\partial^2}{\partial x^2}\ln\varepsilon = -\frac{2}{x_0^2} \quad , \quad \frac{\partial^2}{\partial y^2}\ln\varepsilon = -\frac{2}{y_0^2} \quad . \tag{2.5.50}$$

Inserting this into the theorem (2.5.48) we obtain directly the result

$$\Delta\beta_{TE}^2 = -\frac{1}{y_0^2} \quad ; \quad \Delta\beta_{TM}^2 = -\frac{1}{x_0^2} \quad , \tag{2.5.51}$$

which agrees with the results obtained by other methods [2.4, 40, 75]. For the birefringence of the parabolic channel we get the estimate

$$\beta_{TE}^2 - \beta_{TM}^2 = \frac{1}{x_0^2} - \frac{1}{y_0^2} = \left(\frac{\lambda}{\pi n_f} \right)^2 \left(\frac{1}{w_x^4} - \frac{1}{w_y^4} \right) \quad . \tag{2.5.52}$$

2.5.7 The Effective-Index Method

Because of its immediate intuitive appeal, the effective-index approach has been used since the beginnings of integrated optics. It has helped in the understanding of structures such as guided-wave prisms, lenses and gratings. It has been proposed for the approximate analysis of channel guides by *Knox* and *Toulios* [2.64], and has produced results which were in close agreement with more exact computer results as well as experimental results for a considerable number of practical guide structures. This includes its application to ridge guides [2.65], buried channel guides [2.66] and diffused channel guides [2.66].

The effective-index approach starts with a birds-eye view of a planar film guide, with the film in the y-z plane. The guide is viewed in the x-direction. For a uniform planar guide, the viewer sees a uniform effective index N independent of y and z. When small variations are introduced either in the guide thickness or the refractive indices, the viewer will see an effective index pattern $N(y, z)$. When this pattern has the shape of a familiar bulk optical component, analogy arguments are used to understand its characteristics. For a channel guide the viewer sees the pattern reminiscent of a planar film guide with the film in the x-z plane. This is used to predict the modal fields and the propagation constants of the channel guide.

In the following, we discuss the application of the effective-index method to step-index channel guides. For a detailed treatment of diffused channel guides, the reader is referred to *Hocker* and *Burns* [2.66]. Our discussion aims to provide broad applicability while it uses the rib-guide structure as an illustrative example (and gives, in parenthesis, numerical results for the specific case of $\lambda = 0.8\,\mu$m, $n_f = 2.234$, $n_s = 2.214$, $n_c = 1$, $h = 1.8\,\mu$m, $l = 1\,\mu$m, $w = 2\,\mu$m, a rib guide made of Ti:LiNbO$_3$). The normalized guide parameters N, V and b, introduced in Sect. 2.1, will be used throughout to allow for easy scaling of the results. As shown in Fig. 2.28, we use the subscripts f and l to distinguish between the parameters of the channel and the lateral guides, respectively. The figure shows the cross-section (x, y) of the rib guide example together with the top view (z, y) of the channel guide. From this view we see a channel of width w and effective index N_f,

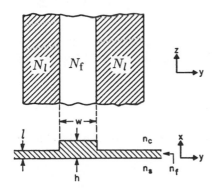

Fig. 2.28. Illustration of the effective-index method showing the top view and the cross section of a rib guide

and a lateral cladding of effective index N_l. The slab guide thickness h is larger than the lateral guide thickness l. Therefore $N_f > N_l$, which leads to the confinement of light in the channel.

The effective-index method proceeds with the following four steps:

1) Determine the normalized thickness V_f and V_l of the channel and lateral guides

$$V_f = kh\sqrt{n_f^2 - n_s^2} \quad , \tag{2.5.53}$$

$$V_l = kl\sqrt{n_f^2 - n_s^2} \quad . \tag{2.5.54}$$

(For our numerical example we get $V_f = 4.2$, $V_l = 2.3$).

2) Use available planar slab-guide results to determine the normalized guide indices b_f and b_l of the two guides (Fig. 2.8 gives $b_f = 0.65$, $b_l = 0.2$ for TE modes and asymmetry $a = \infty$). Determine the corresponding effective indices N_f and N_l from (2.1.17)

$$N_{f,l}^2 = n_s^2 + b_{f,l}(n_f^2 - n_s^2) \tag{2.5.55}$$

(for our example: $N_f = 2.227$, $N_l = 2.218$).

3) The method tells us to regard the slab guide suggested by the top view as equivalent to the actual channel guide. Determine the normalized width V_{eq} of the equivalent slab guide

$$V_{eq} = kw\sqrt{N_f^2 - N_l^2} = kw\sqrt{(n_f^2 - n_s^2)(b_f - b_l)} \quad . \tag{2.5.56}$$

Now, use available slab-guide results such as Fig. 2.8 to determine the normalized guide index b_{eq} of the equivalent guide ($V_{eq} = 3.14$, $b_{eq} = 0.64$ for the example).

4) Determine the effective index N_{eq} of the equivalent guide and postulate its equality to effective index N of the channel guide:

$$N \equiv N_{eq} = N_l^2 + b_{eq}(N_f^2 - N_l^2) \quad . \tag{2.5.57}$$

This concludes the procedure, providing a result for the effective index N and the propagation constant $\beta = kN$ of the channel guide ($N = 2.224$ in the example).

For purposes of easy comparisons and easy scaling it is often useful to cast the results in terms of the normalized channel guide index (2.5.16) which is given by

$$b = \frac{N^2 - n_s^2}{n_f^2 - n_s^2} = \frac{N_l^2 - n_s^2 + b_{eq}(N_f^2 - N_l^2)}{n_f^2 - n_s^2} \quad . \tag{2.5.58}$$

Table 2.4 presents formulas for the key parameters of the six channel guide structures shown in Fig. 2.24 as derived by use of the effective index

Table 2.4. Effective index parameters for channel guides

Channel structure	Guide height V_f, V_l	Eff. index N_f, N_l	$N_f^2 - N_l^2$	Channel guide index b
a) General	$V_f = kh\sqrt{n_f^2 - n_s^2}$ $V_l = kh\sqrt{n_l^2 - n_f^2}$	$N_f^2 = n_s^2 + b_f(n_f^2 - n_s^2)$ $N_l^2 = n_s^2 + b_l(n_l^2 - n_s^2)$	$b_f(n_f^2 - n_s^2) - b_l(n_l^2 - n_s^2)$	$b_f b_{eq} + b_l(1 - b_{eq})a_{ch}$
b) Buried	$V_f = kh\sqrt{n_f^2 - n_s^2}$	$N_f^2 = n_s^2 + b_f(n_f^2 - n_s^2)$ $N_l = n_s$	$b_f(n_f^2 - n_s^2)$	$b_f b_{eq}$
c) Raised	$V_f = kh\sqrt{n_f^2 - n_s^2}$	$N_f^2 = n_s^2 + b_f(n_f^2 - n_s^2)$ $N_l = n_c$	$(n_s^2 - n_c^2) + b_f(n_f^2 - n_s^2)$	$b_f b_{eq} - (1 - b_{eq})a$
d) Rib	$V_f = kh\sqrt{n_f^2 - n_s^2}$ $V_l = kl\sqrt{n_f^2 - n_s^2}$	$N_f^2 = n_s^2 + b_f(n_f^2 - n_s^2)$ $N_l^2 = n_s^2 + b_l(n_f^2 - n_s^2)$	$(b_f - b_l)(n_f^2 - n_s^2)$	$b_f b_{eq} + b_l(1 - b_{eq})$
e) Embedded	$V_f = kh\sqrt{n_f^2 - n_s^2}$	$N_f^2 = n_s^2 + b_f(n_f^2 - n_s^2)$ $N_l = n_s$	$b_f(n_f^2 - n_s^2)$	$b_f b_{eq}$
f) Ridge	$V_f = kh\sqrt{n_f^2 - n_s^2}$ $V_l = kl\sqrt{n_f^2 - n_s^2}$	$N_f^2 = n_{s1}^2 + b_f(n_f^2 - n_{s1}^2)$ $N_l^2 = n_{s2}^2 + b_l(n_{s1}^2 - n_{s2}^2)$	$(1 - b_l)(n_{s1}^2 - n_{s2}^2) + b_f(n_f^2 - n_{s1}^2)$	$b_{eq}(1 + b_f \cdot a_{ridge}) + b_l(1 - b_{eq})$

method. Here h always refers to the height of the channel, and l to the height of the lateral guide. The following comments refer to specifics in the treatment of the six structures (Table 2.4).

a) General Channel. The asymmetry measure a of (2.1.19)

$$a = \frac{n_s^2 - n_c^2}{n_f^2 - n_s^2} \tag{2.5.59}$$

is used for most structures to simplify the expressions and to determine the b parameters from the normalized plots of Fig. 2.8.

A similar simplification results from the introduction of a channel measure a_{ch} defined as

$$a_{ch} = \frac{n_l^2 - n_s^2}{n_f^2 - n_s^2} \quad . \tag{2.5.60}$$

Buried Heterostructure Guides. Buried heterostructures are frequently used for semiconductor junction lasers (Chap. 5). A good model for this channel guide is obtained by extending the height l of the lateral guide to infinity $(l \to \infty)$. In this limit the formulas for the general channel guide parameters simplify to

$$N_l = n_l \quad , \quad b_l = 1 \quad , \quad \text{and} \tag{2.5.61}$$

$$b = b_f b_{eq} + (1 - b_{eq}) \cdot a_{ch} \quad . \tag{2.5.62}$$

b) Buried Channel Guide. In the absence of a guiding structure, the effective index N_l of the lateral guide is set equal to the index of the bulk material $(N_l = n_s)$.

c) Raised Strip Guide. The field shadow argument is used in the lateral region: The substrate index is ignored and N_l is set equal to the bulk index adjacent to the channel $(N_l = n_c)$.

d) Rib Guide. See main text.

e) Embedded Strip Guide. The field shadow argument is used in the lateral region: The cover index is ignored and N_l is set equal to the bulk index adjacent to the channel $(N_l = n_c)$. With this approximation, the results are similar to those obtained for the buried channel. The only exception is the use of the asymmetry measure a in the determination of b_f. This is required to take account of the effect of the cladding index n_c.

f) Ridge Guide. The derivation assumes that $n_f > n_{s1} > n_{s2}$. The normalized channel guide index is defined in terms of the refractive indices of the lateral slab guide

$$b = \frac{N^2 - n_{s2}^2}{n_{s1}^2 - n_{s2}^2} \quad , \tag{2.5.63}$$

and simplicity is achieved by definition of the ridge measure

$$a_{\text{ridge}} = \frac{n_f^2 - n_{s1}^2}{n_{s1}^2 - n_{s2}^2} \quad . \tag{2.5.64}$$

A second assumption is that the second substrate (n_{s2}) is of negligible influence on the properties of the slab guide in the channel region. This assumption is not essential: results for four-layer slab guides can be used for better precision.

Numerical Comparisons for the Buried Channel

For the case of the buried channel, the numerical results of *Goell* shown in Fig. 2.25, allow a direct check of the effective index method. *Hocker* and *Burns* [2.66] have made such a comparison.[1] Figure 2.29 shows their results,

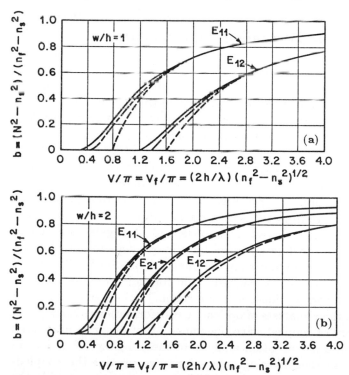

Fig. 2.29a,b. Normalized dispersion curves for a buried channel guide comparing the predictions of the numerical calculations (*dot-dashed lines*), of the effective index method (*solid lines*), and of the field-shadow method (*dashed lines*). Comparisons are shown for the aspect ratios of $w/h = 1$ and $w/h = 2$. (After [2.66])

---

[1] We are grateful to these authors for pointing out a misprint in their figures which is corrected here.

modified for our purposes. The figure shows the normalized dispersion curves $b(V = V_{\mathrm{f}})$ for aspect ratios of $w/h = 1$ and $w/h = 2$ for the case of small index differences. The solid curves show the prediction

$$b = b_{\mathrm{f}} \cdot b_{\mathrm{eq}} \qquad (2.5.65)$$

of the effective index method. The dot-dashed curves show *Goell*'s data. As a further comparison the figure shows also the prediction (2.5.36)

$$b = b_x + b_y - 1 \qquad (2.5.66)$$

of *Marcatili*'s shadow field method depicted by the dashed curves. Data are presented for the fundamental E_{11} mode as well as for the modes E_{12} and E_{21} of the next higher order. The effective index method is seen to give fairly good predictions even near cut-off. For larger aspect ratios we would expect even better predictions because effects at the channel edges are then of lesser importance.

2.6 Coupled-Mode Formalism and Periodic Waveguides

Many phenomena occurring in physics or engineering can be viewed as coupled-mode processes. Examples for this include the diffraction of x-rays in crystals [2.67], the directional couplers of microwave technology [2.68, 69], the energy exchange between electron beams and slow-wave structures in traveling-wave tubes [2.70, 71], and the scattering of light by acoustic waves and by hologram gratings [2.72]. The coupled-mode formalism has also proved a very powerful tool in integrated optics, where it has helped in the understanding and analysis of a large variety of important phenomena and devices. These range from the scattering loss due to waveguide irregularities and the behavior of grating couplers and corrugated waveguide filters to distributed feedback lasers, electro-optic or magneto-optic TE-to-TM mode converters and nonlinear optical interactions. *Snyder* [2.73] and *Marcuse* [2.2] have developed a coupled-mode formalism applicable to a large class of dielectric waveguides including channel guides and fibers, and they have used it in their analyses of optical fiber deformations. *Yariv* [2.74] has given a summary of coupled-wave phenomena that occur in integrated optics and presented a perturbation analysis of these for the TE modes in planar waveguides. In the following subsections, we present the coupled-mode formalism and its derivation in sufficient generality to provide the basis for the treatment of the various coupled-wave phenomena of interest and to permit its application to all practical waveguide structures including channel guides and fibers. We will first consider the excitation of modes by arbitrary sources in the guide, and then apply that formalism to waveguide deformations such as surface irregularities. This is followed by a listing of the standard coupled-wave solutions both for co-directional and contra-directional interactions, and finally we shall treat periodic waveguides and present results for the specific example of a planar waveguide with periodic surface corrugations.

2.6.1 Excitation of Waveguide Modes

We consider a distribution of sources exciting various waveguide modes and represent these sources by the complex amplitude $P(x, y, z)$ of the corresponding induced polarization vector. In the presence of these sources, the complex Maxwell's equations have the form

$$\nabla \times E = -\mathrm{j}\omega\mu H \quad , \tag{2.6.1}$$

$$\nabla \times H = \mathrm{j}\omega\varepsilon E + \mathrm{j}\omega P \quad . \tag{2.6.2}$$

The following steps are now quite analogous to those used in the derivation of the orthogonality relation in Sect. 2.2.5. We consider two different induced polarizations P_1 and P_2 and the fields caused by them, and obtain

$$\nabla \cdot (E_1 \times H_2^* + E_2^* \times H_1) = -\mathrm{j}\omega P_1 \cdot E_2^* + \mathrm{j}\omega P_2^* \cdot E_1 \quad , \tag{2.6.3}$$

which is essentially the Lorentz reciprocity relation. Now we set $P_2 = 0$ and identify the field 2 with a mode of the waveguide. As in Sect. 2.2.5, we integrate over the guide cross-section, use the divergence theorem, and find

$$\int\!\!\!\int_{-\infty}^{+\infty} dx\, dy\, \frac{\partial}{\partial z}(E_1 \times H_2^* + E_2^* \times H_1)_z = -\mathrm{j}\omega \int\!\!\!\int_{-\infty}^{+\infty} dx\, dy\, P_1 \cdot E_2^* \quad . \tag{2.6.4}$$

The next step is to expand the transverse components of the field 1 in terms of the modes of the guide according to (2.2.58, 59)

$$E_{1t} = \sum(a_\nu + b_\nu)E_{t\nu} \quad , \quad H_{1t} = \sum(a_\nu - b_\nu)H_{t\nu} \tag{2.6.5}$$

where we use the $\sum$ notation of Sect. 2.2.6. It should be noted that the coefficients $a_\nu(z)$ and $b_\nu(z)$ have to be regarded as functions of z in the present context. If we choose for the field 2 a forward running mode

$$E_2 = E_\mu \exp(-\mathrm{j}\beta_\mu z) \quad , \quad H_2 = H_\mu \exp(-\mathrm{j}\beta_\mu z) \quad , \tag{2.6.6}$$

insert these fields and the mode expansion into (2.6.4) and apply the orthogonality relation (2.2.51), we then find that the b_ν coefficients drop out and we obtain

$$a_\mu' + \mathrm{j}\beta_\mu a_\mu = -\mathrm{j}\omega \int\!\!\!\int_{-\infty}^{+\infty} dx\, dy\, P \cdot E_\mu^* \quad , \tag{2.6.7}$$

where the prime indicates differentiation with respect to z. Similarly we get

$$b'_\mu - \mathrm{j}\beta_\mu b_\mu = \mathrm{j}\omega \int\!\!\!\int\limits_{-\infty}^{+\infty} dx\, dy\, \boldsymbol{P} \cdot \boldsymbol{E}^*_{-\mu} \tag{2.6.8}$$

if we identify the field 2 with a backward running mode

$$E_2 = \boldsymbol{E}_{-\mu}\exp(\mathrm{j}\beta_\mu z) \quad , \quad H_2 = \boldsymbol{H}_{-\mu}\exp(\mathrm{j}\beta_\mu z) \quad . \tag{2.6.9}$$

The above formulas hold only for propagating modes (real β); for the evanescent modes we have to use the orthogonality relation (2.2.63). It is usually convenient to define the amplitudes $A_\mu(z)$ and $B_\mu(z)$ of the forward and backward propagating modes by

$$a_\mu = A_\mu\exp(-\mathrm{j}\beta_\mu z) \quad , \quad b_\mu = B_\mu\exp(\mathrm{j}\beta_\mu z) \quad . \tag{2.6.10}$$

The change of these amplitudes due to the presence of sources can then be expressed as

$$A'_\mu = -\mathrm{j}\omega \int\!\!\!\int\limits_{-\infty}^{+\infty} dx\, dy\, \boldsymbol{P} \cdot \boldsymbol{E}^*_\mu \exp(\mathrm{j}\beta_\mu z) \quad , \tag{2.6.11}$$

$$B'_\mu = \mathrm{j}\omega \int\!\!\!\int\limits_{-\infty}^{+\infty} dx\, dy\, \boldsymbol{P} \cdot \boldsymbol{E}^*_{-\mu} \exp(-\mathrm{j}\beta_\mu z) \quad . \tag{2.6.12}$$

As expected, there is no change in amplitude when no sources are present. It should be emphasized that the above relations are exact; no assumption about the smallness of the perturbation caused by the sources has been made.

The formalism allows for an arbitrary polarization $\boldsymbol{P}(x, y, z)$ which can be brought about by a variety of physical effects. A standard example is a scalar deformation of the waveguide which is represented by the difference $\Delta\varepsilon(x, y, z)$ of the actual dielectric constant from the nominal distribution $\varepsilon(x, y)$ and which results in an induced polarization

$$\boldsymbol{P} = \Delta\varepsilon\boldsymbol{E} \tag{2.6.13}$$

proportional to the field $\boldsymbol{E}$ in the guide. We shall pursue this case in more detail in the next subsection. To represent loss in the waveguide material, we can use an imaginary-valued $\Delta\varepsilon$. Anisotropy in the guide materials can be represented by a tensor perturbation leading to an induced polarization with the components

$$P_i = \Delta\varepsilon_{ij}E_j \quad , \tag{2.6.14}$$

where we have employed the standard tensor notation which assumes summation over repeated indices. The off-diagonal elements of $\Delta\varepsilon_{ij}$ can cause TE-to-TM mode conversion which has been treated in terms of the coupled-mode formalism by *Yariv* [2.74], and *Sosnowski* and *Boyd* [2.76]. Electro-optic (and other nonlinear optical) effects lead to a $\Delta\varepsilon_{ij}$ of the form

76

$$\Delta\varepsilon_{ij} = \varepsilon_0 X_{ijk} E_k \quad , \tag{2.6.15}$$

where X_{ijk} is the second-order nonlinear susceptibility and E_k a component of the applied electric field [2.77].

2.6.2 Waveguide Deformations

In this subsection, we discuss in more detail the application of the coupled-mode formalism to scalar waveguide deformations. These are represented by the difference $\Delta\varepsilon(x, y, z)$ in the dielectric constant and an induced polarization of the form given in (2.6.13). We stress again that, so far, the formalism is exact, and we will make no approximations in this subsection. The reader should pay attention to the particular way in which the z-components are handled; this is necessary because only the tangential field components are orthogonal and the mode expansion can be applied only to these components. With (2.6.5, 13) we have, therefore,

$$P_t = \Delta\varepsilon E_t = \Delta\varepsilon \sum (a_\nu + b_\nu) E_{t\nu} \quad . \tag{2.6.16}$$

For E_z we obtain from (2.6.2)

$$j\omega(\varepsilon + \Delta\varepsilon) E_z = \nabla_t \times H_t \quad , \tag{2.6.17}$$

which allows us to rewrite expressions for P_z in the following sequence

$$
\begin{aligned}
P_z &= \Delta\varepsilon E_z \\
&= \frac{\Delta\varepsilon}{\varepsilon + \Delta\varepsilon} \frac{1}{j\omega} \nabla_t \times H_t \\
&= \frac{\Delta\varepsilon}{\varepsilon + \Delta\varepsilon} \frac{1}{j\omega} \sum (a_\nu - b_\nu) \nabla_t \times H_{t\nu} \\
&= \frac{\Delta\varepsilon \cdot \varepsilon}{\varepsilon + \Delta\varepsilon} \sum (a_\nu - b_\nu) E_{z\nu} \quad ,
\end{aligned}
\tag{2.6.18}
$$

where we have used the mode expansion (2.6.5) and the modal Maxwell equation (2.2.16). We are now ready to insert the components of P into (2.6.11 and 12) with the result

$$
A'_\mu = -j\omega \int\!\!\!\int_{-\infty}^{+\infty} dx\, dy \sum \Big[(a_\nu + b_\nu) \Delta\varepsilon E_{t\nu} \cdot E^*_{t\mu}
$$

$$
+ (a_\nu - b_\nu) \frac{\Delta\varepsilon \cdot \varepsilon}{\varepsilon + \Delta\varepsilon} E_{z\nu} \cdot E^*_{z\mu} \Big] \exp(j\beta_\mu z) \quad , \tag{2.6.19}
$$

$$
B'_\mu = j\omega \int\!\!\!\int_{-\infty}^{+\infty} dx\, dy \sum \Big[(a_\nu + b_\nu) \Delta\varepsilon E_{t\mu} \cdot E^*_{t\mu}
$$

$$
- (a_\nu - b_\nu) \frac{\Delta\varepsilon \cdot \varepsilon}{\varepsilon + \Delta\varepsilon} E_{z\nu} \cdot E^*_{z\mu} \Big] \exp(-j\beta_\mu z) \quad , \tag{2.6.20}
$$

where we have used the symmetry relations (2.2.31) to express the mode distribution $E_{-\mu}(x, y)$ in terms of the components of the field $E_\mu(x, y)$ of the corresponding forward running mode. To simplify these expressions, we introduce tangential and longitudinal coupling coefficients $K^t_{\nu\mu}(z)$ and $K^z_{\nu\mu}(z)$ defined by

$$K^t_{\nu\mu} = \omega \int\!\!\!\int_{-\infty}^{+\infty} dx\, dy\, \Delta\varepsilon\, E_{t\nu} \cdot E^*_{t\mu} \quad , \tag{2.6.21}$$

$$K^z_{\nu\mu} = \omega \int\!\!\!\int_{-\infty}^{+\infty} dx\, dy\, \frac{\Delta\varepsilon \cdot \varepsilon}{\varepsilon + \Delta\varepsilon} E_{z\nu} E^*_{z\mu} \quad , \tag{2.6.22}$$

leading to real and positive quantities for positive $\Delta\varepsilon$. Using these coupling coefficients and the mode amplitudes of (2.6.10) we can rearrange (2.6.19 and 20) in the final form

$$A'_\mu = -\mathrm{j} \sum \{ A_\nu (K^t_{\nu\mu} + K^z_{\nu\mu}) \exp[-\mathrm{j}(\beta_\nu - \beta_\mu)z] $$
$$+ B_\nu (K^t_{\nu\mu} - K^z_{\nu\mu}) \exp[\mathrm{j}(\beta_\nu + \beta_\mu)z] \} \quad , \tag{2.6.23}$$

$$B'_\mu = \mathrm{j} \sum \{ A_\nu (K^t_{\nu\mu} - K^z_{\nu\mu}) \exp[-\mathrm{j}(\beta_\nu + \beta_\mu)z] $$
$$+ B_\nu (K^t_{\nu\mu} + K^z_{\nu\mu}) \exp[\mathrm{j}(\beta_\nu - \beta_\mu)z] \} \quad . \tag{2.6.24}$$

These two expressions form the basis for the solution of a number of coupled-mode problems. They show the change in the amplitude of each mode (μ) as a function of the deformation $\Delta\varepsilon$, the modal field distribution, and of the amplitudes of all other modes present in the guide. Depending on the particular problem at hand, one can usually make some simplifying assumptions at this stage. A very common and usually good assmuption is that only two guided modes are important and that all other modes can be neglected. This leads to coupled-wave interactions with characteristics that are discussed further in the next subsection. Another common assumption is that $\Delta\varepsilon(x, y, z)$ is only a small perturbation of the dielectric constant $\varepsilon(x, y)$ of the waveguide. Often, this is also a good assumption, but there are configurations of interest where it is not justified. An example is the corrugated glass waveguide used for filter devices where a corrugation of the glass-air interface leads to a large $\Delta\varepsilon$. This is illustrated further in the discussion of Sect. 2.6.4.

The expressions above point out a general difference between co-directional coupling (e.g., coupling between two forward modes A_μ and A_ν) and contra-directional coupling (e.g., between A_μ and B_μ) which occurs in the presence of E_z components. As K^t and K^z have equal sign, the factor $(K^t + K^z)$ indicates stronger coupling for co-directional interactions as compared to contra-directional interactions where we have the factor

$(K^t - K^z)$. This happens because forward and backward modes have E_z components of opposite sign and E_t components of the same sign.

2.6.3 Coupled-Wave Solutions

In a large class of coupled-mode interactions of interest, there are only two guided modes that possess sufficient phase synchronism to allow a significant interchange of energy. One can then neglect all other modes and obtain simple coupled-wave equations that describe the interaction. These equations follow from (2.6.11 and 12) for the general case addressed by this subsection, and, from (2.6.23 and 24) for the case of waveguide deformations which we use as an illustration. The solutions to the coupled-wave equations are well known and have been derived in several ways in the literature cited above. We give here only a brief listing of results. We have to distinguish between two different types, co-directional interactions and contra-directional interactions. Co-directional interactions occur between two forward (or two backward) propagating modes, and contra-directional interactions occur between a forward and a backward running mode. To be more precise, it is the relative direction of the group velocities of the modes that is considered here.

We deal with co-directional interactions first and, referring to (2.6.23), call the amplitudes of the two significant waves $A(= A_\mu)$ and $B(= A_\nu)$. We find that changes in these amplitudes are generally described by differential equations of the form

$$A' = -j\kappa B \exp(-2j\delta z) \quad , \tag{2.6.25}$$

$$B' = -j\kappa A \exp(2j\delta z) \quad , \tag{2.6.26}$$

where κ is the coupling coefficient which, in the simple case considered here, is real and uniform (i.e., independent of z), and δ is a normalized frequency which measures the deviation from synchronism (for which one has $\delta = 0$). In the next subsection, these two parameters are derived explicitly for the example of a corrugated waveguide. By means of the simple substitution

$$A = R \exp(-j\delta z) \quad , \quad B = S \exp(j\delta z) \tag{2.6.27}$$

we can transform the above into coupled-wave equations of the form

$$R' - j\delta R = -j\kappa S \quad , \tag{2.6.28}$$

$$S' + j\delta S = -j\kappa R \quad . \tag{2.6.29}$$

For the boundary conditions $R(0) = 1$, $S(0) = 0$ the solutions of the coupled-wave equations are

$$S(z) = -j\kappa \sin(z \cdot \sqrt{\kappa^2 + \delta^2})/\sqrt{\kappa^2 + \delta^2} \quad , \tag{2.6.30}$$

$$R(z) = \cos(z \cdot \sqrt{\kappa^2 + \delta^2}) + j\delta \sin(z \cdot \sqrt{\kappa^2 + \delta^2})/\sqrt{\kappa^2 + \delta^2} \quad . \tag{2.6.31}$$

For the case of synchronism ($\delta = 0$), this predicts a sinusoidal interchange of energy between R and S,

$$S(z) = -\mathrm{j}\sin(\kappa z) \quad , \quad R(z) = \cos(\kappa z) \quad . \tag{2.6.32}$$

A TE-to-TM mode converter or a directional coupler are examples for this type of interaction.

Next we consider contra-directional interactions with a forward wave of amplitude $A(= A_\mu)$ and a backward wave of amplitude $B(= B_\mu)$. For these, one finds differential equations of the form, see (2.6.23 and 24),

$$A' = -\mathrm{j}\kappa B \exp(2\mathrm{j}\delta z) \tag{2.6.33}$$

$$B' = \mathrm{j}\kappa A \exp(-2\mathrm{j}\delta z) \quad . \tag{2.6.34}$$

The substitution

$$A = R\exp(\mathrm{j}\delta z) \quad , \quad B = S\exp(-\mathrm{j}\delta z) \tag{2.6.35}$$

transforms these into the coupled-wave equations

$$R' + \mathrm{j}\delta R = -\mathrm{j}\kappa S \quad , \tag{2.6.36}$$

$$S' - \mathrm{j}\delta S = \mathrm{j}\kappa R \quad . \tag{2.6.37}$$

For this contra-directional interaction, we have to prescribe the boundary condition for the forward wave R at the beginning $z = 0$ of the interaction region and for the backward wave S at the end $z = L$. With $R(0) = 1$ and $S(L) = 0$, we obtain the solutions

$$S(0) = -\mathrm{j}\kappa/[\sqrt{\kappa^2 - \delta^2}\coth(L \cdot \sqrt{\kappa^2 - \delta^2}) + \mathrm{j}\delta] \tag{2.6.38}$$

$$R(L) = \sqrt{\kappa^2 - \delta^2}/[\sqrt{\kappa^2 - \delta^2}\cosh(L \cdot \sqrt{\kappa^2 - \delta^2})$$
$$+ \mathrm{j}\delta\sinh(L\sqrt{\kappa^2 - \delta^2})] \quad . \tag{2.6.39}$$

For the example of a corrugated waveguide filter, $S(0)$ represents the amplitude of the light reflected as a function of frequency δ, and $R(L)$ is the amplitude of the transmitted light. For synchronism (which corresponds to the center frequency of the filter or $\delta = 0$), the above formulae simplify to

$$S(0) = -\mathrm{j}\tanh(\kappa L) \quad , \quad R(L) = 1/\cosh(\kappa L) \quad . \tag{2.6.40}$$

An illustrative experimental result is given in Fig. 2.30 which is taken from [2.78] by *Flanders* et al. It shows the wavelength dependence of the reflectivity $|S(0)|^2$ of a corrugated waveguide filter as measured with a tunable dye laser (solid curve), and predicted by (2.6.38) (dashed curve).

In the distributed feedback structures considered for lasers, we also have a contra-directional coupled-wave interaction, but with some differences compared to the case considered above. First, there is laser gain

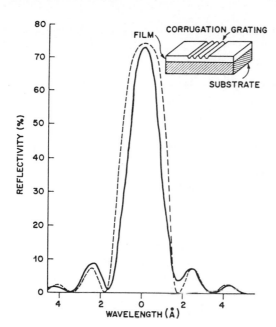

Fig. 2.30. Wavelength response of a corrugated waveguide filter. (After [2.78])

present in the waveguide, and second there is no incident wave, which is represented by the boundary conditions $R(0) = 0$ and $S(L) = 0$. For a detailed discussion of this interaction, we refer to *Kogelnik* and *Shank* [2.79].

2.6.4 Periodic Waveguides

Periodic waveguides are guides with a deformation $\Delta\varepsilon(x,y,z)$ that is periodic in z. These guides are used for a variety of purposes including the construction of filters, grating couplers, and distributed feedback lasers, and for the purpose of mode matching. The physical process that occurs in such a guide is the scattering of light by the periodic structure, which is similar to the light scattering by a diffraction grating. It can be viewed and analyzed as a coupled-mode process. We have already developed above the tools necessary for this analysis and their application is rather straightforward. First, we have to identify two modes which are at least approximately synchronous for a given optical frequency; to these we can apply the coupled-wave solutions given above. What remains to be done is the explicit evaluation of the parameters κ and δ, which we can do with the help of (2.6.21–24).

As an example, we will treat here the case of a corrugated planar slab waveguide, which has been used in experiments with filters and distributed feedback lasers. Figure 2.31 shows a sketch of such a corrugated guide with a periodically varying film thickness

$$h(z) = h_0 + \Delta h \cos(Kz) \quad , \tag{2.6.41}$$

where K is a grating constant related to the corrugation period Λ by

$$K = 2\pi/\Lambda \quad . \tag{2.6.42}$$

81

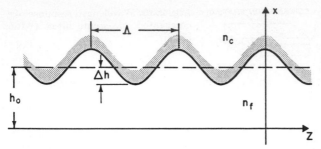

Fig. 2.31. Side view of a corrugated slab waveguide. Here h_0 is the average film thickness, Δh the amplitude of the corrugation, and Λ the period

Effective Index Method

Before using the rigorous coupled-wave formalism, we will first sketch the derivation of the coupling coefficients of corrugated waveguides by means of the effective-index method [2.82], which we have used in Sect. 2.5 to analyse channel waveguides. If viewed from above (along the x axis), the corrugated slab guide is seen as having an effective index $N(z)$ which varies as

$$N(z) = N + \Delta N \cos(\kappa z) \tag{2.6.43}$$

in the z-direction. In this view, the corrugated guide looks like the cross-section of a holographic Bragg grating. From the theory of these gratings [2.72], we know that the coupling coefficient describing the strength of the Bragg diffraction is given by

$$\kappa = \frac{\pi}{\lambda} \Delta N \quad . \tag{2.6.44}$$

By analogy we apply this to the corrugated waveguide, where the effective-index variation is caused by the variation in film thickness, and get

$$\kappa = \frac{\pi}{\lambda} \frac{dN}{dh} \Delta h \quad . \tag{2.6.45}$$

The quantity dN/dh can be derived from the dispersion relation of the guide.

For *TE modes* we start from the normalized dispersion relation (2.1.20) and find after differentiating that

$$2(1 - b) \frac{dh}{db} = h_{\text{eff}} \quad , \tag{2.6.46}$$

where h_{eff} is the effective guide thickness defined in (2.1.33) and b is the normalized guide index as explained in Sect. 2.1. We use this result to rewrite dN/dh as follows

$$\frac{dN}{dh} = \frac{dN}{db} \cdot \frac{db}{dh} = \frac{(n_f^2 - n_s^2)(1 - b)}{N\, h_{\text{eff}}} = \frac{n_f^2 - N^2}{N\, h_{\text{eff}}} \quad . \tag{2.6.47}$$

82

With this, we can express the coupling coefficient as

$$\kappa_{\mathrm{TE}} = \frac{\pi}{\lambda} \frac{\Delta h}{h_{\mathrm{eff}}} \frac{n_{\mathrm{f}}^2 - N^2}{N} \quad . \tag{2.6.48}$$

Similar manipulations yield an expression for the coupling coefficient of the *TM modes*, i.e.

$$\kappa_{\mathrm{TM}} = \frac{\pi}{\lambda} \frac{\Delta h}{h_{\mathrm{eff}}} \frac{n_{\mathrm{f}}^2 - N^2}{N} p \quad , \tag{2.6.49}$$

where p is a reduction factor defined by

$$p = \frac{(N/n_{\mathrm{f}})^2 - (N/n_{\mathrm{c}})^2 + 1}{(N/n_{\mathrm{f}})^2 + (N/n_{\mathrm{c}})^2 - 1} \quad . \tag{2.6.50}$$

Coupled-Mode Formalism

To derive the same coupling coefficients with the more general coupled-mode formalism, we consider the perturbations caused by the corrugation of the guide and use the modal fields of the slab guide tabulated in Sect. 2.3.

In terms of the refractive indices n_{f} and n_{c} of film and cover, the corrugation produces a perturbation $\Delta \varepsilon$ which can be written as

$$\begin{aligned}
\Delta \varepsilon &= \varepsilon_0 (n_{\mathrm{f}}^2 - n_{\mathrm{c}}^2) \quad &\text{for} \quad h(z) > h_0 \quad , \\
\Delta \varepsilon &= -\varepsilon_0 (n_{\mathrm{f}}^2 - n_{\mathrm{c}}^2) \quad &\text{for} \quad h(z) < h_0 \quad .
\end{aligned} \tag{2.6.51}$$

We insert this into (2.6.21 and 22) to determine the coupling coefficients, restricting the discussion to "backward" scattering from a forward propagating mode (μ) to a backward propagating mode of the same mode number $(-\mu)$.

For TE modes we get $K_{\mu,-\mu}^z = 0$, and

$$\begin{aligned}
K_{\mu,-\mu}^t &= \omega \int\!\!\!\int_{-\infty}^{+\infty} dx \, \Delta \varepsilon E_y^2 \approx E_{\mathrm{c}}^2 \int\!\!\!\int_{-\infty}^{+\infty} dx \, \Delta \varepsilon \\
&= \omega \varepsilon_0 E_{\mathrm{c}}^2 (n_{\mathrm{f}}^2 - n_{\mathrm{c}}^2) \Delta h \cos(Kz) = \kappa (e^{jKz} + e^{-jKz}) \quad , \tag{2.6.52}
\end{aligned}$$

where we have assumed that Δh is small enough so that we can replace $E_y(x)$ by the constant field value E_{c} assumed by the mode at the film-cladding interface. The relation of $K_{\mu,-\mu}^t(z)$ to the coupling coefficient κ of the coupled-wave equations is also indicated. From (2.3.11, 12) we can get the normalized value of E_{c}, and determine κ again as

$$\kappa = \frac{\pi}{\lambda} \frac{\Delta h}{h_{\mathrm{eff}}} \frac{n_{\mathrm{f}}^2 - N^2}{N} \quad , \tag{2.6.53}$$

where N is the effective index and h_{eff} the effective guide thickness as used

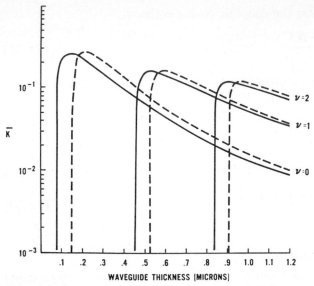

Fig. 2.32. Normalized coupling coefficient $\overline{\kappa} = \lambda\kappa/(2\pi\Delta h)$ as a function of waveguide thickness for a GaAlAs waveguide with $n_f = 3.59$, $n_s = 3.414$ and $n_c = 1$ (*solid curves*) and $n_c = 3.294$ (*dashed curves*) respectively for the three lowest mode orders. (After [2.80])

before. This can be shown to agree with the formula given by *Marcuse* [2.2] who also gives results for coupling between modes of unequal mode number. One remarkable thing about this formula is that neither of the subscripts c or s for cladding or substrate appear. This indicates that we get exactly the same coupling coefficient κ when the corrugation is made on the substrate-film interface instead of the film-cover interface. The reason for this may be found in (2.3.11) which indicates that a smaller value of $(n_f^2 - n_s^2)$ is exactly balanced by a larger value of the field strength E_s at the film-substrate interface.

Figure 2.32, taken from *Shank* et al. [2.80], shows the dependence of κ on film thickness h for the example of a GaAlAs waveguide with $n_f = 3.59$, $n_s = 3.414$ and $n_c = 1$ (solid curves) and $n_c = 3.294$ (dashed curves). The normalized quantity $\overline{\kappa} = \lambda\kappa/(2\pi\Delta h)$ is used as the ordinate. We note that the κ of each mode assumes a maximum value fairly close to cut-off.

To determine the normalized frequency δ, we compare the exponentials in (2.6.23 and 33) and find

$$2\delta = 2\beta_\mu - K \quad . \tag{2.6.54}$$

The scattering is largest at the center frequency where $\delta = 0$, which corresponds to a center wavelength λ_0 and a propagation constant $\beta_0 = 2\pi/\lambda_0$ given by the Bragg condition

$$K = 2\beta_0 \quad , \quad \lambda_0 = 2N\Lambda \quad . \tag{2.6.55}$$

Referring to these quantities, we can rewrite δ in the form

$$\delta = \beta_\mu - \beta_0 = \Delta\beta \approx \frac{d\beta}{d\omega}\Delta\omega = \Delta\omega/v_{\mathrm{g}} \quad , \tag{2.6.56}$$

where $\Delta\omega$ is the radian frequency deviation from the center frequency, and v_{g} is the group velocity of (2.2.80).

For the *TM modes* there are difficulties which arise in the application of the coupled-mode formalism to our particular example of a corrugated waveguide[2]. The root of the problem is thought to be a violation of the boundary conditions of the perturbed guide. This problem has been discussed in greater detail in [2.83–85].

In addition to these references, the literature contains discussions of several alternate methods for the analysis of corrugated guides. Examples are the treatment of a variety of grating profiles in [2.86], the treatment of gratings of rectangular tooth shape [2.87], the application of Rouard's method [2.88], and the analysis of corrugated multi-layer guides [2.89–92].

2.6.5 TE-to-TM Mode Conversion

We have mentioned before that the conversion from TE-to-TM modes can be regarded as another case of a coupled-wave interaction [2.74, 76] In this final subsection, we sketch briefly how this mode conversion is treated with the coupled-mode formalism presented in this section. We consider TE-to-TM mode conversion due to a tensor perturbation $\Delta\varepsilon_{ij}(x)$ of the form

$$\Delta\varepsilon_{ij} = \begin{pmatrix} \Delta\varepsilon_1 & \eta & 0 \\ \eta & \Delta\varepsilon_2 & 0 \\ 0 & 0 & \Delta\varepsilon_3 \end{pmatrix} \quad . \tag{2.6.57}$$

This can be caused, for example, by application of a dc electric field to a planar electro-optic waveguide. Only two modes are assumed present in the guide, a TE mode with a field distribution $\boldsymbol{E}_{\mathrm{E}}(x)$ and a TM mode with the field $\boldsymbol{E}_{\mathrm{M}}(z)$. The corresponding propagation constants are β_{E} and β_{M}, and the complex amplitudes are $A_{\mathrm{E}}(z)$ and $A_{\mathrm{M}}(z)$. We write for the transverse component $\boldsymbol{E}_{\mathrm{t}}(x, z)$ of the total field in the guide

$$\boldsymbol{E}_{\mathrm{t}} = A_{\mathrm{E}} \cdot \boldsymbol{E}_{\mathrm{Et}}\exp(-\mathrm{j}\beta_{\mathrm{E}}z) + A_{\mathrm{M}} \cdot \boldsymbol{E}_{\mathrm{Mt}}\exp(-\mathrm{j}\beta_{\mathrm{M}}z) \quad . \tag{2.6.58}$$

Only the TM mode contributes to the longitudinal component E_z, which, in analogy to (2.6.18), is written as

$$E_z = \frac{\varepsilon}{\varepsilon + \Delta\varepsilon_3}A_{\mathrm{M}}E_{\mathrm{M}z}\exp(-\mathrm{j}\beta_{\mathrm{M}}z) \approx A_{\mathrm{M}}E_{\mathrm{M}z}\exp(-\mathrm{j}\beta_{\mathrm{M}}z) \quad . \tag{2.6.59}$$

---

[2] The author is indebted to D.G. Hall for illuminating discussions on this issue.

With this, we can use (2.6.14) to calculate the induced polarization and obtain, after insertion into (2.6.11), for the changes of the complex amplitudes

$$A_E' = -j\omega \int\limits_{-\infty}^{+\infty} dx \, E_{Ei}^* \Delta\varepsilon_{ij} E_j \exp(j\beta_E z) \quad , \tag{2.6.60}$$

$$A_M' = -j\omega \int\limits_{-\infty}^{+\infty} dx \, E_{Mi}^* \Delta\varepsilon_{ij} E_j \exp(j\beta_M z) \quad . \tag{2.6.61}$$

If we combine this with (2.6.57–59), we can write these equations in the form

$$A_E' = -j\Delta\beta_E A_E - j\kappa A_M \exp[-j(\beta_M - \beta_E)] \quad , \tag{2.6.62}$$

$$A_M' = -j\Delta\beta_M A_M - j\kappa A_E \exp[j(\beta_M - \beta_E)] \quad , \tag{2.6.63}$$

where we have defined the coupling constant as

$$\kappa = \omega \int\limits_{+\infty}^{-\infty} dx \, \eta E_{Mx} E_{Ey}^{*-} = \omega \int\limits_{-\infty}^{+\infty} dx \, \eta E_{Mx}^* E_{Ey} \quad , \tag{2.6.64}$$

and the induced changes in the propagation constant as

$$\Delta\beta_E = \omega \int\limits_{-\infty}^{+\infty} dx \, \Delta\varepsilon_2 E_{Ey} E_{Ey}^* \quad , \tag{2.6.65}$$

$$\Delta\beta_M = \omega \int\limits_{-\infty}^{+\infty} dx (\Delta\varepsilon_1 E_{Mx} E_{Mx}^* + \Delta\varepsilon_3 E_{Mz} E_{Mz}^*) \quad . \tag{2.6.66}$$

It is easy to see that (2.6.62 and 63) can be transformed into the standard form (2.6.28 and 29) of the coupled-wave equations by using (2.6.27) and a normalized frequency deviation δ of the form

$$2\delta = \beta_M - \beta_E + \Delta\beta_M - \Delta\beta_E \tag{2.6.67}$$

which, again, indicates the degree of deviation from synchronism. The mode conversion problem is now cast in a form that allows us to apply directly the coupled-wave solutions of Sect. 2.6.3. The overlap integrals of (2.6.64–66) have to be evaluated for each specific case; examples have been given by *Yariv* [2.74], and *Sosnowski* and *Boyd* [2.76].

References

2.1   N.S. Kapany, J.J. Burke: *Optical Waveguides* (Academic, New York 1974)
2.2   D. Marcuse: *Theory of Dielectric Optical Waveguides* (Academic, New York 1974)
2.3   D. Marcuse: *Light Transmission Optics* (Van Nostrand Reinhold, New York 1972)
2.4   M.S. Sodha, A.K. Ghatak: *Inhomogeneous Optical Waveguides* (Plenum, New York 1977)
2.5   H.G. Unger: *Planar Optical Waveguides and Fibers* (Clarendon, Oxford 1977)
2.6   M.J. Adams: *An Introduction to Optical Waveguides* (Wiley, Chichester 1981)
2.7   T. Okoshi: *Optical Fibers* (Academic, New York 1982)
2.8   A.W. Snyder, J.D. Love: *Optical Waveguide Theory* (Chapman and Hall, London 1983)
2.9   H.A. Haus: *Waves and Fields in Optoelectronics* (Prentice-Hall, Englewood Cliffs, NJ 1984)
2.10 W.W. Anderson: IEEE J. **QE-1**, 228 (1965)
2.11 A. Reisinger: Appl Opt. **12**, 1015 (1973)
2.12 I.P. Kaminow, W.L. Mammel, H.P. Weber: Appl. Opt. **13**, 396 (1974)
2.13 D.F Nelson, J. McKenna: J. Appl. Phys. **38**, 4057 (1967)
2.14 S. Yamamoto, Y. Koyamada, T. Makimoto: J. Appl. Phys. **43**, 5090 (1972).
2.15 V. Ramaswamy: Appl. Opt. **13**, 1363 (1974)
2.16 V. Ramaswamy: J. Opt. Soc. Am. **64**, 1313 (1974)
2.17 P.K. Tien: Appl. Opt. **10**, 2395 (1971)
2.18 S.J. Maurer, L.B. Felsen: Proc. IEEE **55**, 1718 (1967)
2.19 H.K.V. Lotsch: Optik **27**, 239 (1968)
2.20 H. Kogelnik, V. Ramaswamy: Appl Opt. **13**, 1857 (1974)
2.21 K Artmann: Ann. Physik **2**, 87 (1948)
      H.K.V. Lotsch: Optik **32**, 116, 189, 299, 553 (1970/71)
2.22 H. Kogelnik, T.P. Sosnowski, H.P. Weber: IEEE J. **QE-9**, 795 (1973)
2.23 J.J. Burke: Opt. Sci. Newslett. **5**, 31 (Univ. Arizona, 1971)
2.24 J. McKenna: Bell Syst. Tech. J. **46**, 1491 (1967)
2.25 R.B. Adler: Proc. IRE **40**, 339 (1952)
2.26 W.P. Allis, S.J. Buchsbaum, A. Bers: *Waves in Anisotropic Plasmas* (Wiley, New York 1962)
2.27 H. Kogelnik, H.P. Weber: J. Opt. Soc. Am. **64**, 174 (1974)
2.28 L.D. Landau, E.M. Lifshitz: *Electrodynamics of Continuous Media* (Pergamon, Oxford 1960)
2.29 L.D. Landau, E.M. Lifshitz: *Quantum Mechanics* (Pergamon, Oxford 1958)
2.30 A.D. Berk: IRE Trans. **AP-4**, 104 (1956)
2.31 R.F. Harrington: *Time-Harmonic Electromagnetic Fields* (McGraw-Hill, New York 1961)
2.32 K. Kurokawa: IRE Trans. **MTT-10**, 314 (1962)
2.33 K. Moroshita, N. Kumagai: IEEE Trans. **MTT-25**, 34 (1977)
2.34 M. Matsuhara: J. Opt. Soc. Am. **63**, 1514 (1973)
2.35 H.F. Taylor: IEEE J. **QE-12**, 748 (1976)
2.36 S.K. Korotky, W.J. Minford, L.L. Buhl, M.D. Divino, R.C. Alferness: IEEE J. **QE-18**, 1976 (1982)
2.37 M. Geshiro, M. Ohtaka, M. Matsuhara, N. Kumagai: IEEE J. **QE-14**, 259 (1978)
2.38 H.A. Haus, W.P. Huang, S. Kawakami, N.A. Whitaker: IEEE J. Lightwave Technology **LT-5**, 16 (1987)
2.39 S. Akiba, H.A. Haus: Appl. Opt. **21**, 804 (1982)
2.40 M. Ohtaka, M. Ohtaka, M. Matsuhara, N. Kumagai: IEEE J. **QE-10**, 647 (1974)
2.41 Y. Yamamoto, T. Kamiya, H. Yanai: IEEE J. **QE-11**, 729 (1975)
2.42 J.N. Polky, G.I. Mitchell: J. Opt. Soc. Am. **64**, 274 (1974)
2.43 G.E. Smith: IEEE J. **QE-4**, 288 (1968)
2.44 V.V. Cherny, G.A. Juravlev, A.I. Kirpa, I.L. Rylov, V.P. Ijoy: IEEE J. **QE-15**, 1401 (1979)
2.45 H.C. Casey Jr., M.B. Panish: *Heterostructure Lasers A*, (Academic, New York 1978)

2.46   H. Kressel, J.K. Butler: *Semiconductor Lasers and Heterojunction LEDS* (Academic, New York 1977), p. 137
2.47   M. Born, E. Wolf: *Principles of Optics* (Pergamon, New York 1959), p. 50
2.48   F. Abeles: Ann. Physique **5**, 596 (1950)
2.49   H. Kogelnik, T. Li: Appl. Opt. **5**, 1550 (1966)
2.50   G. Poschl, E. Teller: Z. Physik **83**, 143 (1933)
2.51   J.P. Gordon: Bell Syst. Tech. J. **45**, 321 (1966)
2.52   E.M. Conwell: Appl. Phys. Lett. **23**, 328 (1973)
2.53   J.R. Carruthers, I.P. Kaminow, L.W. Stulz: Appl. Opt. **13**, 2333 (1974)
2.54   R.D. Standley, V. Ramaswamy: Appl. Phys. Lett. **25**, 711 (1974)
2.55   H.A.Haus, R.V. Schmidt: Appl. Opt. **15**, 774 (1976)
2.56   B.S. Jeffreys: *Quantum Theory*, Vol. I, ed. by D.R. Bates (Academic, New York 1961)
2.57   L.B. Felsen, N. Marcuvitz: *Radiation and Scattering of Waves* (Prentice Hall, Englewood Cliffs, NJ 1973)
2.58   J.E. Goell: Bell Syst. Tech. J. **48**, 2133 (1969)
2.59   W. Schlosser, H.G. Unger: *Advances in Mircowaves* (Academic, New York 1966)
2.60   W.O. Schlosser: A.E.Ü. **18**, 403 (1964)
2.61   K. Ogusu: IEEE Trans. **MTT-25**, 874 (1977)
2.62   E.A.J. Marcatili: Bell Syst. Tech. J. **48**, 2071 (1969)
2.63   A. Kumar, K. Thyagarajan, A.K. Ghatak: Opt. Lett. **8**, 63 (1983)
2.64   R.M. Knox, P.P. Toulios: Proc. MRI Symp. Submillimeter Waves (Polytechnic Press, Brooklyn, 1970), p. 497
2.65   V. Ramaswamy: Bell Syst. Tech. J. **53**, 697 (1974)
2.66   G.B. Hocker, W.K. Burns: Appl. Opt. **16**, 113 (1977)
2.67   W.H. Zachariasen: *Theory of X-Ray Diffraction in Crystals* (Wiley, New York 1945)
2.68   S.E. Miller: Bell Syst. Tech. J. **33**, 661 (1954)
2.69   J.R. Pierce: J. Appl. Phys. **25**, 179 (1954)
2.70   W.H. Louisell: *Coupled Mode and Parametric Electronics* (Wiley, New York 1960)
2.71   C.F. Quate, C.D.W. Wilkinson, D.K. Winslow: Proc. IEEE **53**, 1604 (1965)
2.72   H. Kogelnik: Bell Syst. Tech. J. **48**, 2909 (1969)
2.73   A.W. Snyder: J. Opt. Soc. Am. **62**, 1267 (1972)
2.74   A. Yariv: IEEE J. **QE-9**, 919 (1973)
2.75   D. Marcuse, IEEE J. **QE-9**, 958 (1973)
2.76   T.P. Sosnowski, G.D. Boyd: IEEE J. **QE-10**, 306 (1974)
2.77   I.P. Kaminow: *An Introduction to Electrooptic Devices* (Academic, New York 1974)
2.78   D.C. Flanders, H. Kogelnik, R.V. Schmidt, C.V. Shank: Appl. Phys. Lett. **24**, 194 (1974)
2.79   H. Kogelnik, C.V. Shank: J. Appl. Phys. **43**, 2327 (1972)
2.80   C.V. Shank, R.V. Schmidt, B.I. Miller: Appl. Phys. Lett. **25**, 200 (1974)
2.81   S.M. Saad: IEEE Trans. **MTT-33**, 894 (1985)
2.82   P.G. Verly, R. Tremblay, J.W.Y. Lit: J. Opt. Soc. Am. **70**, 964 and 1218 (1980)
2.83   W. Streifer, D.R. Scifres, R.D. Burnham: IEEE J. **QE-12**, 74 (1976)
2.84   G.I. Stegeman, D. Sarid, J.J. Burke, D.G. Hall: J. Opt. Soc. Am. **71**, 1497 (1981)
2.85   R.W. Gruhlke, D.G. Hall: Appl. Opt. **23**, 127 (1984)
2.86   W. Streifer, D.R. Scifres, R.D. Burnham: IEEE J. **QE-11**, 867 (1975)
2.87   A. Hardy: IEEE J. **QE-20**, 1132 (1984)
2.88   L.A. Weller-Brophy, D.G. Hall: J. Opt. Soc. Am. **A-4**, 60 (1987)
2.89   Y. Yamamoto, T. Kamiya, H. Yanai: IEEE J. **QE-14**, 245 (1978)
2.90   S.T. Peng, T. Tamir, H.L. Bertoni: IEEE Trans. **MTT-23**, 123 (1975)
2.91   R. Petit: *Electromagnetic Theory of Gratings*, Topics Current Phys., Vol. 22 (Springer, Berlin, Heidelberg 1980)
2.92   K.C. Chang, V. Shah, T. Tamir: J. Opt. Soc. Am. **70**, 804 (1980)

3. Waveguide Transitions and Junctions

By W.K. Burns and A.F. Milton

With 43 Figures

Most useful waveguide devices use a series of waveguide transitions or junctions where the cross section of the waveguide structure changes in the direction of propagation. For example, transitions are typically placed on both sides of a phase shift region to form an electrooptical switch. The performance of the switch in terms of extinction ratio and crosstalk depends critically on how well those transitions perform their intended function. To understand and optimize device performance it is therefore useful to be able to predict the transmission characteristics of waveguide transitions and junctions.

These characteristics are, of course, completely defined by Maxwell's equations. However, finding a computer-generated solution to Maxwell's equations for each particular geometry of interest is a tedious proposition which usually does not provide design guidance. It is therefore instructive to look for an approach that uses aggregate properties of the junction geometry (its degree of symmetry, etc.) to predict performance. Such high-order formalisms are well-known for the case of microwave waveguides. The chief difference in the case of optical waveguides concerns the possibility of designing transitions that take place over hundreds of wavelengths (i.e., adiabatic transitions) which would not be practical at microwave wavelengths.

Given the importance of adiabatic or near-adiabatic transitions in the optical regime, the local-normal-mode formalism described below, first introduced by *Louisell,* turns out to be an extremely powerful prediction tool. Coupling between local normal modes is caused by variations of waveguide parameters in the direction of propagation. As we shall see, using this formalism, the transmission properties of most waveguide transitions can be shown to depend on two or three aggregate parameters which make most devices easy to understand and design. Under certain approximations, analytic solutions have even been obtained to describe the amount of power transferred between these modes. These solutions can be used to predict the performance of important transition structures.

3.1 Waveguide Modes and Coupled-Mode Theory

3.1.1 Normal Modes of Coupled Waveguides

If we consider a simple three-layer waveguide structure and derive a solution of Maxwell's equations for the guided modes of the structure, we obtain electric-field profiles, as shown in Fig. 3.1. These solutions are called the normal modes of the waveguide, as discussed in Sect. 2.2.2. In Fig. 3.1, the geometry and optical wavelength are assumed such that the structure supports two normal modes. We denote the first or symmetric mode as ψ_i and the second or antisymmetric mode as ψ_j.

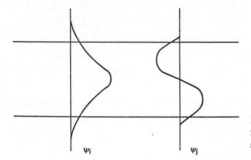

Fig. 3.1. Electric field profiles for the first two normal modes of a three-layer structure

ψ_i ψ_j

More generally, we will be dealing with two waveguides which can each independently support a single mode but which are sufficiently close to each other so that the optical fields of the two guides overlap in some region. The waveguides are then said to be coupled because transfer of optical power between them can occur. We first define the modes of the guides when they are uncoupled, i.e., the separation between the guides approaches infinity so that optical coupling between them does not occur (Fig. 3.2). Then, if the waveguides are denoted a and b, the uncoupled mode field profiles are denoted by $\phi_a \equiv \psi_{ia}$ and $\phi_b \equiv \psi_{ib}$, and their propagation constants by β_a and β_b, respectively. These are the normal modes of each guide considered separately and are also called the modes of the uncoupled waveguides.

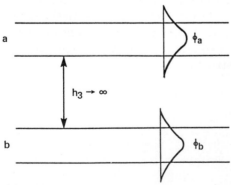

a

$h_3 \to \infty$

b

Fig. 3.2. Uncoupled mode field profiles for waveguides at large separation

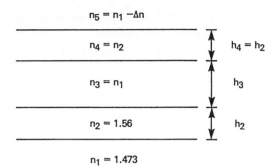

$n_5 = n_1 - \Delta n$

$n_4 = n_2$ $h_4 = h_2$

$n_3 = n_1$ h_3

$n_2 = 1.56$ h_2

$n_1 = 1.473$

Fig. 3.3. Two coupled waveguides (regions 2 and 4) represented as a five-layer structure. The index difference Δn describes the index asymmetry between the outer layers of the structure

Next, consider the modes of the total structure when the waveguides are not at a large separation, i.e., the modes of the coupled waveguides. One approach is to obtain from Maxwell's equations the exact solution for the normal modes. For that purpose, the two coupled waveguides, with separation h_3, are considered as a five-layer structure, as shown in Fig. 3.3, and analytic or computer solutions can be obtained. Such solutions [3.1] for the structure of Fig. 3.3 are shown in Fig. 3.4 as a function of waveguide separation and the index asymmetry $\Delta n = n_1 - n_5$ between the outer layers of the structure. When the waveguide separation is zero, we obtain the two normal-mode solutions of Fig. 3.1, because doubling the width of a single-mode guide will generally create a double moded guide. When the waveguide separation is large and the asymmetry Δn is not equal to zero,

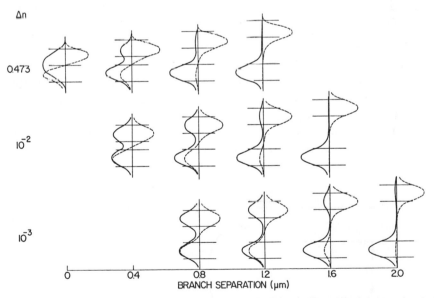

Fig. 3.4. Plots of mode electric-field distributions in model waveguide of Fig. 3.3 versus branch separation for index asymmetry Δn of 0.473, 10^{-2}, and 10^{-3}. The solid curve is the first-order mode (i), and the dashed curve is the second-order mode (j). Each mode has unity power

91

solutions are obtained that look like the uncoupled waveguide solutions of Fig. 3.2, i.e., each normal mode is predominantly associated with one of the waveguides and assumes the shape of the first or symmetrical mode of that waveguide. Note, however, that the normal modes ψ_i and ψ_j of the structure can now be associated with one or the other of the coupled waveguides. The shape change that the normal modes undergo as the waveguide separation increases from a strongly coupled to an uncoupled condition is termed modal evolution. For perfectly symmetric coupled waveguides, with $\Delta n = 0$, the normal modes do not evolve to a condition of having most of their power associated with a single waveguide at large waveguide separation. Rather, they maintain a condition of evenly dividing their power between the waveguides at all waveguide separations.

3.1.2 Coupled-Mode Theory Representation

A second approach [3.2–4] to the representation of the modes of coupled waveguides is the use of coupled-mode theory, as described in Sect. 2.6. Coupled-mode theory is basically a perturbation approach which uses the modes of the uncoupled waveguides (Fig. 3.2) to describe optical power transfer between two coupled waveguides. It can also be used to provide an approximate representation for the normal modes of the coupled structure. In Fig. 3.5, we assume that the guides a and b have a separation h_3, and K is the coupling coefficient between them. The quantity K is an exponentially decaying function of the separation between the waveguides, which is given by

$$|K| = Fe^{-\gamma_3 h_3} \quad , \tag{3.1.1}$$

where F is a constant, and γ_3 is the transverse momentum component of the electric field in the region between the waveguides. We define the difference in the separated-guide propagation constants $\Delta\beta = \beta_a - \beta_b$, as a measure of the momentum mismatch between the waveguides. The difference $\Delta\beta$ is also referred to as the asynchronism. We then introduce the coupled-waveguide parameter X

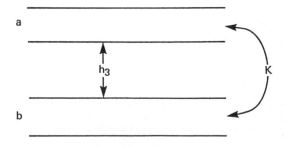

Fig. 3.5. Two coupled waveguides with separation h_3 and coupling coefficient K

$$X \equiv \frac{\Delta\beta}{2|K|} \qquad (3.1.2)$$

which is the ratio of the momentum mismatch or guide asynchronism to twice the magnitude of the coupling coefficient.

To describe power transfer between two coupled waveguides, using coupled-mode theory, we assume that each waveguide mode retains the transverse shape of the uncoupled mode ϕ_a or ϕ_b, but that the coupled mode acquires a mode amplitude which will vary with position along the waveguide. If a and b are used to denote the coupled-mode amplitudes, the coupled modes may then be written $a\phi_a$ and $b\phi_b$. The coupled-mode equations for the waveguide system of Fig. 3.5 are

$$\frac{da}{dz} - j\beta_a a - j|K|b = 0 \quad , \qquad (3.1.3)$$

$$\frac{db}{dz} - j\beta_b b - j|K|a = 0 \quad , \qquad (3.1.4)$$

where the amplitude coefficients $a(z)$ and $b(z)$ are functions of z, the direction of propagation. Solutions to these equations have been given in Sect. 2.6.3.

Coupled-mode theory can also be used to provide an approximate representation for the normal modes in a coupled-waveguide system. In contrast to the amplitudes of the modes of the coupled waveguides, normal modes propagate in a coupled-waveguide system without change in amplitude. The normal modes can be expressed as a linear combination of the uncoupled modes ϕ_a and ϕ_b [3.5]:

$$\psi_i \simeq d\phi_a + e\phi_b \quad , \qquad (3.1.5)$$

$$\psi_j \simeq -e\phi_a + d\phi_b \quad . \qquad (3.1.6)$$

Normalization is provided by requiring $d^2 + e^2 = 1$. Since the normal modes must be orthogonal, we have $\int \psi_i \psi_j^* \, dx = 0$ where, for simplicity, we assume a two-dimensional (x, z) geometry so that x is the coordinate transverse to the waveguide. This requires

$$\int_{-\infty}^{\infty} \phi_a \phi_a^* dx \simeq 1 \quad , \qquad (3.1.7)$$

$$\int_{-\infty}^{\infty} \phi_b \phi_b^* dx \simeq 1 \quad , \qquad (3.1.8)$$

$$\int_{-\infty}^{\infty} \phi_a \phi_b^* dx \simeq 0 \quad , \qquad (3.1.9)$$

$$\int_{-\infty}^{\infty} \phi_b \phi_a^* dx \simeq 0 \quad . \tag{3.1.10}$$

The constants e and d are related to the coupled-waveguide parameter X by

$$f = \frac{e}{d} = -X + (X^2 + 1)^{1/2} \quad , \tag{3.1.11}$$

which when combined with the normalization requirement given above yields

$$d = \left[\frac{1}{2} \left(1 + \frac{X}{(X^2 + 1)^{1/2}} \right) \right]^{1/2} \quad , \tag{3.1.12}$$

$$e = \left[\frac{1}{2} \left(1 - \frac{X}{(X^2 + 1)^{1/2}} \right) \right]^{1/2} \quad . \tag{3.1.13}$$

We define the ratio e/d as f, which is plotted in Fig. 3.6. The normal mode propagation constants are given by

$$\beta_i = \overline{\beta} + |K|(X^2 + 1)^{1/2} \quad , \tag{3.1.14}$$

$$\beta_j = \overline{\beta} - |K|(X^2 + 1)^{1/2} \quad , \tag{3.1.15}$$

where $\overline{\beta} = (\beta_a + \beta_b)/2$ is the average propagation constant.

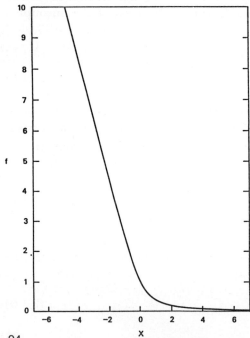

Fig. 3.6. Plot of $f = e/d$ versus the parameter $X = \Delta\beta/2|K|$

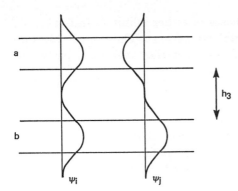

Fig. 3.7. Coupled mode theory representation of the normal mode field profiles for two coupled waveguides when $\Delta\beta = 0$

As an example, the case of $\Delta\beta = 0$, or synchronous case, is plotted in Fig. 3.7. In this case, $X = 0$, so $e = d$. The normal modes are made up of linear combinations of the uncoupled modes, with equal amplitude. As discussed previously, this particular case is independent of the guide separation h_3, so that the normal modes have equal power in each waveguide for all guide separations. For $\Delta\beta \neq 0$ we see that $|K| \to 0$ as the guide separation h_3 becomes infinitely large. Then $X \to \pm\infty$, and, from Fig. 3.6, $f = e/d$ approaches 0 or ∞. This process describes the modal evolution discussed above in which the normal mode power eventually accumulates in one or the other of the two waveguides (Fig. 3.4) as the distance between them increases.

Although the exact normal mode solution of any coupled-waveguide system can always be obtained numerically, the process is tedious and non-predictive. The advantage of the coupled-mode-theory representation of the normal modes of a coupled-waveguide system is that it provides an analytic representation with a direct dependence on the waveguide parameters. For this reason, we will make extensive use of this approach in our study of waveguide transitions. However, this approach is an approximation and it behooves us to keep in mind the implicit assumptions of a perturbation-theory approach. First, we are assuming weak coupling so that K is small or, from (3.1.1), that $\gamma_3 h_3 > 1$. Second, we are assuming that the waveguides are not too dissimilar, i.e., that $\Delta\beta$ is not too large. These limitations will become more apparent and better defined as specific examples of waveguide transitions are discussed.

3.2 Fast and Slow Transitions

So far we have discussed normal modes in fixed structures, i.e., structures whose cross section does not vary in the direction of optical propagation. In this section, we will begin to consider waveguide transitions, in which the waveguide structure does vary in the direction of mode propagation, but will restrict ourselves to the limiting cases of either infinitely fast (abrupt) transitions or infinitely slow (adiabatic) transitions. These cases have an obvious intuitive meaning, and we will see that they take on a particular mathematical significance as well.

3.2.1 Local Normal Modes

Implicit in our discussion of normal modes in Sect. 3.1.1 was the understanding that the normal modes of a structure are, by definition, orthogonal, i.e., they satisfy the orthogonality condition discussed in Sect. 2.5.5, namely

$$\int_{-\infty}^{\infty} \psi_i(x)\psi_j^*(x)dx = 0 , \qquad (3.2.1)$$

where x is the coordinate transverse to the waveguide. In waveguide structures in which the waveguide parameters vary in the direction of propagation, (3.2.1) is no longer true and we cannot strictly speak of the normal modes of the structure. Instead, we introduce the concept of a "locally" normal mode [3.6–8] as suggested in Fig. 3.8. At a waveguide position z_0, we consider the waveguide parameters at that position, and solve for the normal modes of that structure as if it did not vary with z. These normal mode solutions are then said to be the local normal modes of the actual varying waveguide structure at the position z_0. The local normal mode representation will now become a function of z.

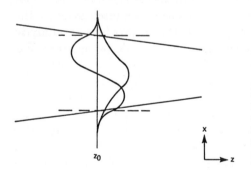

Fig. 3.8. To define a "local" normal mode, the slowly varying waveguide structure (solid line) is replaced at z_0 by a structure constant with z (dashed line), and the normal mode solutions at z_0 are obtained

In contrast to the normal modes of a fixed structure, the amplitudes of the local normal modes in a varying waveguide structure are not necessarily constant. Power transfer between the local normal modes of a varying structure occurs, and we will see that the magnitude of that power transfer depends on the rate of change of the geometry of the structure. The methods for the calculation of the power transfer between local normal modes are the central topic of this chapter.

In our treatment we will generally not account for mode coupling to radiation modes in a waveguide transition. The problem of mode coupling to radiation modes has been treated by *Marcuse* [3.6, 9].

3.2.2 Adiabatic Transition

We define a slow or adiabatic waveguide transition as a transition between two waveguide structures that takes place gradually with propagation distance z so that negligible power transfer occurs between the normal modes, as they propagate from one structure to the other. An example is provided by the separating waveguides of Fig. 3.4. If the waveguide separation occurs sufficiently slowly with z, power injected initially in a given local normal mode will stay in that mode throughout the transition. The local normal mode may change its shape in the process of modal evolution, but coupling to the other local normal modes is assumed not to occur. Power put into the first-order local normal mode (highest β) will end up in the first-order local normal mode, etc.

3.2.3 Abrupt Transition

The opposite extreme is a fast or abrupt transition in which a transition between two waveguide structures is made so abruptly that the maximum amount of power transfer between the local normal modes occurs, consistent with the geometry of the two waveguide structures. The limiting case of an abrupt transition is one that occurs at a plane z_0, along the direction of propagation of the modes. Plane, or step transitions are illustrated in Fig. 3.9. Power transfer between local normal modes at a step transition can be readily calculated from a consideration of boundary conditions at the transition [3.5].

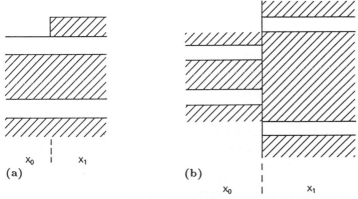

Fig. 3.9. Abrupt transitions where the change in waveguide geometry occurs at a plane along the mode propagation direction. The transition can be two dimensional with the planar waveguides (**a**), or three dimensional with the channel waveguides (**b**)

Since the local normal modes of a waveguide structure are characterized by the waveguide parameter X, see (3.1.5,6 and 11), we can characterize any abrupt transition in terms of the parameter X before (X_0) and after (X_1) the transition. We use the subscripts 0 and 1 to define the waveguide regions on each side of the transition, as shown in Fig. 3.9, where mode propagation is assumed from left to right. The transition can be either two-dimensional, with planar waveguides as in Fig. 3.9a, or three-dimensional, with channel waveguides as in Fig. 3.9b. We assume only an input mode i, with unity power incident on the left, and that the transition will cause a power transfer P from mode i to mode j. The input mode i is represented by

$$\psi_{i0} = d_0 \phi_a + e_0 \phi_b \tag{3.2.2}$$

and to the right of the boundary the output modes are

$$(1 - P)^{1/2} \psi_{i1} = (1 - P)^{1/2} d_1 \phi_a + (1 - P)^{1/2} e_1 \phi_b \quad , \tag{3.2.3}$$

$$P^{1/2} \psi_{j1} = -P^{1/2} e_1 \phi_a + P^{1/2} d_1 \phi_b \quad , \tag{3.2.4}$$

where we have neglected loss to radiation modes and conserved power between the output modes. The requirement of continuous transverse electric fields across the boundary (equating the coefficients of ϕ_a and ϕ_b on each side) leads to

$$d_0 = (1 - P)^{1/2} d_1 - P^{1/2} e_1 \quad , \tag{3.2.5}$$

$$e_0 = (1 - P)^{1/2} e_1 + P^{1/2} d_1 \quad . \tag{3.2.6}$$

Substituting $f_0 = e_0/d_0$ and $f_1 = e_1/d_1$ gives the desired relationship for power transfer

$$P = \frac{(f_0 - f_1)^2}{(f_0 - f_1)^2 + (1 + f_1 f_0)^2} \quad . \tag{3.2.7}$$

The power transfer P is plotted vs X_1 for various values of X_0 in Fig. 3.10.

3.2.4 Tapered Velocity Coupler

With the understanding gained so far on local normal modes and modal evolution, it is possible to understand the operation of complex devices involving waveguide transitions. In Fig. 3.11, we show a tapered velocity coupler [3.10], which consists of two guiding layers with constant separation between them, but with a linear decrease in thickness in the top layer. The uncoupled-mode effective indices become identical in the center of the coupler. The value of X goes from a large positive value (at the left of Fig. 3.11), through zero, to a large negative value (at the right of Fig. 3.11).

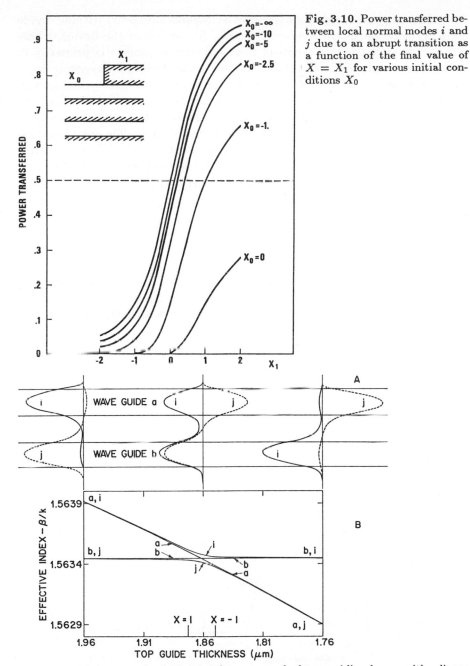

Fig. 3.10. Power transferred between local normal modes i and j due to an abrupt transition as a function of the final value of $X = X_1$ for various initial conditions X_0

Fig. 3.11. (a) A tapered velocity coupler composed of two guiding layers with a linear decrease in the thickness of the top guiding layer. The first (i) and second (j) order local normal-mode amplitudes are plotted at various positions along the coupler. (b) Effective index (β/k) for the various modes as a function of position along the tapered velocity coupler of (a). Curves labeled i and j represent the local normal-modes; a and b represent the uncoupled modes of the two three-layer waveguides taken separately

This device operates solely in the slow or adiabatic regime, without mode coupling between the local normal modes. At the left of the device, where $\Delta\beta$ and X are large and positive, the local normal mode i is primarily in waveguide a, and the local normal mode j is primarily in waveguide b. At the device center with $X = 0$ the local normal modes are symmetrically divided between the guides. However, at the right of the device, where $\Delta\beta$ and X are large and negative, the local normal mode i is now associated with guide b, and the local normal mode j with guide a. Thus, by injecting power into one waveguide at either end, primarily one local normal mode is excited which travels through the device without mode coupling, but emerges from the other waveguide at the other end of the device. Coupling between waveguides is achieved through the change of shape, i.e., modal evolution, of the local normal mode.

3.2.5 3 dB Coupler

A somewhat more complicated situation is shown in the 3 dB coupler of Fig. 3.12 [3.5]. Here the waveguides are identical and the guide separation constant, but a superstrate (layer 5) is introduced abruptly to make the

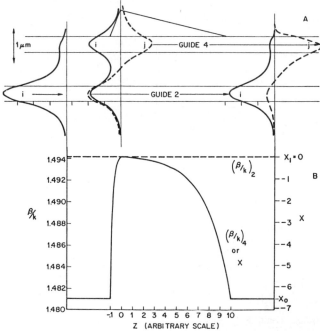

Fig. 3.12. (a) A cross section of a planar modal evolution 3-dB coupler with the electric field amplitudes of the local normal modes i and j at three positions along the propagation direction z. The input mode i has unity power, and 3-dB coupling is shown. (b) The corresponding values of $X = \Delta\beta/2|K|$ and the uncoupled mode effective indices are shown. To the left of $z = 0$, the z scale is exaggerated to show the fast taper

waveguides synchronous ($\Delta\beta = 0$), and removed slowly. At either end of the device, X is large and negative so that the local normal modes are associated primarily with one waveguide or the other. Under the superstrate, $X = 0$ by (3.1.2) and the power in each local normal mode is evenly divided between the waveguides. With optical input from the left in mode i as shown, the mode sees the superstrate introduced abruptly and power transfer between the local normal modes occurs. If the change in X at the boundary is from $-\infty$ to 0, we see from Fig. 3.10 that this power transfer from mode i to mode j is just 1/2 or 3 dB. The relative phase between i and j, as shown in Fig. 3.12 is obvious from a consideration of the boundary conditions used to derive (3.2.7), i.e., the mode fields cancel in guide 4 and add in guide 2. The superstrate is now removed slowly so that adiabatic modal evolution occurs as the modes propagate to the right of the device. At the end of the device the local normal modes are associated with individual guides so that the input power in guide 2 is evenly divided at the output between the waveguides. The devices acts as a power splitter or 3 dB coupler. The reader may convince himself that the coupler operates identically in either direction.

Three-dimensional configurations of this device using channel waveguides (Fig. 3.13) with fast and slow branches operate analogously, in agreement with the discussion given for the planar device. This coupler has been called a modal evolution 3 dB coupler [3.5], and an optical-waveguide hybrid coupler [3.11].

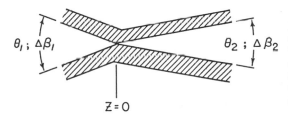

Fig. 3.13. Sketch of a two-dimensional, channel modal evolution 3-dB coupler. A single region of mode conversion ($z < 0$) acts as a power divider and is followed by an adiabatic region that acts as a mode splitter

3.2.6 Directional Coupler

Next we consider the operation of the system of two coupled waveguides shown in Fig. 3.14a. This device, known as a directional coupler [3.12], consists of an abruptly converging branch, an adiabatic section of parallel coupled waveguides, and an abruptly diverging branch. We can assume that the input arms are uncoupled so that $|K| = 0$ and $|X| = \infty$ at the input, but that the guides are nearly synchronous so that $\Delta\beta \simeq 0$ and $X = 0$ in the coupled region. Then, for an input of unity power in guide 1, the local normal modes i and j are excited with equal power and phase, as shown in Fig. 3.14b, at the start of the coupled waveguide section ($z = 0$). The mode amplitudes are $a_i = a_j = 1/2$. This is the assumption of a perfect power divider. In the adiabatic section, the local normal modes propagate

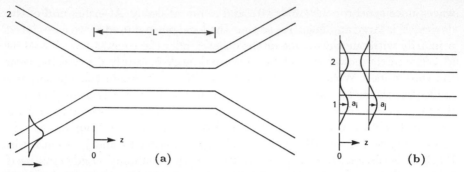

Fig. 3.14. Model of a directional coupler of length L (a). Input in waveguide 1 in (a) results in excitation of normal modes i and j at $z = 0$ as shown in (b)

with different propagation constants β_i, β_j and interfere as they propagate to the end of the section at $z = L$. The mode amplitudes in each guide may be expressed as

$$\psi_1 = \tfrac{1}{2}(e^{j\beta_i z} + e^{j\beta_j z}) \quad , \tag{3.2.8}$$

$$\psi_2 = \tfrac{1}{2}(e^{j\beta_i z} - e^{j\beta_j z}) \quad , \tag{3.2.9}$$

and the mode powers are

$$P_1 = \psi_1 \psi_1^* = \tfrac{1}{2}(1 + \cos \Delta\beta_{ij} z) \quad , \tag{3.2.10}$$

$$P_2 = \psi_2 \psi_2^* = \tfrac{1}{2}(1 - \cos \Delta\beta_{ij} z) \quad , \tag{3.2.11}$$

where $\Delta\beta_{ij} = \beta_i - \beta_j$ is the difference in propagation constants of the normal modes. Complete crossover occurs at the end of the interference section when $P_2(L) = 1$ or $L = \pi/\Delta\beta_{ij}$. From (3.1.14, 15) we have $\Delta\beta_{ij} = 2|K|$ when $X = 0$, so the crossover condition for synchronous operation may also be expressed as $L = \pi/2|K|$. At $z = L$ we again assume a perfect power divider so that the power distribution at $z = L$ is reproduced at the output of the diverging branch. Equations (3.2.10, 11) may then be taken as the output powers of the device. Obviously, the operation of a directional coupler becomes more complicated if the waveguides are not synchronous, or if the initial or final branches are not perfect power dividers.

3.3 Mode Coupling Between Local Normal Modes

So far we have considered the limiting cases of very fast and very slow transitions. To deal with general transitions, we must begin to introduce the appropriate mathematical formulism required to treat them. We have seen that coupled-waveguide structures that do not vary in the direction of propagation can be treated by normal modes which are orthogonal. In waveguide structures that do vary in the direction of propagation, we have introduced the concept of local normal modes. We will see that the operation of waveguide transitions can be described by the coupling between local normal modes, and that such coupling is calculated from an overlap integral between local normal modes across a transition.

3.3.1 Coupled-Amplitude Equations

We start by considering the coupled waveguides shown in Fig. 3.15 where a small abrupt step in the guide separation h_3 occurs at some point along the propagation direction z [3.1]. The waveguide modes of interest are assumed to be incident on the step, and we compute the transmitted mode amplitudes by requiring the transverse-field components to be continuous at the step.

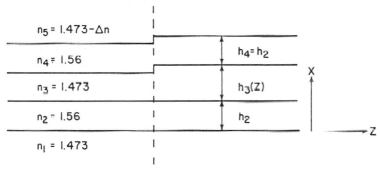

Fig. 3.15. Model of one of the steps used to approximate a continuous taper slope. Note the geometric asymmetry assumed to approximate a planar geometry

We restrict our attention to two transverse electric guided modes, i and j. Guided-mode electric fields are written as

$$E_y = a(z)\mathcal{E}(x,z)e^{j\alpha(z)} \quad , \quad \text{where} \tag{3.3.1}$$

$$\alpha(z) = \beta z + \phi \quad . \tag{3.3.2}$$

Here $a(z)$ is the real amplitude at a point z, $\mathcal{E}(x,z)$ is the real x-dependent field distribution, β is the mode wave vector, and ϕ is a phase constant. We will assume that the effect of reflected radiation into unguided modes E^{ref} is small, and can ultimately be neglected. The transmitted radiation into unguided modes E^{trans} is treated exactly. Boundary conditions for TE modes require E_y and H_x to be continuous across a step at z. We therefore obtain the equations:

$$E_y \,:\, a_{i0}\mathcal{E}_{i0}e^{j\alpha_{i0}} + a_{j0}\mathcal{E}_{j0}e^{j\alpha_{j0}} + a_{i1}^{\text{R}}\mathcal{E}_{i0}e^{-j\alpha_{i0}^{\text{R}}} + a_{j0}^{\text{R}}\mathcal{E}_{j0}e^{-j\alpha_{j0}^{\text{R}}} + E^{\text{ref}}$$
$$= a_{i1}\mathcal{E}_{i1}e^{j\alpha_{i1}} + a_{j1}\mathcal{E}_{j1}e^{j\alpha_{j1}} + E^{\text{trans}} \quad , \tag{3.3.3}$$

$$H_x : \beta_{i0}a_{i0}\mathcal{E}_{i0}e^{j\alpha_{i0}} + \beta_{j0}a_{j0}\mathcal{E}_{j0}e^{j\alpha_{j0}} - \beta_{i0}a_{i0}^{\mathrm{R}}\mathcal{E}_{i0}e^{-j\alpha_{i0}^{\mathrm{R}}}$$
$$- \beta_{j0}a_{j0}^{\mathrm{R}}\mathcal{E}_{j0}e^{-j\alpha_{j0}^{\mathrm{R}}} + H^{\mathrm{ref}}$$
$$= \beta_{i1}a_{i1}\mathcal{E}_{i1}e^{j\alpha_{i1}} + \beta_{j1}a_{j1}\mathcal{E}_{j1}e^{j\alpha_{j1}} + H^{\mathrm{trans}} \quad , \tag{3.3.4}$$

where the subscripts 0 and 1 are used to denote the incident and transmitted fields across the step, and a^{R} is the amplitude of reflected guided modes. Multiplication by a transmitted mode field and integration over x eliminates the transmitted unguided radiation modes, because of orthogonality.

Our goal now is to obtain iterative equations for a_{j1} and α_{j1} from (3.3.3, 4). If we consider the cross reflection between modes i and j to be small, multiplication of these equations by $\int \mathcal{E}_{i1}dx$, and subtraction of the resulting equations, gives an approximate equation for a_{i0}^{R}, which takes the form

$$a_{i0}^{\mathrm{R}}e^{j\alpha_{i0}^{\mathrm{R}}} \simeq \left(\frac{\beta_{i0} - \beta_{i1}}{\beta_{i0} + \beta_{i1}}\right)a_{i0}e^{j\alpha_{i0}} + 0(a_{j0}\int \mathcal{E}_{j0}\mathcal{E}_{i1}dx) \quad . \tag{3.3.5}$$

Now multiply (3.3.3 and 4) by $\int \mathcal{E}_{j1}dx$ and subtract out the term in a_{j0}^{R}. Using (3.3.5) in the resulting equation gives

$$a_{j1}I_{j1,j1}e^{j\alpha_{j1}} \simeq a_{i0}I_{i0,j1}e^{j\alpha_{i0}} + a_{j0}I_{j0,j1}e^{j\alpha_{j0}}$$
$$+ 0(a_{j0}I_{j0,i1}I_{i0,j1}) \quad , \tag{3.3.6}$$

where we have neglected small differences in propagation constant both across the step and between the modes, and

$$I_{\gamma,\delta} = \int \mathcal{E}_\gamma \mathcal{E}_\delta dx \quad , \quad \gamma, \delta = i, j \quad . \tag{3.3.7}$$

Equation (3.3.6) is conveniently normalized by introducing a mode amplitude (a^s) which corresponds to a mode power of unity. At any point z, a^s is related to mode power (P) by

$$\sqrt{P_\gamma} = 1 = a_\gamma^s\sqrt{\frac{\beta_\gamma}{2k_0}I_{\gamma,\gamma}} \quad , \quad \gamma = i, j \tag{3.3.8}$$

where k_0 is the free-space propagation constant. By dividing the terms of (3.3.6) by appropriate factors of the right-hand side of (3.3.8), and neglecting terms which contain the square of an overlap integral between different modes on opposite sides of the step, we obtain

$$A_{j1}e^{j\alpha_{j1}} = c_{ij}A_{i0}e^{j\alpha_{i0}} + c_{jj}A_{j0}e^{j\alpha_{j0}} \quad . \tag{3.3.9}$$

Here $A_\gamma = a_\gamma/a_\gamma^s$ is the ratio of the mode amplitude in the presence of mode conversion to that mode amplitude which corresponds to unity power. The coefficient c_{ij} is given by

$$c_{ij} \simeq \frac{I_{i0,j1}}{\sqrt{I_{i0,i0}I_{j1,j1}}} \quad ,$$

(3.3.10)

and c_{jj} is obtained by substituting j for i in (3.3.10).

3.3.2 Differential Form of Coupled-Amplitude Equations

In (3.3.9, 10) we have the coupled-amplitude equations that describe mode coupling between local normal modes at a single step. One approach, which we will use later on, is to divide the waveguide transition of interest into many small steps and iteratively apply these equations to determine the output mode amplitudes [3.9]. What we will do here is transform these equations into differential equations and show that these equations have analytic solutions for transitions shaped in a particular way [3.13]. These solutions will prove to be useful in our later investigation of horns and branches.

From (3.3.9) and its counterpart for A_{i1} we have coupled amplitude equations of the form

$$A_{j1} = \sum_{\gamma} c_{\gamma i} A_{\gamma 0} \quad , \quad \gamma = i, j \quad ,$$

(3.3.11)

$$A_{i1} = \sum_{\gamma} c_{\gamma j} A_{\gamma 0} \quad , \quad \gamma = i, j \quad ,$$

(3.3.12)

which relate the transmitted local normal mode amplitudes on side 1 of a small step to the local normal mode amplitudes incident on side 0. The amplitudes are normalized so that $|A| = 1$ corresponds to unity power. We generalize to a "step" in which an unspecified parameter P varies from P to $P + \delta P$ across the step. Thus, P may be the channel width in an expansion horn or the waveguide separation in a separating waveguide structure. We expect that c_{ij} will be proportional to the change δP across the step, and remove this dependence by defining a new coefficient

$$C_{ij} = \lim_{\delta P \to 0} \left(\frac{c_{ij}}{\delta P} \right) \quad .$$

(3.3.13)

For a small step $c_{jj} = c_{ii} \simeq 1$, $c_{ij} = -c_{ji} = C_{ij}\delta P$, and (3.3.11, 12) reduce to

$$A_{j1} - A_{j0} = C_{ij}\delta P A_{i0} \quad ,$$

(3.3.14)

$$A_{i1} - A_{i0} = -C_{ij}\delta P A_{j0} \quad .$$

(3.3.15)

If we define propagation between the steps at which P changes by $A_i = |A_i|\exp[j(\beta_i z + \phi_i)]$, we obtain the differential equations describing coupling between the local normal modes, namely

105

$$\frac{dA_j}{dz} = C_{ij}\frac{dP}{dz}A_i + jB_jA_j \quad , \tag{3.3.16}$$

$$\frac{dA_i}{dz} = -C_{ij}\frac{dP}{dz}A_j + j\beta_iA_i \quad , \tag{3.3.17}$$

where β_i and β_j are the propagation constants for the local normal modes.

If $C_{ij}dP/dz = 0$, no power transfer between local normal modes occurs and the evolution of local normal mode i is described by a phase factor $\exp[j\int_0^z \beta_i dz']$. For convenience, we consider reduced mode amplitudes a_i and a_j such that

$$a_i = A_i\exp\left[-j\int_0^z \beta_i dz'\right] \quad , \tag{3.3.18}$$

$$a_j = A_j\exp\left[-j\int_0^z \beta_j dz'\right] \quad . \tag{3.3.19}$$

With this notation, (3.3.16, 17) become

$$\frac{da_j}{dz} = C_{ij}\frac{dP}{dz}a_i\exp\left[j\int_0^z (\beta_i - \beta_j)dz'\right] \quad , \tag{3.3.20}$$

$$\frac{da_i}{dz} = -C_{ij}\frac{dP}{dz}a_j\exp\left[-j\int_0^z (\beta_i - \beta_j)dz'\right] \quad . \tag{3.3.21}$$

The variation of P with z is determined by the shape of the transition. Equations (3.3.20) and (3.3.21) have an analytic solution only for selected shapes. For a shape such that

$$\frac{dP}{dz} = \gamma\left(\frac{\Delta\beta_{ij}}{C_{ij}}\right) \quad , \tag{3.3.22}$$

where $\Delta\beta_{ij} = \beta_i - \beta_j$ and γ is an arbitrary constant, (3.3.20, 21) reduce to

$$\frac{da_j}{du} = \gamma e^{ju}a_i \quad , \tag{3.3.23}$$

$$\frac{da_i}{du} = -\gamma e^{-ju}a_j \quad , \quad \text{where} \tag{3.3.24}$$

$$u = \int_0^z \Delta\beta_{ij}dz' = \int_{P_0}^P \frac{C_{ij}}{\gamma}dP' \quad , \tag{3.3.25}$$

and $P(z) = P_0$ at $z = 0$. Equations (3.3.23, 24) are in the form of standard

106

coupled mode equations and have the analytic solution

$$a_j = a_{i0} \frac{2\gamma}{(4\gamma^2 + 1)^{1/2}} e^{ju/2} \sin\left[\tfrac{1}{2}(4\gamma^2 + 1)^{1/2} u\right] \quad , \tag{3.3.26}$$

$$a_i = a_{i0} e^{-ju/2} \left\{ \cos\left[\tfrac{1}{2}(4\gamma^2 + 1)^{1/2} u\right] \right.$$
$$\left. + \frac{j}{(4\gamma^2 + 1)^{1/2}} \sin\left[\tfrac{1}{2}(4\gamma^2 + 1)^{1/2} u\right] \right\} \quad , \tag{3.3.27}$$

for the initial conditions $a_i = a_{i0}$ and $a_j = 0$ at $z = 0$.

The quantity γ may be considered as a measure of the strength of the coupling between the local normal modes. For $\gamma < 1$ the amount of power converted to mode j oscillates with z; however, for small γ, it has a maximum value of

$$\frac{P_j^{\max}}{P_{i0}} = \frac{(a_j a_j^*)_{\max}}{(a_{i0} a_{i0}^*)} = \frac{4\gamma^2}{4\gamma^2 + 1} \quad . \tag{3.3.28}$$

For large γ the amount of power conversion is controlled by the limiting value of $\sin\tfrac{1}{2}[(4\gamma^2 + 1)^{1/2} u]$. Adiabatic structures should thus be designed with $\gamma \ll 1$ while abrupt structures need $\gamma > 1$. The approach considered here is valid if the principal coupling occurs between mode i and mode j, and coupling to other modes can be neglected.

The shape described by (3.3.22) holds $(dP/dz)(C_{ij}/\Delta\beta_{ij})$ constant throughout the length of the transition. This corresponds to the case of constant hypercoupling coefficient as defined by *Louisell* [3.8]. The pressure for mode conversion between local normal modes is the same everywhere. If we consider averages over mode interference effects, this should produce a transition which is as short as it can be for a given amount of allowed mode conversion.

3.3.3 Coupled-Mode Theory Representation of C_{ij}

By now we see that coupling between local normal modes in a transition structure is governed by the coupling coefficient between them, C_{ij}. This coupling coefficient can be obtained from an overlap integral between the local normal modes at a small step, or, since we have analytic representations for the local normal modes from coupled-mode theory, we can use these representations to calculate an analytic expression for C_{ij} [3.14]. We will do this here for a branching structure (Fig. 3.16) where the separation between the two guiding regions is increasing across the step. From (3.3.10 and 13) we have

$$c_{ij} = C_{ij} \Delta W_3 \simeq \int_{-\infty}^{\infty} \psi_{i0} \psi_{j1} dx \quad , \tag{3.3.29}$$

where ΔW_3 is the increase in the separation between the two guiding regions

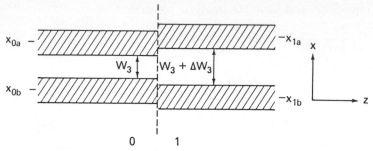

Fig. 3.16. One step of a step transition model for a waveguide branch with a symmetric shape ($x_{1a} - x_{0a} = x_{0b} - x_{1b}$)

across the step, ψ_{i0} refers to ψ_i before the step on side 0, and ψ_{j1} refers to ψ_j after the increase in W_3 on side 1. In the approximation of coupled-mode theory, we can write from (3.1.5, 6) that

$$\psi_{i0} = d_0\phi_a(x - x_{0a}) + e_0\phi_b(x - x_{0b}) \quad , \tag{3.3.30}$$

$$\psi_{j1} = -e_1\phi_a(x - x_{1a}) + d_1\phi_b(x - x_{1b}) \quad , \tag{3.3.31}$$

where x_{0a} is the center of the waveguiding region a on side 0, x_{0b} is the center of the waveguiding region b on side 0, x_{1a} is the center of the waveguiding region a on side 1, and x_{1b} is the center of the waveguiding region b on side 1. Here d_0 and e_0 refer to side 0, whereas e_1 and d_1 refer to side 1. With (3.3.30, 31), C_{ij} can be written as

$$C_{ij}\Delta W_3 = -d_0e_1 \int_{-\infty}^{+\infty} \phi_a(x - x_{0a})\phi_a(x - x_{1a})dx$$

$$+ e_0d_1 \int_{-\infty}^{+\infty} \phi_b(x - x_{0b})\phi_b(x - x_{1b})dx \quad , \tag{3.3.32}$$

assuming that, for well separated guiding regions, we have

$$\int_{-\infty}^{+\infty} \phi_a(x - x_{0a})\phi_b(x - x_{1b})dx$$

$$= \int_{-\infty}^{+\infty} \phi_b(x - x_{0b})\phi_a(x - x_{1a})dx = 0 \quad . \tag{3.3.33}$$

If

$$\int_{-\infty}^{+\infty} \phi_a(x - x_{0a})\phi_a(x - x_{1a})dx$$

$$= \int_{-\infty}^{+\infty} \phi_b(x - x_{0b})\phi_b(x - x_{1b})dx \approx 1 \quad , \tag{3.3.34}$$

then

$$C_{ij}\Delta W_3 = e_0 d_1 - d_0 e_1 \quad . \tag{3.3.35}$$

If we use $f = e/d$, (3.3.35) becomes

$$C_{ij}\Delta W_3 = d_0 d_1 (f_0 - f_1)$$
$$\simeq \frac{-1}{1+f^2} \frac{\partial f}{\partial W_3} \Delta W_3 \quad , \tag{3.3.36}$$

and, using (3.1.1, 2, and 11), we have

$$\frac{\partial f(W_3)}{\partial W_3} = \frac{\gamma_3 \Delta\beta}{2|K|} \left(\frac{\Delta\beta}{\Delta\beta^2 + 4|K|^2} - 1 \right) \quad , \tag{3.3.37}$$

so that

$$C_{ij} = \frac{\gamma_3 X}{2(X^2+1)} \quad . \tag{3.3.38}$$

Equation (3.3.34) is only valid if the step is symmetric (i.e., if $x_{1a} - x_{0a} = x_{0b} - x_{1b}$), which implies a symmetric shape for the branch. If the branch shape is not symmetric, there will be a contribution to C_{ij} which does not depend on a change in the ratio e/d, and which is not included in (3.3.38). For example, the step depicted in Fig. 3.15 is not a symmetric step, and would not lead to a symmetrically shaped branch.

3.4 Two-Arm Branches

In this section, we begin to treat real transition structures that may be intermediate between their fast and slow limits. We would also like to be able to develop criteria that quantify what the fast and slow limits are for a given structure. We will use the example of a separating waveguide, which may be a waveguide branch, as in Fig. 3.17a, or the separating region of two closely coupled waveguides, as in Fig. 3.17b. It is important to note that this analysis applies to the transitions which occur at each end of a directional coupler which are usually assumed to be abrupt.

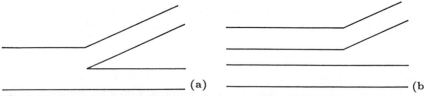

Fig. 3.17. Two types of separating optical wavegudies. Propagation in both structures can be described in an equivalent manner if the parallel sections are sufficiently close in (b) for the normal modes to have power in both guiding regions

3.4.1 Step Approximation for a Waveguide Branch

We consider the step model of Fig. 3.15 where Δn is an index asymmetry parameter. Power transfer between local normal modes at the step is governed by (3.3.9) and its counterpart for mode i. The mode coupling coefficient c_{ij} is given by (3.3.10). The step model has been used to approximate linear branches with various taper slopes and index asymmetries [3.1]. Computer calculations were used to compute the mode propagation constants and electric fields between steps, to evaluate the mode coupling coefficients, and to perform the iterations describing changes in mode amplitude at each step.

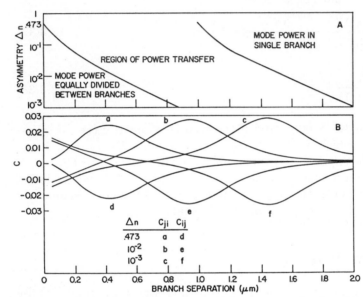

Fig. 3.18. (a) Branch separations at which power transfers from both branches to one as a function of branching waveguide index asymmetry for the model waveguide. **(b)** Plots of mode-conversion coefficients versus branch separation for various index asymmetries. The mode-conversion coefficients are large in the region of power transfers where modal evolution is taking place

Plots of mode electric-field distributions versus increasing branch separation have already been shown in Fig. 3.4. In Fig. 3.18, we show the mode conversion coefficient c_{ij} versus branch separation for each index asymmetry. In this case, the coefficient c_{ij} is initially positive and then peaks at a negative value. The sign of c_{ij} is a consequence of the initial phases assumed for the modes in Fig. 3.4. The initial positive contribution for $\Delta n = 10^{-2}$ or 10^{-3} is a consequence of the geometric asymmetry of the branch structure assumed in Fig. 3.15 (Sect. 3.3.3). We also show in Fig. 3.18, for each index asymmetry, the region of modal evolution during which the mode power evolves from being evenly divided between the branches to being predomi-

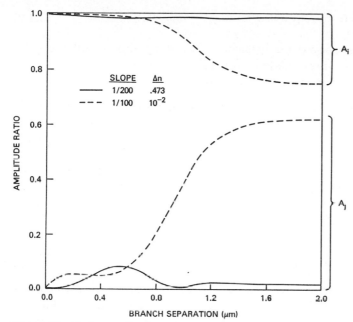

Fig. 3.10. Mode amplitude ratios versus branch separation for mode conversion from mode i to j in the model branching waveguide

nantly in a single branch. This region of modal evolution occurs more slowly, as measured by the amount of branch separation, as the branch becomes more symmetric. The region of modal evolution also coincides with the peak in c_{ij}, for each index asymmetry, implying that the overlap integrals between the modes i and j are largest when the mode shapes are undergoing the greatest change.

In Fig. 3.19, we show the calculated mode amplitudes as a function of branch separation, for two cases of taper slope and index asymmetry. In each case, mode i is the input mode. In the more asymmetric case ($\Delta n = 0.473$, slope $= 1/200$) essentially no cumulative mode conversion to mode j occurs, whereas in the more symmetric branch ($\Delta n = 10^{-2}$, slope $= 1/100$) substantial mode conversion to mode j does occur. In general, increased mode conversion from mode i to mode j occurs as the branches become more symmetric (small Δn) and as the branch tapers become more steep (abrupt limit).

This behavior can be understood by examining (3.3.9) when the condition

$$(\beta_{i0} - \beta_{j1})z \simeq 0 \tag{3.4.1}$$

is satisfied over the length of the branch. For a small step $\beta_{j0} \approx \beta_{j1}$, so that we also have

$$(\beta_{j0} - \beta_{j1})z \simeq 0 \quad . \tag{3.4.2}$$

Equation (3.4.1) is obeyed for branches with steep slopes and for near-symmetric branches whose modes are nearly degenerate. Equation (3.3.9) can then be expressed in real form

$$A_{j1} = c_{ij}A_{i0}\cos(\phi_{i0} - \phi_{j1}) + c_{jj}A_{j0}\cos(\phi_{j0} - \phi_{j1}) \quad , \tag{3.4.3}$$

$$0 = c_{ij}A_{i0}\sin(\phi_{i0} - \phi_{j1}) + c_{jj}A_{j0}\sin(\phi_{j0} - \phi_{j1}) \quad . \tag{3.4.4}$$

If we assume a phase $\phi_{i0} = 0$ for mode i, and a step small enough so that the phase of mode j changes very little across a step, i.e., $\phi_{j0} - \phi_{j1} \simeq 0$, then (3.4.3, 4) are satisfied by taking $\phi_{j1} = 0$ or π. A distinction can be made by expanding (3.4.4) to give

$$\tan\phi_{j1} = \frac{c_{ij}A_{i0}\sin\phi_{i0} + c_{jj}A_{j0}\sin\phi_{j0}}{c_{ij}A_{i0}\cos\phi_{i0} + c_{jj}A_{j0}\cos\phi_{j0}} \quad . \tag{3.4.5}$$

If c_{ij} were always negative, ϕ_{j1} would be locked at π and cumulative mode conversion would occur over the whole length of the taper. If c_{ij} is initially positive and then turns negative ($\Delta n = 10^{-2}, 10^{-3}$), $\phi_{j1} = 0$ until the denominator of (3.4.5) turns negative at

$$c_{ij} = -c_{jj}\frac{A_{j0}}{A_{i0}} \quad , \tag{3.4.6}$$

after which $\phi_{j1} = \pi$. After $\phi_{j1} = \pi$, cumulative mode conversion occurs.

The phase behavior for these two cases is shown in Fig. 3.20, where c_{ij} and $\cos(\alpha_{i0} - \alpha_{j1})$ are plotted versus branch separation. In the case of large asymmetry and small slope ($\Delta n = 0.473$, slope $= 1/200$), the phase term rapidly oscillates in the region where c_{ij} is large and no accumulation of converted mode amplitude is possible. In the case of small asymmetry and larger slope ($\Delta n = 10^{-2}$, slope $= 1/100$), the phase difference $\alpha_{i0} - \alpha_{j1}$ is locked to π through the peak of c_{ij}, and cumulative mode conversion occurs. We conclude that the converted mode field to mode j must be in phase with the field already in that mode for cumulative mode conversion to occur. This phase matching requirement is similar to other mode conversion processes.

The electric field profiles for these two cases of asymmetry and taper slope is shown in Fig. 3.21. In the case without mode conversion (Fig. 3.21a), the mode field evolves until it is in the lower arm of the branch. This case

Fig. 3.21. Examples of a branching waveguide operated as a mode splitter (a) and as a power divider (b). Mode electric field is plotted for various waveguide cross sections, with indices as shown in Fig. 3.15. In (a) 98 % of the input power remains in mode i. In (b) output powers are 55 % for mode i and 36 % for mode j

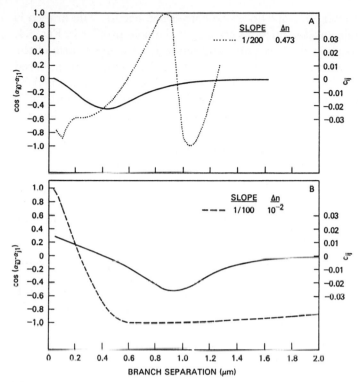

Fig. 3.20. Plots of the phase term and mode conversion coefficient for various cases of slope and index asymmetry versus branch separation. (a) slope $= 1/200$, $\Delta n = 0.473$. (b) slope $= 1/100$, $\Delta n = 10^{-2}$

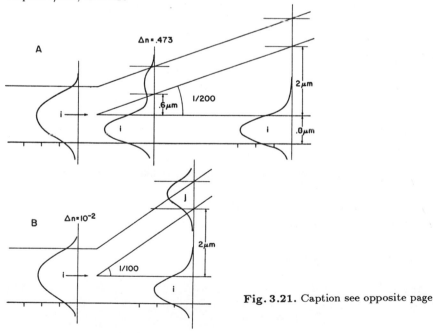

Fig. 3.21. Caption see opposite page

is clearly slow or adiabatic, since no mode conversion occurs. The branch is referred to as a mode splitter since, as shown by the field profiles in Fig. 3.4, normal mode i ends up in the lower arm and mode j in the upper arm of the branch. In the case with mode conversion (Fig. 3.21b), power is transferred from the input mode i to mode j. The phase of mode j is determined by the phase of the input mode i, and the branch operates as if the input mode is spatially divided between the branches. In this mode of operation the branch is referred to as a power divider.

The considerations used in the discussion (3.4.1–5) lead us to conclude that, as long as

$$(\beta_{i0} - \beta_{j1})\Delta z \ll \tfrac{\pi}{2} \quad , \tag{3.4.7}$$

considerable mode conversion will occur and the device will act as a power divider; on the other hand, if

$$(\beta_{i0} - \beta_{j1})\Delta z \gg \tfrac{\pi}{2} \quad , \tag{3.4.8}$$

cumulative mode conversion will not occur and the device will act as a mode splitter. Here Δz represents the width of the region where c_{ij} is appreciable. It is thus possible for the width of the peak in c_{ij} to influence the amount of cumulative mode conversion; however, the position of the peak will be unimportant. For a linear taper

$$\Delta z = \frac{\Delta h_3}{\theta} \quad , \tag{3.4.9}$$

where Δh_3 is the width of the peak in c_{ij} in units of branch separation (measured at the half maximum points), and θ is the taper slope for the separating guides. Clearly, only the slope of the taper in the region of appreciable c_{ij} is important.

Next we can use coupled-mode theory to provide a more useful representation of (3.4.7,8). Using (3.3.13 and 38), the coupled-mode theory representation for c_{ij} is

$$c_{ij} = \frac{\gamma_3 \delta h_3 X}{2(X^2 + 1)} \quad . \tag{3.4.10}$$

From this expression, we can determine that c_{ij} peaks at $X = 1$, or in terms of branch separation, at

$$(h_3)_{\text{peak}} = -\frac{1}{\gamma_3}\ln\left(\frac{\Delta\beta}{2F}\right) \quad . \tag{3.4.11}$$

The half-width for c_{ij} will be

$$\Delta h_3 = \frac{2.6}{\gamma_3} \quad . \quad . \tag{3.4.12}$$

The coupled-mode theory representation of $\Delta\beta_{ij} = \beta_i - \beta_j$ is obtained by subtracting (3.1.14, 15)

$$\Delta\beta_{ij} = 2|K|(X^2 + 1)^{1/2} \quad , \tag{3.4.13}$$

which at $X = 1$ becomes

$$\Delta\beta_{ij} = \sqrt{2}\Delta\beta \quad . \tag{3.4.14}$$

If we approximate $\beta_{i0} - \beta_{j1} \simeq \Delta\beta_{ij}$ the criteria for the fast (power divider) or slow (mode splitter) branches (3.4.7, 8) becomes

$$\frac{\Delta\beta}{\theta\gamma_3} \gtrless 0.43 \quad , \tag{3.4.15}$$

where we have used (3.4.9, 12, and 14). In (3.4.15) the branch is a mode splitter if the upper limit holds and a power divider if the lower limit holds. Equation (3.4.15) thus provides a criterion to determine the mode of operation of a branch in terms of its parameters. In Fig. 3.22 this criterion is shown with the mode converted amplitude from several of the branches of [3.1] plotted vs $\Delta\beta/\theta\gamma_3$. Plotted in this way, the data lies on a single curve. We note that (3.4.15) only holds for $\Delta\beta \neq 0$.

3.4.2 Analytic Solution for Shaped Branches

If we are willing to consider shaped branches, use can be made [3.14] of the analytic solution to the coupled normal mode equations of Sect. 3.3.2. The

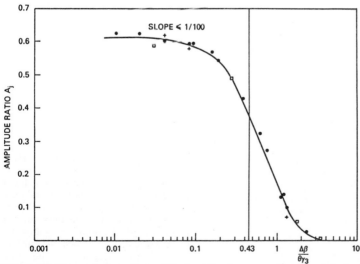

Fig. 3.22. Converted mode amplitude ratio A_j versus the calculated parameter $\Delta\beta/\Theta\gamma_3$, which is indicative of mode-conversion magnitude. For abscissa values > 0.43, the separating waveguide acts as a mode splitter; < 0.43, as a power divider

waveguide parameter that varies is the branch arm separation W_3 and then (3.3.22), which defines the shape of the structure, becomes

$$\frac{dW_3}{dz} = \gamma\left(\frac{\Delta\beta_{ij}}{C_{ij}}\right) \quad . \tag{3.4.16}$$

Coupled-mode theory approximations for $\Delta\beta_{ij}$ and C_{ij} have been given in (3.4.13 and 38). We consider a branch or separating waveguide where θ is the local full branch angle. Then (3.4.16) becomes

$$\tan\left(\frac{\theta}{2}\right) = \frac{1}{2}\frac{dW_3}{dz} = \gamma\frac{\Delta\beta}{\gamma_3}\frac{(X^2+1)^{3/2}}{X^2} \quad , \tag{3.4.17}$$

where by using (3.1.1), the waveguide parameter X can be expressed as

$$X = X_0 e^{\gamma_3(W_3 - W_{30})} \quad . \tag{3.4.18}$$

Here X_0 is the value of X when $z = 0$ and $W_3 = W_{30}$. Using (3.4.18), (3.4.17) can be integrated to yield the required branch shape, which we express as

$$\cos(\tan^{-1}X_0) - \cos(\tan^{-1}X) = 2\gamma\Delta\beta z \quad . \tag{3.4.19}$$

The parameter X_0 is small for nearly symmetric branches and increases with $\Delta\beta$, which reflects the branch asymmetry, and with W_{30}, the initial branch separation of a separating waveguide. We plot the branch shape as defined by (3.4.18, 19) in Fig. 3.23 for different values of X_0. The minimum value of θ in (3.4.17) is

$$\theta_{\min} = 3^{3/2}\frac{\gamma\Delta\beta}{\gamma_3} \quad , \tag{3.4.20}$$

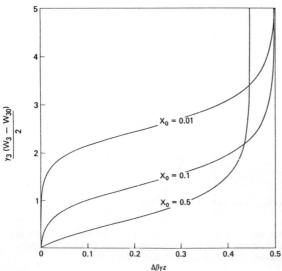

Fig. 3.23. A normalized plot of branch arm separation W_3 versus direction of propagation z for a shaped structure with initial condition X_0. The shape is symmetrc about a center line between the two guiding regions. γ is an arbitrary constant which determines both the steepness of the transition and the amount of power transfer

which occurs at $X = \sqrt{2}$, where the coupling effects are strongest ($C_{ij}/\Delta\beta_{ij}$ is a maximum). Then $W_3 \to \infty$ and θ approaches π when

$$z_{\max} = \frac{1}{2\gamma\Delta\beta(X_0^2 + 1)^{1/2}} \quad , \tag{3.4.21}$$

which means that, for $X_0 \ll 1$, the length of the structure is approximately $(2\gamma\Delta\beta)^{-1}$. Smaller values of γ correspond therefore to longer branches. Equation (3.3.25) for u can be evaluated as

$$u = \frac{1}{2\gamma}(\tan^{-1}X - \tan^{-1}X_0) \quad . \tag{3.4.22}$$

Finally, the cumulative power conversion from mode i to mode j is given by (3.3.26) and (3.3.18, 19), which yield

$$\frac{|A_j|^2}{|A_{i0}|^2} = \frac{4\gamma^2}{4\gamma^2 + 1} \sin^2\left[\frac{u}{2}(4\gamma^2 + 1)^{1/2}\right] \quad , \tag{3.4.23}$$

which is a function both of γ and the maximum value of u.

Equations (3.4.22, 23) describe in detail the power transfer between local normal modes in any branch or separating waveguide structure whose shape is defined by (3.4.19). Curves for $|A_j|^2/|A_{i0}|^2$ for structures with typical values of X_0 and γ are plotted as a function of propagation distance in the branch in Fig. 3.24, assuming $|A_j| = 0$ at $z = 0$. The quantity $\tan^{-1}X$

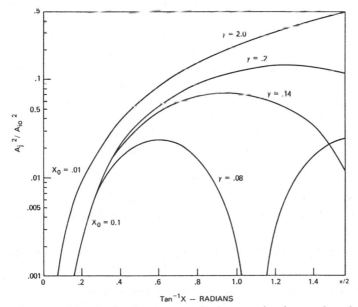

Fig. 3.24. Magnitude of power transfer between local normal modes in shaped separating waveguides as a function of position in the structure. Position is described by the waveguide parameter $X = \Delta\beta/2|K|$. Power transfer is shown for several values of γ, which is proportional to the steepness of the structure, and initial condition X_0

117

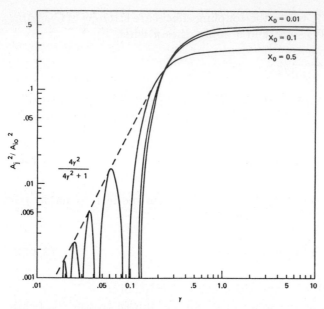

Fig. 3.25. Power transfer between local normal modes at the end of a shaped separating structure as a function of γ, for initial condition X_0. The oscillations in power transfer that occur at small γ are plotted for $X_0 = 0.5$

is used as a normalized parameter to describe position in the branch so that $W_3 = \infty$ at the end of the branch and $\tan^{-1}X = \pi/2$. In Fig. 3.25 we plot $|A_j|^2/|A_{i0}|^2$ at the end of the branch ($W_3 = \infty$) as a function of γ for several values of X_0. For branches with values of $\gamma \geq 1$, power buildup in the jth mode approaches 50% of the input power as $X_0 \to 0$. for values of $\gamma \leq 0.1$, oscillations occur and power is traded back and forth between mode i and mode j as the modes propagate through the branch. For the branch shapes considered here, the maximum value of this oscillation does not change with propagation distance and is given simply by $4\gamma^2/(4\gamma^2+1)$. This limiting value of power transfer for small γ is shown by the dashed curve in Fig. 3.25. For small γ, the taper becomes more gradual, leading to long structures with little power transfer between local normal modes.

For abrupt structures where $\gamma \to \infty$, (3.4.23) describes the power transfer which should depend upon the initial and final values of X, but not upon the shape of the structure in between. As $\gamma \to \infty$, we obtain from (3.4.22, 23) that

$$\lim_{\gamma \to \infty} \frac{|A_j|^2}{|A_{i0}|^2} = \sin^2\left[\frac{1}{2}(\tan^{-1}X - \tan^{-1}X_0)\right] \quad, \tag{3.4.24}$$

which, at the end of the structure (where $X \to \infty$), reduces to

$$\frac{|A_j|^2}{|A_{i0}|^2} = \frac{1}{2}\left(1 - \frac{X_0}{(X_0^2 + 1)^{1/2}}\right) \quad. \tag{3.4.25}$$

118

Equation (3.4.25) predicts the power transfer between local normal modes for an abrupt transition of any shape between $X = X_0$ and $X = \infty$. It is identical to the result obtained in (3.2.7) where only a single overlap integral was needed for the calculation.

It is instructive to compare the shaped branch with its analytical solution to the linear branch where θ does not vary with z. In the shaped structure, γ is a measure of the strength of the coupling between local normal modes, as shown by (3.4.23), and is held constant in the structure. In a linear branch, $\gamma \sim \theta\, C_{ij}/\Delta\beta_{ij}$ becomes z dependent and a plot of γ versus z shows where mode coupling takes place in the branch. In (3.4.20), we showed that the shaped structure has a minimum value of θ, which is independent of X_0 for any given structure. We can compare the shaped structure with a linear branch for which $\theta = \theta_{\min}$. By using (3.4.20) we then find that the shaped structure has a parameter

$$\frac{\Delta\beta}{\theta_{\min}\gamma_3} = \frac{1}{3^{3/2}\gamma} \quad . \tag{3.4.26}$$

The linear branch will have a maximum value of γ at $X = \sqrt{2}$ which, using (3.4.8), corresponds to

$$W_3 = \frac{1}{\gamma_3}\ln\left(\frac{\sqrt{2}}{X_0}\right) + W_{30} \quad , \tag{3.4.27}$$

which is the same value of X where $\theta_{\min}$ occurs for the shaped structure.

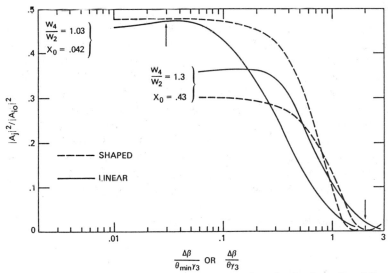

Fig. 3.26. Power transfer between local normal modes in shaped and linear branches is plotted versus the waveguide parameter $\Delta\beta/\Theta\gamma_3$, where the branch angle Θ for the linear branch is taken equal to the minimum angle $\Theta_{\min}$ of the corresponding shaped branch. Power transfer for the shaped branches is calculated from equations given in the text

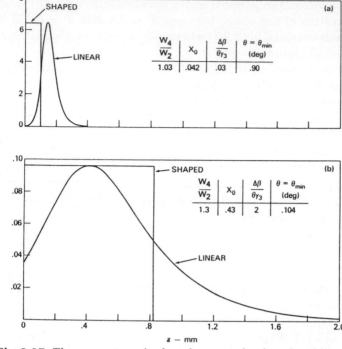

Fig. 3.27. The parameter γ is plotted versus z for shaped and linear branches specified by the vertical arrows in Fig. 3.26. The structures in (**a**) act as power dividers and in (**b**) as mode splitters. This plot shows where the power transfer occurs in the linear branches and demonstrates the length advantages of the shaped branches

We make these comparisons in Figs. 3.26 and 27 where we compare two linear branching structures taken from the calculations of *Yajima* [3.15] with structures identical in indices and layer thicknesses but shaped according to (3.4.18, 19). For each structure $\Delta\beta$, γ_3, and X_0 are given in Table 3.1 identified by the arm thickness ratio W_4/W_2. In Fig. 3.26 we plot $|A_j|^2/|A_{i0}|^2$ at the end of the structure versus the appropriate $\Delta\beta/\theta\gamma_3$, taking the data for the linear branch directly from [3.15], and calculating the result for the shaped structure from (3.4.22, 23, and 26). Next we choose a specific value θ (or θ_{min}) for each structure by choosing $\Delta\beta/\theta\gamma_3 = 0.03$ for the power divider ($W_4/W_2 = 1.03$), and $\Delta\beta/\theta\gamma_3 = 2$ for the mode splitter ($W_4/W_2 = 1.3$), and plot γ versus z for the four resulting structures in Fig. 3.27a, b. Here γ

Table 3.1. Waveguide parameters for branching structures

W_4/W_2	$\Delta\beta$ $[10^{-3}\ \mu\text{m}^{-1}]$	γ_3 $[\mu\text{m}^{-1}]$	X_0
1.03	0.75	1.59	0.042
1.3	5.8	1.61	0.43

120

is obtained from (3.4.17) for the linear structures and from (3.4.26) for the shaped structures.

Although the power transfer curves in Fig. 3.26 have the same general shape, two trends appear. For the symmetric structure (small X_0) the analytic solution overestimates the amount of power transfer in the corresponding linear branch. This occurs because the linear branch will have smaller θ at each end than the shaped structure and thus less overall mode conversion. The second trend is that, for the asymmetric structure (larger X_0), the analytic solution underestimates the amount of mode conversion in an abrupt branch. This effect is related to the failure of (3.3.38), which is based on coupled mode theory, to properly account for C_{ij} in the region where $W_3 < 1/\gamma_3$. Figure 3.27 clearly demonstrates the geometric efficiency of the shaped structure. Then γ, which is a measure of the strength of local normal mode coupling, is constant with length, whereas for linear structures γ peaks at a value of W_3 defined by (3.4.27). The shaped structures have a resulting theoretical length advantage over the linear structures. This result would be most useful in the design of mode splitters, which must be long to avoid appreciable power transfer.

Limitations to the analytic solution for a shaped structure arise in two areas: 1) the local normal mode coupling constant C_{ij}, which is used to obtain the analytic solution given in (3.4.22, 23), is obtained from approximate coupled mode equations; and 2) we have only considered coupling between the two lowest order local normal modes, neglecting both higher-order local normal modes and radiation modes. The neglect of radiation modes means that their effects must be separately considered in order to ensure that excessive losses do not occur. This consideration will limit the maximum value of θ at the branch apex and at large arm separation. These limitations are discussed in more detail in [3.14].

3.4.3 Experimental Results

The theory given above has been tested experimentally in a series of channel waveguide branches [3.16], as shown in Fig. 3.28. These branches were

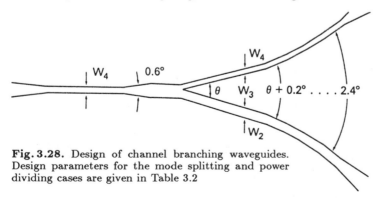

Fig. 3.28. Design of channel branching waveguides. Design parameters for the mode splitting and power dividing cases are given in Table 3.2

Table 3.2. Experimental branching waveguide parameters

Sample	W_2, W_4 [μm]	$\Delta W = W_2 - W_4$ [μm]	$\Delta\beta$ [$10^{-3}\ \mu m^{-1}$]	γ_3 [μm^{-1}]	X_0	$P(X_0)$
10-5, 10-6	2,3	1	0.87	0.16	0.16	
Planar mode splitter	2,3	1	10.8	1.87	0.71	
10-9	3,3	0.033	0.067	0.21	0.012	0.494
10-7	4.8,4.8	0.053	0.12	0.48	0.014	0.493
Planar power divider	3,3.1	0.1	0.68	1.93	0.054	0.473

fabricated in Ti diffused LiNbO$_3$. The initial branching angles for mode splitters varied from 0.4° to 0.8°, and for power dividers from 0.8° to 2.0°. The waveguide parameters for these branches and for two similar planar branches that were computer modeled are shown in Table 3.2. The calculations leading to these parameters have been detailed in [3.16]. Here X_0 is the initial value of X at zero branch arm separation given by

$$X_0 = \frac{\Delta\beta}{2|K(0)|} \quad , \qquad (3.4.28)$$

where $|K(0)| \equiv F$ is the coupling coefficient at zero branch arm separation. For the power dividers, power transfer between the local normal modes $P(X_0)$ in the abrupt limit depends only on the initial condition X_0, as the final value of X is infinite. The power $P(X_0)$ is then given by (3.4.25). The experimental measurements of power transfer in the branches vs the parameter $\Delta\beta/\theta\gamma_3$ is shown in Fig. 3.29 along with the computer calculations for the similar planar branches. Experimentally, power transfer was taken to be

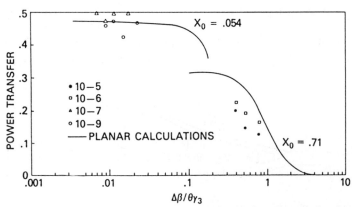

Fig. 3.29. Power transfer in power dividing and mode-splitting branches vs. parameter $\Delta\beta/\Theta\gamma_3$. Experimental points are Ti : LiNbO$_3$ channel waveguide branches, and solid curves are calculated for similar planar branches

the ratio of the power in the arm with the minimum output to the sum of the output power from both arms. The computed planar branches were intended to show the trend of power transfer behavior and were not identical to the experimental channel-waveguide branches. However, the experimental branch results are seen to generally follow these theoretical models and the calculated values of power transfer $P(X_0)$ for the power dividers were in good agreement with three of the experimental devices.

Radiation loss in waveguide branches has been measured experimentally by several researchers [3.16–18], and has also been treated theoretically [3.19–22]. Radiation loss can be significant and is of practical concern in designing branching waveguide structures [3.22].

Several attempts have been made to add electrodes to waveguide branches fabricated on electrooptic material to switch the optical output from one branch arm to the other. Accounts of these experiments are given in [3.23, 24]. A device closely related to the active waveguide branch is an electro-optic switch using crossing channel waveguides at a small angle. This device, also known as an X switch, was originally proposed to work on the principle of total internal reflection [3.25]. However, *Neyer* [3.26] has shown that this switch is better described using a model of interference between the local normal modes, similar to a directional coupler. Using this model, he compared [3.27] the operation of a directional coupler, a directional coupler of zero gap, h_3, and an X switch.

3.4.4 Superposition of Solutions

For a single local normal mode incident on the structure, the behavior of a given branch can be understood in terms of the local normal modes of the structure. We will now consider more complicated cases, in which more than one local normal mode is incident on a branch, or in which propagation is reversed in a branch with arbitrary inputs into the branch arms. These cases can be treated simply by superposition of the solutions already obtained.

We consider waveguide branches that support the two lowest-order local normal modes of amplitude A_i and A_j with propagation constants β_i and β_j. Power transfer between these modes in the branch is described by the coupled-mode equations (3.3.16, 17) with the parameter P replaced by W_3, the separation between the guiding regions. These equations are homogeneous and linear and it follows that if (A_i, A_j) and (A'_i, A'_j) are solutions then $(A_i + A'_i, A_j + A'_j)$ is also a solution. This result, of course, does not depend on the accuracy of the coupled-mode equations given above, but is a consequence of the linearity and resulting superposition properties of Maxwell's equations.

The case of a branch with two local normal modes incident is then quite straightforward. One simply obtains the solution for each local normal mode independently, and superimposes these solutions at the output side

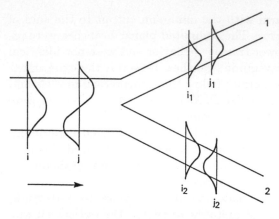

Fig. 3.30. Superposition of local normal mode solutions for incidence on a ideal power dividing branch with $\Delta\beta = 0$

of the branch, at large branch arm separation. An example is shown in Fig. 3.30 where we assume local normal modes i and j with equal power $(P_i = P_j = 1)$ and phases as shown on an ideal power dividing branch with $\Delta\beta = 0$. Considering each local normal mode independently, the output mode amplitudes are shown in the figure where $\psi_{i_1} = \psi_{i_2} = 1/\sqrt{2}$ and $\psi_{j_1} = 1/\sqrt{2}$, $\psi_{j_2} = -1/\sqrt{2}$. We then superimpose the output mode amplitudes in waveguides 1 and 2 to obtain $\psi_1 = 2/\sqrt{2}$ and $\psi_2 = 0$. The output powers are $P_1 = 2$ and $P_2 = 0$, i.e., all the power is in branch arm # 1.

Again assuming an ideal power dividing branch with $\Delta\beta = 0$, we can consider the reverse case of power incident on a branch from the branch arms as in Fig. 3.31a. We assume unequal powers incident with amplitude a_1 in guide # 1 and a_2 in guide # 2, but equal phases $(a_1 > a_2)$. We introduce a new set of mode amplitudes a and b defined by $a_1 = a + b$ and $a_2 = a - b$, and decompose the original set of modes into a new set with equal in-phase (a) and equal out-of-phase (b) amplitudes, as shown in Fig. 3.31b. The in-phase mode amplitudes will combine to form local normal

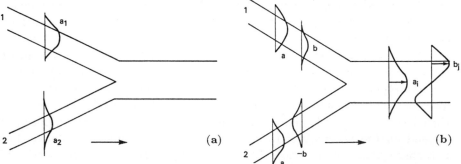

Fig. 3.31. Unequal amplitudes a_1 and a_2 incident on an ideal power dividing branch with $\Delta\beta = 0$ from the branch arms **(a)**. In **(b)** the original modes are decomposed into a new set with in-phase (a) and out-of-phase (b) amplitudes. The in-phase mode amplitudes combine to form normal mode i in the output waveguide, and the out-of-phase mode amplitudes combine to form normal mode j in the output waveguide

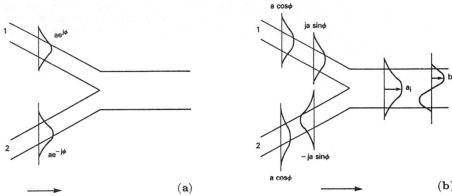

Fig. 3.32. Case of equal amplitudes, with an arbitrary phase difference, incident on a power dividing branch from the branch arms (**a**). In (**b**) the original mode amplitudes are decomposed to form an in-phase and an out-of-phase set, which are treated as in Fig. 3.31

mode i in the output waveguide and the out-of-phase mode amplitudes will combine to form local normal mode j. Conservation of power will show that the output amplitudes of modes i and j are related to the input amplitudes by $a_i = \sqrt{2}\,a$ and $b_j = b$.

Often we encounter the case illustrated in Fig. 3.32a where the power incident on the branch from the branch arms is equal but the amplitudes have an arbitrary phase difference. We use the identity $\exp(\pm j\phi) = \cos\phi \pm j\sin\phi$ to decompose each mode amplitude into its real and imaginary parts as in Fig. 3.32b. Again the in-phase and out-of-phase mode amplitudes combine to form local normal modes i and j, respectively, in the output waveguide. Conservation of power yields $a_i = \sqrt{2}\,a\cos\phi$ and $b_j = j(a\sin\phi)$, so that the local normal modes i and j are 90° out of phase. The output powers are

$$P_i = 2a^2\cos^2\phi = a^2(1 + \cos 2\phi) \quad , \tag{3.4.29}$$

$$P_j = 2a^2\sin^2\phi \quad . \tag{3.4.30}$$

Equation (3.4.29) is the usual output equation for a Mach-Zehnder interferometer, where the output waveguide is usually made single mode so that the local normal mode j is cut-off and radiates into the substrate.

3.5 Waveguide Horns

Horns are used in integrated optics [3.28–30] to change the dimensions of channel waveguides and to couple from planar guides to channel guides, as shown in Fig. 3.33. To provide a low loss transition, the coupling horn must operate adiabatically, i.e., the lowest-order local normal mode of the

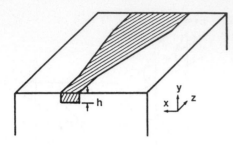

Fig. 3.33. Three-dimensional view of an embedded optical channel waveguide horn. The shaded volume has the highest index of refraction

structure must propagate through the structure without cumulative power transfer to higher-order local normal modes. We will now explicitly consider channel waveguides and note that, due to the symmetry of a horn structure, we will be concerned with mode coupling between the first and third local normal modes, rather than the first and second ones.

In this section, we will first derive a more general expression for c_{ij} which is appropriate for channel waveguides. We will then derive approximate expressions for $\Delta\beta_{ij}$ and C_{ij} valid in a wide channel where the modes are far from cut-off. Finally we will employ the analytic solution of (3.3.2) and show that it leads to an expansion horn with a parabolic shape. Our discussion will closely follow that of [3.30].

3.5.1 Mode-Conversion Coefficient c_{ij} for Channel Waveguides

We provide here a general derivation of the mode-conversion coefficient c_{ij} which describes power transfer between two guided modes at a small step. Two modes, i and j, are assumed with propagation in the z direction. The input and output sides of the step are denoted by the subscripts 0 and 1. For channel waveguide propagation, hybrid modes with TE and TM components are assumed, so that we allow the general field components

$$\boldsymbol{E} = E_x \hat{i} + E_y \hat{j} + E_z \hat{k} \quad , \tag{3.5.1}$$

$$\boldsymbol{H} = H_x \hat{i} + E_y \hat{j} + H_z \hat{k} \quad , \tag{3.5.2}$$

where $\hat{i}$, $\hat{j}$, and $\hat{k}$ are unit vectors in the x, y, and z directions, respectively.

Continuity of the transverse field components at the step is required as expressed by the set of equations:

$$E_{xi0} + E_{xj0} = E_{xi1} + E_{xj1} \quad , \tag{3.5.3}$$

$$E_{yi0} + E_{yj0} = E_{yi1} + E_{yj1} \quad , \tag{3.5.4}$$

$$H_{xi0} + H_{xj0} = H_{xi1} + H_{xj1} \quad , \tag{3.5.5}$$

$$H_{yi0} + H_{yj0} = H_{yi1} + H_{yj1} \quad , \tag{3.5.6}$$

126

where reflected and radiation modes have been ignored. We will make extensive use of the expressions of orthonormality

$$\int_{-\infty}^{\infty} \int_{-\infty}^{\infty} (\boldsymbol{E}_i \times \boldsymbol{H}_j^*)_z\, dx\, dy = 0 \quad , \tag{3.5.7}$$

and power flow in the z direction

$$P_\gamma = \frac{1}{2} \int_{-\infty}^{\infty} \int_{-\infty}^{\infty} (\boldsymbol{E}_\gamma \times \boldsymbol{H}_\gamma^*)_z\, dx\, dy \quad , \quad (\gamma = i, j) \quad . \tag{3.5.8}$$

We multiply (3.5.3) by H_{yj1}^* and (3.5.4) by H_{xj1}^* and subtract, and multiply (3.5.5) by E_{yj1}^* and (3.5.6) by E_{xj1}^* and subtract. We then add the two resulting equations to obtain

$$4P_{j1} = \int_{-\infty}^{\infty} \int_{-\infty}^{\infty} [(\boldsymbol{E}_{j1}^* \times \boldsymbol{H}_{i0})_z + (\boldsymbol{E}_{i0} \times \boldsymbol{H}_{j1}^*)_z$$
$$+ (\boldsymbol{E}_{j1}^* \times \boldsymbol{H}_{j0})_z + (\boldsymbol{E}_{j0} \times \boldsymbol{H}_{j1}^*)_z]dx\, dy \quad , \tag{3.5.9}$$

which represents the power in mode j on the output side of the step.

We now express the field components as

$$E = a\vec{\mathcal{E}}(x,y)e^{j\alpha(z)} \quad , \tag{3.5.10}$$

$$H = a\vec{\mathcal{H}}(x,y)e^{j\alpha(z)} \quad , \tag{3.5.11}$$

where a is the real mode amplitude, $\vec{\mathcal{E}}$ and $\vec{\mathcal{H}}$ are vectors representing the transverse field distribution, and $\alpha(z)$ is the mode phase factor $(\beta z + \phi)$. We define the integral $I_{\gamma\delta}$

$$I_{\gamma,\delta} = \int_{-\infty}^{\infty} \int_{-\infty}^{\infty} (\vec{\mathcal{E}}_\gamma \times \vec{\mathcal{H}}_\delta^*)_z dx\, dy \quad (\gamma, \delta = i, j) \quad . \tag{3.5.12}$$

Then (3.5.8) can be written

$$P_\gamma = \tfrac{1}{2}a_\gamma^2 I_{\gamma,\gamma} \quad (\gamma = i, j) \quad , \tag{3.5.13}$$

and the mode amplitude a_γ^s introduced

$$a_\gamma^s = \left(\frac{2}{I_{\gamma,\gamma}}\right)^{1/2} \quad (\gamma = i, j) \quad , \tag{3.5.14}$$

which corresponds to unity power flow in the mode. Finally, we can express (3.5.9) as

$$A_{j1}e^{j\alpha_{j1}} = c_{ij}A_{i0}e^{j\alpha_{i0}} + c_{jj}A_{j0}e^{j\alpha_{j0}} \quad , \tag{3.5.15}$$

127

where $A_\gamma = a_\gamma/a_\gamma^s$ and

$$c_{ij} = \frac{I_{i0,j1} + I_{j1,i0}^*}{2(I_{i0,i0}I_{j1,j1})^{1/2}} \quad .$$

(3.5.16)

Similarly, c_{jj} is obtained from (3.5.16) by replacing j for i. Result (3.5.16) is a more general representation of the derivation of (3.3.10), which assumed planar waveguide modes.

Finding an appropriate analytical expression for c_{ij} is somewhat difficult since an overlap integral is involved; however, we can simplify the problem by using an analytical approach developed by Marcuse to demonstrate modal orthogonality [3.31]. In [3.30], Maxwell's equations and the divergence theorem are used to derive a relationship between the fields of two guided modes ($i0$ and $j1$) at a step. The result is

$$(\beta_{i0} - \beta_{j1}) \int_{-\infty}^{\infty} \int_{-\infty}^{\infty} [(\boldsymbol{E}_{i0} \times \boldsymbol{H}_{j1}^*)_z + (\boldsymbol{E}_{j1}^* \times \boldsymbol{H}_{i0})_z]dx\,dy$$

$$= k_0 \int_{-\infty}^{\infty} \int_{-\infty}^{\infty} (\varepsilon_0 - \varepsilon_1)\boldsymbol{E}_{i0} \cdot \boldsymbol{E}_{j1}^* dx\,dy \quad .$$

(3.5.17)

Here ε_0 and ε_1 represent the dielectric constants on each side of the step, which are functions of x and y, and k_0 is the free-space propagation constant. Then (3.5.17) can be used to calculate the overlap integral in c_{ij}. Combining (3.5.17) with (3.5.16), and using the definitions in (3.5.10–12) gives

$$c_{ij} = \frac{k_0 \int_{-\infty}^{\infty} \int_{-\infty}^{\infty} (\varepsilon_0 - \varepsilon_1)\vec{\mathcal{E}}_{i0} \cdot \vec{\mathcal{E}}_{j1}^* dx\,dy}{2(\beta_{i0} - \beta_{j1})\left[\int_{-\infty}^{\infty} \int_{-\infty}^{\infty} (\vec{\mathcal{E}}_{i0} \times \vec{\mathcal{H}}_{i0}^*)_z dx\,dy \int_{-\infty}^{\infty} \int_{-\infty}^{\infty} (\vec{\mathcal{E}}_{j1} \times \vec{\mathcal{H}}_{j1}^*)_z dx\,dy \right]^{1/2}} \quad .$$

(3.5.18)

We will use (3.5.18) to approximate c_{ij} below.

3.5.2 Approximation for $\Delta\beta_{ij}$

Our approach to approximate the wave-vector difference $\Delta\beta_{ij}$ will be to first develop an expression for a planar guide. This result can then be extended to the channel case. For TM modes in a planar guide with cladding layers of similar index, the dispersion relationship can be approximated away from the cutoff by (Sects. 2.1.2–4)

$$h_{\text{eff}} = (m + 1)\pi/(n_f^2 k_0^2 - \beta_m^2)^{1/2} \quad ,$$

(3.5.19)

128

where h_{eff} is the effective waveguide thickness

$$h_{\text{eff}} = h + \left(\frac{n_s}{n_f}\right)^2 \frac{2}{k_0(n_f^2 - n_s^2)^{1/2}} \quad . \tag{3.5.20}$$

Here, n_f is the index of the guiding layer and n_s is the index of the surrounding layers, β_m is the wavevector of the mth mode in the waveguide, k_0 is the free space propagation constant, and h is the waveguide thickness. This relationship will be approximately valid whenever $\beta_m \to k_0 n_f$, i.e., whenever the mode is not close to cutoff. With this dispersion relationship, we can obtain

$$\Delta\beta_{0m} = \beta_0 - \beta_m = \pi^2(m^2 + 2m)/[h^2(\beta_0 + \beta_m)] \tag{3.5.21}$$

which, for $m = 2$, reduces to

$$\Delta\beta_{02} = 2\pi\lambda_{\text{g}}/h^2 \tag{3.5.22}$$

under the assumption that $\beta_0 + \beta_m \simeq 2\beta_0$. Here $\lambda_{\text{g}} = 2\pi/\beta_0$ is the modal wavelength in the guide.

These planar TM results are directly applicable to a channel structure with E^x excitation (electric field parallel to the x direction) if we take $\Delta\beta_{0m}$ to refer to the wave-vector difference beteen the $E^x_{1,q}$ mode and the $E^x_{1+m,q}$ mode in *Marcatili*'s notation [3.32]. The index n_s refers to the regions alongside the channel guide in the x direction, i.e., on either side of the horn for our structures. Now $W \equiv h$ refers to the width of the channel in the x direction. If W is large enough so that the modes do not extend significantly beyond the guiding region in the x direction, our approximations will be justified and we can use (3.5.21) to express the wave-vector difference for channel modes of different orders in the x direction but of the same order in the y direction.

3.5.3 Approximation for C_{ij}

If we consider a small symmetric step, such as the one shown in Fig. 3.34, in a horn structure with a channel waveguide of width W on side 0 and

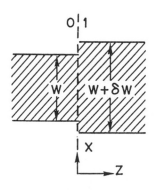

Fig. 3.34. A single step where the waveguide width changes from W to $W + \delta W$

129

width $W + \delta W$ on side 1 of the step, the difference $\varepsilon_0 - \varepsilon_1$ will be finite only near the edge of the two channels where $-W/2 - \delta W/2 < x < -W/2$ and $W/2 < x < W/2 + \delta W/2$. Here n_f is the index of refraction of the channel and n_s is the index of the cladding on each side of the channel. Then if the channels are rectangular with a constant depth h, the numerator in (3.5.18) can be simplified to give

$$
k_0 \frac{\delta W}{2}(n_s^2 - n_f^2) \int_{-h/2}^{h/2} \left[\vec{\mathcal{E}}_{i0}\left(\frac{W}{2}, y\right) \cdot \vec{\mathcal{E}}_{j1}^*\left(\frac{W}{2}, y\right) \right.
$$
$$
\left. + \vec{\mathcal{E}}_{i0}\left(-\frac{W}{2}, y\right) \cdot \vec{\mathcal{E}}_{j1}^*\left(-\frac{W}{2}, y\right) \right] dy \quad . \tag{3.5.23}
$$

Here the center of our coordinate system is in the middle of the guiding region.

We now consider the $E_{p,q}^x$ modes to evaluate the expression (3.5.23). Due to orthogonality, the step will cause no significant coupling between $E_{p,q}^x$ and $E_{m,n}^x$ modes unless $q = n$, so we can confine our attention to E_{p1}^x modes. In addition, an examination of (3.5.23) shows that, for a symmetric step, coupling will occur only between modes that are both even in the x direction or both odd in the x direction, but not between even and odd modes. This means that the most important power transfer from the lowest order E_{11}^x mode will be to the E_{31}^x mode, so that for our calculation we will consider E_{i0} to be an E_{11}^x mode and E_{j1} to be an E_{31}^x mode. For $W/2 < |x| < W/2 + \delta W/2$ and $-h/2 < y < h/2$, the transverse field distributions become

$$
\mathcal{E}_{i0}^x = \left(\frac{n_f}{n_s}\right)^2 M_{1i} \cos\left(\kappa_{xi}\frac{W}{2}\right)\cos(\kappa_{yi}y + \alpha_i) \quad ,
$$
$$
\mathcal{E}_{j1}^x = M_{1j} \cos\left(\kappa_{xj}\frac{W}{2}\right)\cos(\kappa_{yj}y + \alpha_j) \quad . \tag{3.5.24}
$$

In (3.5.24), where M_{1i} and M_{1j} are real amplitudes, we have used the relationship between H^y and E^x:

$$
H_\gamma^y = \frac{\varepsilon_0 n_\gamma^2 \beta}{k_0^2 n_\gamma^2 - \kappa_{x\gamma}^2} E_\gamma^x \quad (\gamma = i0, j1) \tag{3.5.25}
$$

to write $\mathcal{E}_{i0}^x$ in terms of parameters valid in the guiding region. For modes of the same order in y, $\kappa_{yi} = \kappa_{yj} = \kappa_y$ and $\alpha_i = \alpha_j = \alpha$. Both $\mathcal{E}_{i0}^z$ and $\mathcal{E}_{j1}^z$ are much smaller so that we can neglect the z component in the scalar product in (3.5.23). The dispersion relationship tells us that

$$
\tan \kappa_x \frac{W}{2} = j\left(\frac{n_f^2}{n_s^2}\right)\frac{\gamma_x}{\kappa_x} \quad , \tag{3.5.26}
$$

where γ_x is the x component of momentum in the cladding,

$$\gamma_x^2 \equiv \kappa_x^2 - k_0^2(n_f^2 - n_s^2) \quad , \tag{3.5.27}$$

so that

$$\cos \kappa_x \frac{W}{2} = \frac{n_s^2 \kappa_x}{(n_s^4 \kappa_x^2 - n_f^4 \gamma_x^2)^{1/2}} \quad . \tag{3.5.28}$$

If we consider a channel guide with a small index difference $[(n_f/n_s)^2 \simeq 1]$, we can write for $-h/2 < y < h/2$ that

$$\mathcal{E}_{i0}^x \left(\pm \frac{W}{2}, y\right) = M_{1i} \frac{\kappa_{xi}}{k_0(n_f^2 - n_s^2)^{1/2}} \cos(\kappa_y y + \alpha) \quad ,$$

$$\mathcal{E}_{j1}^x \left(\pm \frac{W}{2}, y\right) = M_{1j} \frac{\kappa_{xj}}{k_0(n_f^2 - n_s^2)^{1/2}} \cos(\kappa_y y + \alpha) \quad . \tag{3.5.29}$$

Then (3.5.23) can be reduced to

$$-\delta W \frac{\kappa_{xi}\kappa_{xj}}{k_0} M_{1i} M_{1j} \int_{-h/2}^{h/2} \cos^2(\kappa_y y + \alpha)dy \quad . \tag{3.5.30}$$

If the modes are largely confined to the guiding region, we can neglect those parts of the integral in the denominator of (3.5.18) that fall outside the guiding region. Then that denominator can be evaluated using (3.5.25) to give

$$\frac{(\beta_{i0} - \beta_{j1})k_0 n_f^2 (\beta_i \beta_j)^{1/2} W M_{1i} M_{1j}}{(k_0^2 n_f^2 - \kappa_{xi}^2)^{1/2} (k_0^2 n_f^2 - \kappa_{xj}^2)^{1/2}} \int_{-h/2}^{h/2} \cos^2(\kappa_y y + \alpha)dy \quad . \tag{3.5.31}$$

We assume a wide guide, well away from cutoff in the x direction, so that $\kappa_{xi}, \kappa_{xj} \ll k_0 n_f$. Then c_{ij} is given by dividing (3.5.30) by (3.5.31), which reduces to

$$c_{ij} = -\frac{\kappa_{xi}\kappa_{xj}}{(\beta_{i0} - \beta_{j1})(\beta_i \beta_j)^{1/2}} \frac{\delta W}{W} \quad . \tag{3.5.32}$$

With our assumptions, κ_{xi} and κ_{xj} are approximately given by

$$\kappa_{xi} \simeq \pi/W \quad , \tag{3.5.33}$$

$$\kappa_{xj} \simeq 3\pi/W \quad . \tag{3.5.34}$$

Finally, using (3.5.22) for $\Delta\beta_{ij}$ we arrive at simple approximations for c_{ij} and thus C_{ij}, which are valid for large W, namely

131

$$c_{ij} = -\frac{3}{4}\frac{\delta W}{W} \quad , \tag{3.5.35}$$

$$C_{ij} = -\frac{3}{4W} \quad . \tag{3.5.36}$$

Both C_{ij} and $\Delta\beta_{ij}$ decrease as W increases, but as $\Delta\beta_{ij}$ decreases faster the ratio $|C_{ij}/\Delta\beta_{ij}|$ increases. This will result in an increasing coupling problem between the modes as W increases in a linear horn with constant θ.

3.5.4 Parabolic Solution

We can now proceed to obtain a solution to the coupled-amplitude equations given in Sect. 3.3.2. Such a solution can only be obtained analytically for particular horn shapes and its generality will be limited by the assumptions made in the previous two sections. The parameter P used in Sect. 3.3.2 becomes the horn width W. We are interested in a quantitative estimate of power transfer between the E_{11}^x mode (i) and the E_{31}^x mode (j) and we ignore all other modes.

For an analytic solution of (3.3.20, 21), the horn shape required is

$$\frac{dW}{dz} = \gamma\frac{\Delta\beta_{ij}}{C_{ij}} \quad , \tag{3.5.37}$$

with γ being a constant. We substitute (3.5.22) for $\Delta\beta_{ij}$ and (3.5.36) for C_{ij} into (3.5.27) and obtain the local horn angle θ

$$\theta \equiv \frac{1}{2}\frac{dW}{dz} = -\gamma\frac{4\pi}{3}\frac{\lambda_g}{W} \quad . \tag{3.5.38}$$

The shape of this horn will be parabolic with

$$W = (2\alpha\lambda_g z + W_0^2)^{1/2} \quad , \tag{3.5.39}$$

where W_0 is the horn width at $z = 0$ and $\alpha = -8\pi\gamma/3$ (Fig. 3.35).

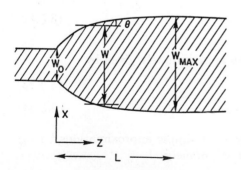

Fig. 3.35. Top view of a parabolic channel waveguide horn

Power transferred into the E_{31}^x mode from an initial launch of power P_0 into the E_{11}^x mode will oscillate as a function of distance with a maximum power transfer given by (3.3.28), where

$$\frac{P_{j\max}}{P_0} = \frac{4\gamma^2}{4\gamma^2 + 1} = \frac{(3\alpha/4\pi)^2}{(3\alpha/4\pi)^2 + 1} \quad , \tag{3.5.40}$$

which is just $(3\alpha/4\pi)^2$ for $\alpha \ll 4\pi/3$. A horn designed with $\alpha = 1$ should therefore conserve at least 94 % of the power in the lowest-order mode.

For the parabolic solution, (3.3.25) becomes

$$u = \int_{W_0}^{W} \frac{C_{ij}}{\gamma} dW \quad . \tag{3.5.41}$$

In the limit of wide W, this reduces to

$$u = \frac{2\pi}{\alpha} \ln \frac{W}{W_0} \quad . \tag{3.5.42}$$

Using (3.5.42) for u with the exact solution (3.3.26), fractional power transferred from mode i to mode j is plotted in Fig. 3.36 as a function of W/W_0, for increasing W and several values of α. The actual amount of power transferred depends upon the particular length of the horn-shaped waveguide section, according to (3.3.26). The power that remains in the lowest-order mode P_i is just $(1 - P_j)$ in this two-mode approximation.

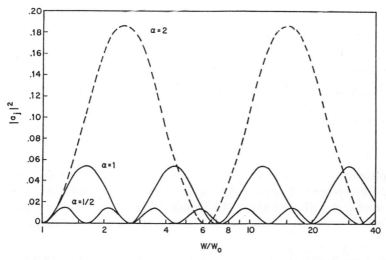

Fig. 3.36. Plot of power in mode j for several values of α in a parabolic horn with increasing width W. The approximation used is of questionable validity for $\alpha > 1$

For a given waveguide expansion from W_0 to W_{max}, the length of a parabolic horn will depend on the particular value of α which is chosen. According to (3.3.26), the amount of power left in the lowest-order mode after passing through the horn will depend upon the length of the horn since, for a given expansion, both α and the final value of u are functions of horn length. Figure 3.37 plots the fraction of the initial power remaining in the lowest-order mode (P_i/P_0) after passing through a parabolic horn which expands from a width of $3\,\mu m$ to a width of $50\,\mu m$ (for $\lambda_{\mathrm{g}} = 0.287\,\mu m$) as the design length is varied. For the exact solution, the fraction of power remaining in the lowest-order mode is not a monotonically increasing function of horn length. If the right length is chosen, mode interference effects can lead to less accumulated power transfer out of the lowest-order mode and to higher horn efficiency. As can be seen from Fig. 3.37, the envelope derived in (3.5.40) does, however, set a lower bound on horn efficiency.

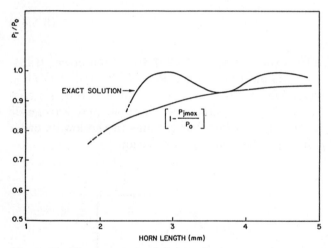

Fig. 3.37. The power remaining in the lowest order mode P_i after passage through a parabolic horn which expands from a channel width of $3\,\mu m$ to a channel width of $50\,\mu m$ as the length of the horn is varied with $\lambda_{\mathrm{g}} = 0.287\,\mu m$. P_0 is the power initially in the lowest order mode and P_j is the power converted to the E^x_{31} mode. The exact solution as given by (3.3.26) is shown as in the approximation given by (3.5.40)

With a parabolic shape $\gamma = 2C_{ij}\theta/\Delta\beta_{ij}$ remains constant throughout the horn, and the strength of mode coupling between the E^x_{11} mode and the E^x_{31} mode is evenly distributed along the length. This is not true for other horn shapes where γ will be a function of u; however, for all horn shapes

$$\int_0^{u_{\mathrm{max}}} \gamma\, du = \frac{-3}{4} \ln \frac{W_{\mathrm{max}}}{W_0} \quad , \tag{3.5.43}$$

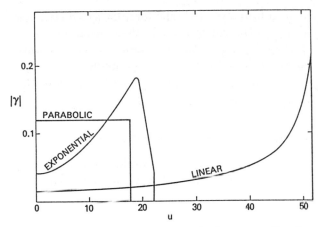

Fig. 3.38. The coupling constant γ is shown as a function of u for three different horn shapes. All three horns expand from 3 to 50 μm within a length of 4325 μm with $\lambda_g = 0.287\,\mu$m. The shape of the parabolic horn is given by $W = \sqrt{(3)^2 + 2\lambda_g z}$. The shape of the linear horn is given by $W = 3 + 2(0.00543)z$. The shape of the exponential horn is given by $W = 50[1 - \exp(-z/1537)] + 3$

so that the integrated value of γ over u is only a function of horn expansion. The value of $|\gamma|$ as a function of u is plotted for three horn shapes in Fig. 3.38 where, for all three cases, the horn expands from 3 to 50 μm within the same length. The mode interference effects make comparisons between different horn shapes difficult. However, in order to minimize cumulative power transfer out of the lowest-mode order, it is desirable to use horn shapes where the average value of $|\gamma|$ is not large over an interval where u changes by a factor of π. Since towards the wide end of the horn the value of γ for a linear horn exceeds the value of γ for a parabolic horn of similar expansion and length, it is clear that a parabolic horn shape will be preferred. Comparisons of these results with the computer calculations for other horn shapes [3.28, 29] show this to be the case.

Aside from theoretical arguments that the parabolic horn should be the most efficient structure in terms of mode conversion, its practical advantage lies in the existence of the analytical solution of (3.3.28). This allows the design of expansion horns without the requirement of computer modeling for each case. Experimental measurements with parabolic shaped expansion horns have demonstrated horn transmissions of $\sim 90\,\%$ for waveguide expansions from 4–8 μm to 30 μm, in good agreement with the theory developed here [3.33].

3.6 Branches with Three Arms

So far, we have treated two-arm branches with the familiar symmetric and antisymmetric normal modes. In a three-arm branch, the normal modes are not so obvious and, indeed, they depend on the symmetry of the structure. We will first derive the normal modes for a three waveguide system and then use our previous results for abrupt junctions and apply superposition to analyze a 3×2 branching waveguide coupler [3.34].

3.6.1 Normal Modes of Three Coupled Waveguides

In order to treat a three-arm branch, we will first develop representations for the local normal modes of the three coupled waveguides of Fig. 3.39.

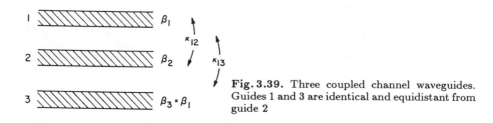

Fig. 3.39. Three coupled channel waveguides. Guides 1 and 3 are identical and equidistant from guide 2

We assume the outer guides to be identical and equidistant from the center guide. Then K_{12} $(= K_{23})$ is the nearest neighbor coupling coefficient and K_{13} is the coupling coefficient between the two outer guides. Coupled-mode equations describing this system are an extension of (3.1.3,4), namely

$$\frac{da_1}{dz} - j\beta_1 a_1 - j|K_{12}|a_2 - j|K_{13}|a_3 = 0 \quad , \tag{3.6.1}$$

$$\frac{da_2}{dz} - j\beta_2 a_2 - j|K_{12}|a_1 - j|K_{12}|a_3 = 0 \quad , \tag{3.6.2}$$

$$\frac{da_3}{dz} - j\beta_3 a_3 - j|K_{13}|a_1 - j|K_{12}|a_2 = 0 \quad , \tag{3.6.3}$$

where a_1 is the amplitude of the mode of the coupled waveguide and β_1 is the mode propagation constant for guide 1, etc. When $\beta_3 = \beta_1$, the propagation equations for this three-guide system can be reduced to the equations of the usual two-guide system by the substitution [3.35]

$$a(z) = a_2(z) \quad , \tag{3.6.4}$$

$$b(z) = \frac{1}{\sqrt{2}}[a_1(z) + a_3(z)] \quad , \tag{3.6.5}$$

to obtain

136

$$\frac{da}{dz} - j\beta_a a - j|K|b = 0 \quad , \tag{3.6.6}$$

$$\frac{db}{dz} - j\beta_b b - j|K|a = 0 \quad . \tag{3.6.7}$$

Equations (3.3.6, 7) are the coupled-mode equations for a set of two coupled waveguides, a and b, where

$$\beta_a = \beta_2 \quad , \tag{3.6.8}$$

$$\beta_b = \beta_1 + |K_{13}| \quad , \tag{3.6.9}$$

$$|K| = \sqrt{2}|K_{12}| \quad . \tag{3.6.10}$$

One local normal mode of our three coupled waveguides is obtained from the particular solution $a(z) = b(z) = 0$. Substitution into (3.6.1–3) yields the antisymmetric local normal mode, which we denote by the subscript j, i.e.,

$$A_j(z) = \begin{pmatrix} 1/\sqrt{2} \\ 0 \\ -1/\sqrt{2} \end{pmatrix} e^{j\beta_j z} \quad , \qquad \text{where} \tag{3.6.11}$$

$$\beta_j = \beta_1 - |K_{13}| \quad . \tag{3.6.12}$$

The column vector in (3.6.11) represents the amplitudes of the transverse field distributions in each channel

$$\begin{pmatrix} a_1 \\ a_2 \\ a_3 \end{pmatrix}$$

and has been normalized to unity power. The two local normal modes for the reduced two-guide system are obtained from the standard normal mode transformation of (3.6.6, 7) as given in Sect. 3.1.2. However, we note that the antisymmetric mode j of the reduced two-guide system will become a symmetric mode, denoted k, of the three-guide system. The result for the local normal mode amplitudes a_i and a_k is

$$a_i(z) = \begin{pmatrix} d \\ e \end{pmatrix} e^{j\beta_i z} \quad , \tag{3.6.13}$$

$$a_k(z) = \begin{pmatrix} -e \\ d \end{pmatrix} e^{j\beta_k z} \quad , \tag{3.6.14}$$

where the column vector represents the amplitudes of ϕ_a and ϕ_b

$$\begin{pmatrix} a \\ b \end{pmatrix} \quad .$$

The normal mode propagation constants are

$$\beta_i = \bar{\beta} + |K|(X^2 + 1)^{1/2} \quad , \tag{3.6.15}$$

$$\beta_k = \bar{\beta} - |K|(X^2 + 1)^{1/2} \quad , \tag{3.6.16}$$

where, as defined in Sect. 3.1.2,

$$\bar{\beta} = \tfrac{1}{2}(\beta_a + \beta_b) \quad , \tag{3.6.17}$$

$$X = \frac{\Delta\beta}{2|K|} \quad , \tag{3.6.18}$$

$$\Delta\beta = \beta_a - \beta_b \quad , \tag{3.6.19}$$

and

$$d = \left[\frac{1}{2} \left(1 + \frac{X}{(X^2 + 1)^{1/2}} \right) \right]^{1/2} \quad , \tag{3.6.20}$$

$$e = \left[\frac{1}{2} \left(1 - \frac{X}{(X^2 + 1)^{1/2}} \right) \right]^{1/2} \quad , \tag{3.6.21}$$

so that $d^2 + e^2 = 1$.

By transforming the column vectors in (3.6.13, 14) from the reduced two-guide system to the original three-guide system, we obtain the symmetric local normal modes of the three-guide system. We employ (3.6.4, 5) and maintain the constraint of unity power to obtain

$$A_i(z) = \begin{pmatrix} e/\sqrt{2} \\ d \\ e/\sqrt{2} \end{pmatrix} e^{j\beta_i z} \quad , \tag{3.6.22}$$

$$A_k(z) = \begin{pmatrix} d/\sqrt{2} \\ -e \\ d/\sqrt{2} \end{pmatrix} e^{j\beta_k z} \quad . \tag{3.6.23}$$

These local normal modes have the property $a_1 = a_3$, as expected from the symmetry of the system.

Power distribution between the center and outer guides in the local normal modes i and k depends on the parameter X which, from (3.6.18, 19) and (3.6.8–10), can be written as

$$X = \frac{\overline{\Delta\beta} - |K_{13}|}{2\sqrt{2}|K_{12}|} \quad , \tag{3.6.24}$$

where $\overline{\Delta\beta} = \beta_2 - \beta_1$ is the difference in propagation constants for the three coupled guides. For example, if $X = 0$, then $d = e = 1/\sqrt{2}$. At large guide

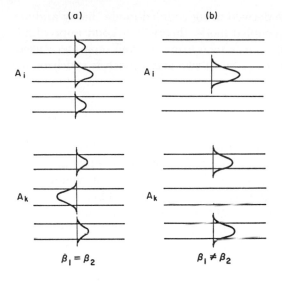

Fig. 3.40. Shape of the symmetric local-normal modes i and k for large guide separation such that the waveguides are uncoupled. (**a**) Three identical waveguides. (**b**) Outer waveguides different from center waveguide

separation, $X \to 0$ if $\overline{\Delta\beta} = 0$. However, if $\overline{\Delta\beta} > 0$, then $X \to \infty$ at large guide separation, and $d = 1$, $e = 0$. These local normal mode distributions are shown in Fig. 3.40. This behavior is the analog of the behavior of the two-arm branch for synchronous ($\Delta\beta = 0$) or nonsynchronous ($\Delta\beta \neq 0$) operation. The difference here is that, at small guide separation, $\Delta\beta$ becomes a function of guide separation through the coupling coefficient K_{13}.

The utility of the reduced two-guide solution is that many of the results of the two-arm branch can be applied to the three-arm branch. In particular, power transfer between local normal modes at an abrupt transition can be obtained from (3.2.7)

$$\frac{|A_{k1}|^2}{|A_{i0}|^2} = \frac{(f_0 - f_1)^2}{(f_0 - f_1)^2 + (1 + f_1 f_0)^2} \tag{3.6.25}$$

for mode i incident on the transition, where $f = e/d$ (3.1.11), and 0, 1 refer to opposite sides of the transition. We also note that, because of the symmetry of the structure, overlap integrals between the antisymmetric mode j and the symmetric modes i and k are always zero. Thus, there is never any power transfer at a transition between j and i or k.

For $\overline{\Delta\beta} > 0$ we typically have a transition between X_0 at small or zero guide separation and $X = \infty$ at large guide separation. For input in i, (3.6.25) for this case reduces to (3.4.25), given by

$$\frac{|A_{k1}|^2}{|A_{i0}|^2} = \frac{1}{2}\left(1 - \frac{X_0}{(X_0^2 + 1)^{1/2}}\right) \tag{3.6.26}$$

which, as a shorthand notation, is identical to e^2 evaluated at X_0 (3.6.21).

Equations (3.6.25, 26) were derived using coupled-mode theory and are accurate in the regime where coupled-mode theory is a good approximation. This circumstance occurs when branch arms are well separated (weak coupling) or, in (3.6.26), when the reduced two-arm branch is synchronous (small X_0).

3.6.2 3 × 2 Waveguide Coupler

We will use the local normal mode formalism developed in Sect. 3.6.1 for three coupled waveguides to analyze the 3 × 2 waveguide coupler shown in Fig. 3.41. In a fiber gyroscope application [3.36], source input would be to the center guide on the left of the coupler, the gyroscope fiber loop would be attached to the two guides on the right, and output signals would emit from the outer guides on the left. We will consider the branching waveguide behavior in the abrupt or power dividing limit because it will allow us to superpose modes in a branch while maintaining their relative phase, and because it will allow us to use the analytic expression for power transfer in (3.6.26). We will also assume the more general case of small difference in propagation constant between the center and outer guides ($\overline{\Delta\beta}\neq0$), although the problem could be worked out either way.

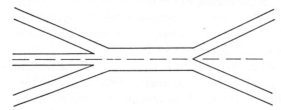

Fig. 3.41. (3 × 2) branching waveguide coupler

In Fig. 3.42 we assume $W_1 = W_3 \simeq W_2$ and $W_4 = W_5$, where W_i is the width of channel i. Each channel is assumed to be single mode so that the central channel of width $2W_4$ supports two normal modes, and the channel at position B of width $2W_1 + W_2$ supports three normal modes. Between B and C, the channel width narrows from a three-mode guide to a two-mode guide.

For $\overline{\Delta\beta} > 0$ the local normal modes at large separation (position A) are shown in Fig. 3.40b. Input of unity power into guide 2 will excite only local normal mode i, which will travel in the branch and, from (3.6.26), convert e^2 of power to mode k by the time it reaches position B (Fig. 3.42a). The total mode amplitude at B, $\psi(B)$, is given by

$$\psi(B) = (1 - e^2)^{1/2}\psi_i - e\psi_k \quad , \tag{3.6.27}$$

where e is evaluated at B and ψ_i and ψ_k represent the transverse field distribution of the symmetric local normal mode amplitudes at B. In (3.6.27) we

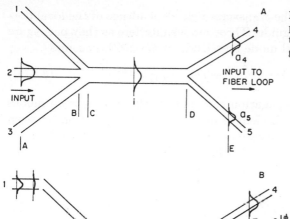

Fig. 3.42. Input (a) and output (b) mode distributions for the (3 × 2) branching waveguide coupler

have inserted the proper phase (π) between ψ_i and ψ_k, which is apparent from the assumption of power division and the definitions of (3.6.22, 23). This choice allows the mode amplitudes in the center guide to have the same phase. We assume adiabatic guide narrowing between B and C so that mode k cuts off with power loss e^2 and mode i evolves without power loss to the mode i of the two-mode guide at C. Mode i then propagates to position D where we assume a symmetric ($\beta_4 = \beta_5$) power dividing two-arm branch leading to each end of the gyro fiber loop. After branch arm separation, mode amplitudes of the uncoupled waveguides in each guiding region are $a_4 = a_5 = [(1 - e^2)/2]^{1/2}$. These modes are then input into each end of a fiber loop which forms the gyroscope coil. After traversing the loop each mode experiences a non-reciprocal phase shift of magnitude ϕ, corresponding to the total Sagnac phase shift of 2ϕ imposed by the rotating loop. The returning mode amplitudes at position E are then given by (Fig. 3.42b)

$$\begin{pmatrix} a_4 \\ a_5 \end{pmatrix} = \left(\frac{1 - e^2}{2} \right)^{1/2} \begin{pmatrix} e^{j\phi} \\ e^{-j\phi} \end{pmatrix} \quad . \tag{3.6.28}$$

These modes then enter the branch to form a mixture of the symmetric ($i2$) and antisymmetric ($j2$) modes, $\pi/2$ degrees out of phase, of the two-mode structure at D, as shown in Sect. 3.4.4. The total mode amplitude at D, ψ (D), is then given by

$$\psi(D) = (1 - e^2)^{1/2}(\cos \phi \, \psi_{i2} + j \sin \phi \, \psi_{j2}) \quad . \tag{3.6.29}$$

Here ψ_{i2} and ψ_{j2} represent the transverse field dependence of the local normal modes of the two-mode guide. These modes interfere as they propagate to position C where the total mode amplitude at C, $\psi(C)$, becomes

$$\psi(C) = (1 - e^2)^{1/2}(\cos \phi\, \psi_{i2}e^{j\beta_{i2}L} + j \sin \phi\, \psi_{j2}e^{j\beta_{j2}L}) \quad . \tag{3.6.30}$$

Here β_{i2} and β_{j2} are the propagation constants of the local normal modes between C and D, and L is the length of the interference section between C and D. As these modes propagate out of the branch to position A, mode i will again transfer e^2 of power to k, while mode j remains uncoupled. The output local normal modes at A are then given by

$$\psi_i = (1 - e^2) \cos \phi \begin{pmatrix} 0 \\ 1 \\ 0 \end{pmatrix} e^{j\beta_{i2}L} \quad , \tag{3.6.31}$$

$$\psi_j = (1 - e^2)^{1/2} \sin \phi \begin{pmatrix} 1/\sqrt{2} \\ 0 \\ -1/\sqrt{2} \end{pmatrix} e^{j(\beta_{j2}L - \pi/2)} \quad , \tag{3.6.32}$$

$$\psi_k = e(1 - e^2)^{1/2} \cos \phi \begin{pmatrix} 1/\sqrt{2} \\ 0 \\ 1/\sqrt{2} \end{pmatrix} e^{j\beta_{i2}L} \quad , \tag{3.6.33}$$

where e is to be evaluated at B. The output power in each guide is obtained by superposition of the local normal modes of (3.6.31–33). The result is

$$P_{\substack{1 \\ 3}} = \tfrac{1}{2}(1 - e^2)[1 - (1 - e^2) \cos \phi \pm (e \cos \alpha) \sin 2\phi] \quad , \tag{3.6.34}$$

$$P_2 = (1 - e^2)^2 \cos^2 \phi \quad , \tag{3.6.35}$$

where in (3.6.34) the plus sign is taken for P_1 and the minus for P_3, and $\alpha = (\beta_{j2} - \beta_{i2})L - \pi/2$.

Equations (3.6.34, 35) show that the output powers for the 3×2 waveguide coupler depend on the value of X at B and the length L of the interference section between C and D. For the gyroscope application these choices are made by maximizing the gyroscope sensitivity S which, assuming unity input power, is defined as

$$S = \frac{dP_1}{d2\phi}\bigg|_{\phi=0} = \tfrac{1}{2}(1 - e^2)e \cos \alpha \quad . \tag{3.6.36}$$

Maximum sensitivity is obtained for the parameter values $e^2(B) = 1/3$ and $\alpha = 0$. The 3×2 branching coupler is convenient to analyze in that, with the assumption of an abrupt three-arm branch, all the phase shift between the local normal modes occurs in the interference section of length L. From (3.6.36) we see that sensitivity is maximized by a $\pi/2$ phase shift in this section. Optimization of the branch design is achieved by maximizing the interference term between local normal modes j and k, $\psi_j \psi_k^* + \psi_j^* \psi_k$, in the output guide 1 or 3. Mode i plays no role since it has no amplitude in an output guide. The value of X at B required to accomplish the condition $e^2(B) = 1/3$ is obtained from (3.6.26) as

$$X(B) \equiv \frac{\overline{\Delta\beta} - |K_{13}(B)|}{2\sqrt{2}\,|K_{12}(B)|} = \frac{1}{2\sqrt{2}} \quad , \tag{3.6.37}$$

which in turn requires $\overline{\Delta\beta} = |K_{12}(B)| + |K_{13}(B)|$. Power loss upon input in the coupler is then $1/3$, independent of ϕ. With this optimization, the output powers become

$$P_{\substack{1 \\ 3}} = \tfrac{1}{9}(2 - \cos 2\phi + \sqrt{3}\sin 2\phi) \quad , \tag{3.6.38}$$

$$P_2 = \tfrac{2}{9}(1 + \cos 2\phi) \quad . \tag{3.6.39}$$

The maximized coupler sensitivity is then $S = (3\sqrt{3})^{-1}$. Equations (3.6.38, 39) are plotted in Fig. 3.43. Compared to fiber gyros constructed with the usual 2×2 coupler, gyros constructed with the 3×2 coupler have a maximum sensitivity at zero Sagnac phase shift (rotation rate) [3.36].

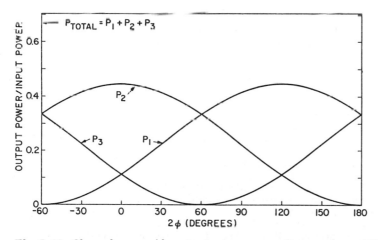

Fig. 3.43. Channel waveguide output power versus Sagnac phase shift as described by (3.6.38 and 39)

3.7 Conclusion

The examples described above for couplers, the two- and three-arm branches and horns demonstrate the power of the coupled local normal mode approach in providing quantitative guidance to device designers. Quantitative guidance is needed because if an abrupt transition is too abrupt it can radiate power and if an adiabatic transition is made too long it will take up too much space and suffer ordinary propagation losses. Later chapters will show how transitions of both types can be combined into additional useful devices.

References

3.1   W.K. Burns, A.F. Milton: IEEE J. **QE-11**, 32–39 (1975)
3.2   S.E. Miller: Bell Syst. Tech. J. **33**, 661–719 (1954)
3.3   W.H. Louisell: *Coupled Mode and Parametric Electronics* (Wiley, New York 1960)
3.4   A. Yariv: IEEE J. **QE-9**, 919–933 (1973)
3.5   W.K. Burns, A.F. Milton, A.B. Lee, E.J. West: Appl. Opt. **15**, 1053–1065 (1976)
3.6   D. Marcuse: *Theory of Dielectric Optical Waveguides* (Academic, New York 1974) Chap. 3
3.7   A.G. Fox: Bell Syst. Tech. J. **34**, 823–852 (1955)
3.8   W.H. Louisell: Bell. Syst. Tech. J. **34**, 853–870 (1955)
3.9   D. Marcuse: Bell Syst. Tech. J. **49**, 273–290 (1970)
3.10 A.F. Milton, W.K. Burns. Appl. Opt. **14**, 1207–1212 (1975)
3.11 M. Izutsu, A. Enokihara, T. Sueta: Opt. Lett. **7**, 549–551 (1982)
3.12 A. Ihaya, H. Furuta, H. Noda: Proc. IEEE **60**, 470 (1972)
3.13 A.F. Milton, W.K. Burns: IEEE Trans. **CAS-26**, 1020–1028 (1979)
3.14 W.K. Burns, A.F. Milton: IEEE J. **QE-16**, 446–454 (1980)
3.15 H. Yajima: IEEE J. **QE-14**, 749–755 (1978)
3.16 W.K. Burns. R.P. Moeller, C.H. Bulmer, H. Yajima: Appl. Opt. **19**, 2890–2896 (1980)
3.17 Y. Murakami, M. Ikeda: Electron. Lett. **17**, 411–433 (1981)
3.18 T.J. Cullen, C.D.W. Wilkinson: Opt. Lett. **10**, 134–136 (1984)
3.19 I. Anderson: Microwaves, Optics and Acoustics **2**, 7–12 (1978)
3.20 H. Sasaki, N. Mikoshiba: Electron. Lett. **17**, 136–138 (1981)
3.21 R. Baets, P.E. Lagasse: Appl. Opt. **21**, 1972–1978 (1982)
3.22 M. Kuznetsov: J. Lightwave Tech. **LT-3**, 674–677 (1985)
3.23 W.K. Burns, A.B. Lee, A.F. Milton: Appl. Phys. Lett. **29**, 790–792 (1976)
3.24 H. Sasaki, I. Anderson: IEEE J. **QE-14**, 883–892 (1978)
3.25 C.S. Tsai, B. Kim, F.R. El-Akkai: IEEE J. **QE-14**, 513–517 (1978)
3.26 A. Neyer: Electron. Lett. **19**, 553–554 (1983)
3.27 A. Neyer, W. Mevenkamp, L. Thylen, B. Lagerstorm: J. Lightwave Tech. **LT-3**, 635–642 (1985)
3.28 R.K. Winn, J.H. Harris: IEEE Trans. **MTT-23**, 92–97 (1975)
3.29 A.R. Nelson: Appl. Opt. **14**, 3012–3015 (1975)
3.30 A.F. Milton, W.K. Burns: IEEE J. **QE-13**, 828–835 (1977)
3.31 D. Marcuse: *Light Transmission Optics* (Van Nostrand Reinhold, New York 1972) pp. 322–324
3.32 E.A.J. Marcatili: Bell Syst. Tech. J. **48**, 2071–2102 (1969)
3.33 W.K. Burns, A.F. Milton, A.B. Lee: Appl. Phys. Lett. **30**, 28–30 (1977)
3.34 W.K. Burns, A.F. Milton: IEEE J. **QE-18**, 1790–1796 (1982)
3.35 A.W. Snyder: J. Opt. Soc. Am. **62**, 1267–1277 (1972)
3.36 S.K. Sheem: Appl. Phys. Lett. **37**, 869–871 (1980)

4. Titanium-Diffused Lithium Niobate Waveguide Devices

R.C. Alferness

With 39 Figures

Optical waveguide devices based upon lithium niobate substrates are presently in an advanced state. A large variety of devices have been demonstrated; fabrication parameters have been optimized by several laboratories; and waveguide devices have shown, for example, improved performance as external modulators in long-distance high-bit-rate systems. In fact, prototype research-grade devices are now commercially available. Lithium niobate has several important characteristics that make it attractive for waveguide devices, in particular its good electro-optic and acousto-optic coefficients. However, equally important is the fact that lithium niobate is readily available commercially in large substrates (7.5 cm wafers). Furthermore it is relatively easily processed and its material properties have been studied and documented extensively.

In this chapter, we will present an overview of the lithium niobate guided-wave device technology. The emphasis is on electro-optically controlled channel waveguide devices. These devices are particularly applicable to telecommunications and sensing applications. Significant progress has also been achieved in planar devices for signal processing applications which we will not cover [4.1]. While the technology of lithium niobate devices is quite well developed, it is still evolving and our understanding of material and processing interdependence continues to grow. Nevertheless, from a phenomenological point of view, we know how to fabricate "good" waveguides to achieve satisfactory overall device performance with respect to loss, drive voltage and modulation bandwidth. Furthermore, the principal device structures, their relative advantages and drawbacks are now reasonably well understood. It is these aspects of the materials and device technology that are emphasized in this chapter.

4.1 Waveguide Fabrication

Three different techniques have been used to form waveguides in lithium niobate. Initially waveguides were formed by thermal outdiffusion of Li_2O which results in an increased refractive index for the extraordinary index n_e [4.2]. (Lithium niobate is a uniaxial crystal with extraordinary index for light polarized along the z axis.) In addition to being limited to guiding light in only one polarization, the achievable index change is small and therefore provides waveguide modes whose confinement is relatively weak. In addition, channel waveguides cannot be formed conveniently except by etching ridge waveguides. These problems can be overcome using waveguides created by indiffusion of a dopant – almost exclusively titanium – to raise the refractive index [4.3]. More recently, waveguides have been formed by an exchange process similar to that used for glass substrates. Initial results employed Ag-ion exchange using AgNO3 solution [4.4]. More recently, proton exchange using benzoic [4.5] and other acids [4.6] have been employed. As in the case of out-diffused waveguides, only the extraordinary index n_e is increased. However, very large index differences have been achieved using this technique.

Based upon published results, titanium diffused waveguides are currently preferred for general device development. This is the technology that will be emphasized here. Waveguide stability and reduction in the electro-optic effect are currently issues being addressed in proton-exchange waveguides. Nevertheless, electro-optic devices have been fabricated using proton exchange techniques. Furthermore, the high index change achievable and the apparent reduction in photorefractive effects are important advantages. Also, proton exchange has been combined as an auxiliary process with titanium-diffusion waveguides to build some special function devices. Therefore, we also give an overview of the proton exchange process.

4.1.1 Titanium Diffused Waveguides

The fabrication of titanium diffused waveguides is quite straightforward. Although there are slight variations, the typical steps to fabricate channel waveguides by lift-off patterning are shown in Fig. 4.1. The polished crystal is carefully cleaned and spun with photoresist. A mask with the desired waveguide pattern is placed in contact with the crystal which is exposed to uv light. Upon developing to remove the exposed photoresist, a window corresponding to the waveguide pattern is left in the photoresist. Titanium

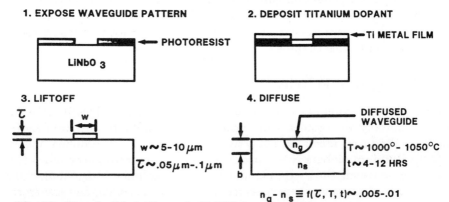

Fig. 4.1. Steps for fabricating strip Ti:LiNbO3 waveguides

is deposited over the entire crystal by rf sputtering, by e-beam deposition or by using a resistively heated evaporator. The crystal is then placed in a photoresist solvent which removes the photoresist and the unwanted titanium, leaving the desired strip of titanium. Instead of liftoff, the patterning can also be achieved by depositing titanium over the entire surface and selectively removing it outside the desired waveguide region.

The crystal is then placed in a diffusion furnace for diffusion at temperatures that range from 980° to 1050° C for typical diffusion times of 4–10 hours. The lower temperature results in an overly long diffusion time, while the upper limit is set by the desire to remain below the Curie temperature ($\sim 1125°$ C) [4.7] to avoid the need to repole the crystal after diffusion. Various ambient conditions have been used for the diffusion. Historically, the heating and diffusion cycles were performed in an atmosphere of flowing argon bubbled through a water column [4.8]. Cool down after diffusion is performed in flowing oxygen to allow reoxidation of the crystal to compensate for oxygen loss during diffusion. The water vapor treatment was initially employed to reduce the photorefractive effect [4.9]. Later, it was realized that this process also reduces Li_2O outdiffusion [4.10, 11] which can cause unwanted planar guiding for the extraordinary polarization. Typically, 80 % relative humidity in the flow gas is sufficient to eliminate unwanted guiding for diffusion temperatures below 1000° C [4.12]. However, the technique has been successfully used at temperatures as high as 1050° C.

Other techniques [4.12] to reduce the unwanted effect of outdiffusion include an overpressure of Li_2O by putting a powdered source, for example Li_2CO_3 [4.13] or $LiNbO_3$ [4.14], upstream from the lithium niobate crystal in an open tube furnace. A similar technique is to use a closed tube furnace without a separate Li_2O source [4.15]. Another approach is to replace the substrate in a crucible made from pressed $LiNbO_3$ powder of composition identical to that of the substrate crystal [4.16]. The ambient atmosphere of Li_2O is determined by the crucible, and the Li_2O content in the waveguide substrate can be maintained. One other approach is to compensate by diffusing outside the strip region a material such as magnesium oxide which lowers the refractive index [4.17]. While none of these techniques totally eliminates outdiffusion, they work sufficiently well to keep unwanted planar guiding from being a problem under most circumstances.

In addition to ease of fabrication, $Ti : LiNbO_3$ waveguides offer the advantage that the important waveguide parameters – waveguide width, effective depth and peak index change – can be independently controlled through the fabrication parameters. From a somewhat oversimplified viewpoint, the waveguide depth d depends upon the diffusion time t, upon the diffusion temperature T and, to a lesser extent on the titanium strip width. The waveguide width is photolithographically defined although increased somewhat by lateral diffusion. Most importantly, the peak waveguide-substrate index change Δn depends, for fixed T and t, upon the titanium thickness

147

and density. These considerations can be made more quantitative by using a simple diffusion model. Assuming complete diffusion into the crystal, the relative titanium concentration as a function of depth, y, in the crystal is [4.18]

$$C(y) = \frac{2}{\sqrt{\pi}} \frac{\tau}{d} \exp\left(\frac{-y}{d}\right)^2 , \qquad (4.1.1)$$

where τ is the deposited titanium thickness and d is the diffusion depth. The latter depends upon the diffusion temperature T, the activation temperature T_0 and bulk diffusivity D_0, with

$$d = 2(Dt)^{1/2} , \qquad (4.1.2)$$

where $D = D_0 \exp(-T_0/T)$. D_0 and T_0 depend upon the lithium niobate composition. For the commonly used congruent $LiNbO_3$ composition (48.6 mole % of Li_2O), representative values are [4.19] $T_0 \simeq 2.5 \times 10^4$ K and $D_0 = 2.5 \times 10^{-4}$ cm^2/s.

The exact mechanism by which titanium doping causes a refractive index change is not fully understood; however, a photoelastic change due to Ti induced strain seems to be a significant factor [4.21]. The refractive index increase, of course, depends upon the Ti concentration, but in a way that is different for the ordinary n_o and extraordinary index n_e. Measured results are shown in Fig. 4.2. The value of n_e depends essentially linearly upon titanium concentration. n_o varies linearly with titanium concentration (with different slope) for small values, but tends to saturate for larger values [4.20].

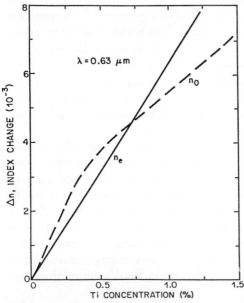

Fig. 4.2. Refractive index change for the ordinary (n_o) and extraordinary (n_e) indices versus titanium concentration for $\lambda = 0.63\,\mu$m. (After [4.20])

Due to lateral (surface) diffusion the peak Ti concentration for strip waveguides also depends upon the titanium strip width. Taking the lateral diffusion into account, the peak refractive index change depends upon the parameter

$$\Delta n_p \alpha \frac{\tau}{d} \ \text{erf}\left(\frac{w}{2d}\right) \quad , \tag{4.1.3}$$

where w is the titanium strip width and erf denotes the error function. It is important to note that the metal thickness τ is a meaningful parameter only for two titanium films of the same density. The titanium density depends upon the deposition technique; it is approximately equal to the bulk value for e-beam deposition and approximately 70 % of that value for resistively evaporated titanium.

4.1.2 Proton Exchange LiNbO₃ Waveguides

While several proton sources have been used to create the increase in the extraordinary refractive index that results from the exchange of protons for lithium ions, benzoic acid [4.5], either pure or diluted in lithium benzoate [4.22], remain the most common. The procedure is quite simple. A metallic mask with openings to define the exchange stripe is delineated on the cleaned lithium niobate substrate. Benzoic acid, a solid at room temperature, is heated (melting point 122° C) to a temperature between 200° and 249° C (its boiling point). The masked LiNbO₃ sample is immersed in the melt. In pure benzoic acid, the index profile is stepped-like with a surface index change $\Delta n_e = 0.12$ for the extraordinary index and a decrease to the ordinary value of $\Delta n_o = -0.04$. The temperature dependent diffusion coefficient is also crystal direction dependent, being smaller for diffusion in the z direction than in the x direction. As an example, for pure benzoic acid melt at a temperature of 235° C, the diffusion coefficients in the x and z directions are ~ 1 and $0.4 \ \mu\text{m}^2/\text{h}$, respectively [4.22]. The high index change and large diffusion coefficient allow single mode waveguides to be formed for either visible or infrared operation with exchange times $\lesssim 1$ h. Planar waveguide losses are $\sim 0.5 \ \text{dB/cm}$ for visible operation [4.5].

Unfortunately, early waveguides formed in a pure benzoic acid melt exhibited degradation in waveguide confinement with time [4.23] as well as with the application of a dc electric field [4.24]. In addition, the electro-optic efficiency of channel waveguide devices fabricated with a pure benzoic acid melt is nearly three times less than in titanium diffused waveguides of similar mode size [4.25]. This apparent reduction of the electro-optic effect in the exchanged region is particularly bothersome, because an important potential advantage of the high-index-change waveguides formed by proton exchange is electro-optic devices with greatly reduced drive voltage. This

149

can be achieved with small electrode gap provided small optical mode size can be achieved, as discussed in Sect. 4.2.2. Some of these deleterious effects have been reduced by using a proton source with extra lithium ions, e.g., benzoic acid diluted with lithium benzoate. Typical dilution level is 0.5 to 1 mole % lithium benzoate. While the peak refractive index change is the same as in pure benzoic acid, the diffusion coefficient for dilute melt fabrication is considerably reduced (e.g., at 235° C the diffusion coefficient for a 1% dilution is reduced by about 10 and 6 for the x cut and z cut, respectively). No refractive index reduction was observed after a 10 month period [4.22].

For some applications, including waveguide to fiber coupling, a much smaller index change is desirable. A reduced index change and a more graded index profile can be obtained by annealing the waveguides after proton exchange. Using this approach, low-loss channel waveguides with excellent fiber coupling characteristics have been demonstrated [4.26]. Indeed, except for the fact that only one polarization is guided, these waveguides can be made with properties − mode size, propagation loss and fiber waveguide coupling efficiency − comparable to those of titanium diffused waveguides.

Currently, proton exchange is the only reliable technique to achieve a large index change in lithium niobate. A large value of Δn is important for grating based devices [4.27], small radius of curvature waveguide bends [4.28] which are required in, for example, ring resonators [4.29], planar lenses [4.30] and highly confined, efficient electro-optic or acousto-optic devices. Proton exchange, with a post annealing, can also be used to change, or eliminate, the birefringence of titanium diffused lithium niobate waveguides [4.31]. In passive applications, this advantage of proton exchange waveguides has been clearly demonstrated. However, active proton exchange waveguide devices with reduced voltage requirements, compared to Ti : LiNbO$_3$ devices, have not yet been demonstrated.

4.1.3 Post-Waveguide Processing

If an electrode is to be placed on top of the waveguide, an intermediate (typically dielectric) buffer layer is needed to reduce current induced heating losses to the TM polarized (light polarized perpendicular to the crystal plane) [4.32]. SiO$_2$ is frequently employed [4.33], typically deposited by atmospheric chemical vapor deposition (CVD) [4.10]. For long-wavelength ($\lambda = 1.3\,\mu$m) operation, a $0.2\,\mu$m thick CVD SiO$_2$ layer eliminates measurable loading loss [4.35]. More recently, optically transparent indium tin oxide has been used [4.34]. Some devices with a buffer layer have shown short-term (on the order of seconds) drift which prevents application of the dc bias voltage required for some device applications [4.36]. This problem has been eliminated in some, but not all cases, by etching away the SiO$_2$ in the electrode gap region [4.37]. Presumably the drift is associated with the finite (and deposition dependent) conductivity of the buffer layer, the lithium niobate or both.

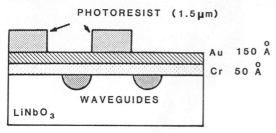

PHOTORESIST (1.5 µm)

Fig. 4.3. Fabrication of thick gold electrodes

Au 150 Å

Cr 50 Å

WAVEGUIDES

LiNbO₃

AFTER PLATING and ETCH

GOLD

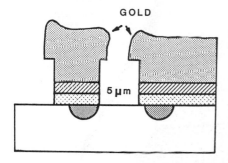

5 µm

The choice of electrode material depends upon the application. For relatively low speed (modulation frequencies below 100 MHz), an evaporated aluminum electrode ($\sim 0.2\,\mu m$ thick), with a flash of chrome for adhesion, is sufficient. Very narrow gap electrodes ($\sim 1\,\mu m$) can be defined by liftoff. Thicker films to reduce electrical loss are essential for high-frequency modulators. In fact, about $1–2\,\mu m$ thick aluminum can be used for multi-gigahertz applications, but patterning the small gap, which is essential for low voltage operation, is difficult. Larger electrode gaps ($\sim 30\,\mu m$) can be conveniently etched chemically [4.39].

The preferred electrode material for very-high-speed devices is gold, which can be electro-plated several micrometers thick with an electrode gap as small as five micrometers [4.40]. The electrode fabrication steps are shown in Fig. 4.3. A thin gold seed layer is deposited over the entire crystal; photoresist is spun and the electrode pattern aligned over the waveguides opening a window in the resist. The crystal is then placed in a plating bath and gold plated to a thickness of $2–3\,\mu m$. Afterwards, the gold seed layer is removed in the gap region. A photomicrograph of a sample electrode is shown in Fig. 4.4. Gold plated in this fashion has conductivity comparable to bulk conductivity.

The crystal end face is prepared for either end-fire lens coupling or fiber butt coupling by careful polishing. Chipping during lapping and polishing is avoided by epoxying a LiNbO₃ cover plate to the top surface at the crystal ends. To avoid rounding during polishing the gap between the two lithium niobate plates should be less than $1\,\mu m$.

→| |← 15 μm

Fig. 4.4. Scanning electron micrograph of gold traveling-wave electrode

To reduce Fresnel reflection and, more importantly, to eliminate this back reflection, which can cause serious laser stability problems in external modulator applications, an antireflection coating may be deposited. A single layer $\lambda/4$ film of yttrium oxide (Y_2O_3) has been used to reduce Fresnel reflections at the fiber-Ti : LiNbO$_3$ interface to -35 dB [3.38]. Y_2O_3 has a refractive index of 1.795 at the wavelength used ($\lambda = 1.55\,\mu$m). An angled endface can also be used to reduce reflection.

Several drift phenomena have been observed in Ti : LiNbO$_3$ devices. A poor buffer layer as described above is the most common cause. However, smaller transient effects have been observed in devices without buffer layers as well [4.40a]. Such transients may be explained by time-dependent anisotropic redistribution of stored charge in lithium niobate after a step voltage change. However, such transient effects, their origin, effect upon device applications [4.40b] as well as potential solutions are currently under investigation.

The impact of environmental conditions upon the performance of Ti : LiNbO$_3$ devices is also currently under study. The pyroelectric effect can

152

cause temperature dependent changes in device performance [4.40c]. These effects can be minimized if not eliminated by judicious choice of crystal orientation or device geometry. Ti : LiNbO₃ device performance has also been shown to be degraded by high ambient water vapor levels [4.40d].

In addition, lithium niobate exhibits a strongly wavelength dependent photorefractive effect for which its refractive index changes under high optical intensity levels. This effect is very strong for visible wavelengths [4.40e] but is dramatically reduced for wavelengths in the 1.3–1.55 μm region [4.40f]. For example, Ti : LiNbO₃ interferometric modulators operated at a wavelength of 1.5 μm have exhibited no photorefractive effects at power levels as high as 75 mW [4.40g].

4.2 Basic Device Considerations

4.2.1 Electro-Optic Effect

The linear electro-optic (Pockels) effect, which is the basis for active waveguide device control, provides a change in refractive index proportional to the applied electric field. The way in which this index change results in optical switching, intensity modulation, filter tuning, etc. depends upon the device configuration which will be considered in detail later. A voltage V applied to the electrodes placed over or alongside the waveguide, as shown in Fig. 4.5, creates an internal electric field of approximate magnitude

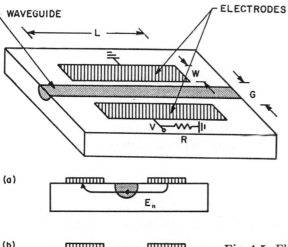

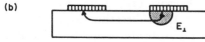

Fig. 4.5. Electrode placement to utilize the electric field component parallel ($E_\parallel$) or perpendicular ($E_\perp$) to the crystal surface

153

$|E| \simeq V/G$, G being the width of the electrode gap. The linear change in the coefficients of the index ellipsoid due to an applied electric field (E_j) along the principal crystal axis is [4.41]

$$\Delta\left(\frac{1}{n_i^2}\right) = \sum_{j=1}^{3} r_{ij} E_j \quad , \quad \text{or} \tag{4.2.1}$$

$$(\Delta n)_i = -\frac{n^3}{2} \sum_{j=1}^{j=3} r_{ij} E_j \quad , \tag{4.2.2}$$

where $i = 1, 2, \ldots, 6$ and r_{ij} is the 6×3 electro-optic tensor. By inserting the electro-optic tensor r_{ij}, the six values of Δn can be written as the elements of a symmetric 3×3 matrix. For lithium niobate

$$\Delta n_{ij} = \frac{-n^3}{2} \begin{pmatrix} -r_{22}E_y + r_{13}E_z & -r_{22}E_x & r_{51}E_x \\ -r_{22}E_x & r_{22}E_y + r_{13}E_z & r_{51}E_y \\ r_{51}E_x & r_{51}E_y & r_{33}E_x \end{pmatrix} , \tag{4.2.3}$$

where n is either the ordinary n_o or extraordinary n_e value.

Utilization of the diagonal elements 11, 22, and 33 of the perturbed refractive index matrix results in an index and, therefore, phase change, for an incident optical field polarized along the crystallographic x, y and z axes, respectively. These diagonal elements effect an index change, essential for switches and modulators, for the optical field aligned (polarized) along the crystallographic j, (4.2.3), axis given an electric field applied in the appropriate direction. For example, an electric field directed along E_z (E_3) causes index change (to the extraordinary index, $j = 3$)

$$\Delta n_{33} = \frac{-n^3}{2} r_{33} E_z \quad . \tag{4.2.4}$$

The electrode orientation relative to the waveguide needed to generate E_z depends, of course, upon the orientation of the crystal used. The orientation is frequently specified by the "cut" − the direction perpendicular to the flat surface on which the waveguide is fabricated.

The off-diagonal elements of (4.2.3), on the other hand, represent electro-optically induced conversion or mixing between orthogonal polarization components, For example,

$$\Delta n_{13} = \frac{-n^3}{2} r_{51} E_x \tag{4.2.5}$$

represents a rotation of the index ellipsoid that causes a coupling proportional to the r_{51} coefficient between the otherwise orthogonal A_1 and A_3 optical fields due to an electric field applied in the x direction (E_x). Utilization of off-diagonal electro-optic elements is necessary to induce polarization

154

change in Ti : LiNbO$_3$ waveguides. It is essential for polarization control devices as well as an interesting class of waveguide filters described in Sects. 4.5 and 4.6, respectively.

4.2.2 Phase Modulator

Perhaps the simplest waveguide electro-optic device is the phase modulator, in which the linear electro-optic induced index change causes a phase shift in the guided light. It is also the essential element of most of the more complex structures to be discussed. In addition, it is the simplest structure for considering the fundamental device parameters, viz. modulation voltage, electrical/optical field overlap and modulation frequency response. We therefore consider these parameters in detail for the phase modulator. Later we carry them over to the more complicated amplitude modulator structures, like the interferometer or directional coupler switch, indicating the structure dependent adjustments where necessary.

Ti : LiNbO$_3$ waveguide phase modulators appropriate to z cut or x or y cut lithium niobate are shown in Fig. 4.5. With electrodes placed on either side of the waveguide, the horizontal electric field $E_\parallel$ is applied while the vertical electrical field $E_\perp$ is employed with one electrode placed directly over the waveguide. In the latter case, an insulating buffer layer is required to eliminate loss to the TM-like polarization, as discussed earlier. In either case, the crystal orientation should typically be chosen to use the largest electro-optic coefficient, r_{33} ($= 30.9 \times 10^{-10}$ cm/V).

Drive Voltage

The local electro-optically induced index change is given by (4.2.4). However, neither the applied electric field nor the optical field is uniform. It is convenient to model the effective applied field inside the waveguide by that of a simple parallel plate capacitor as in the bulk case modulator where the lithium niobate crystal is sandwiched between two electrodes and both optical and electrical fields are assumed to be uniform [4.41]. The correction factor from this simple model is given by an overlap parameter Γ. The effective electro-optically induced index change within a cross section of the optical mode can be written as

$$\Delta n(V) = \frac{-n^3 r}{2} \frac{V}{G} \Gamma \quad , \tag{4.2.6}$$

where G is the interelectrode gap and Γ is the overlap integral between the applied electric field and the optical mode. The quantity Γ is given by

$$\Gamma = \frac{G}{V} \iint\limits_{-\infty}^{\infty} E|A|^2 dA \quad , \tag{4.2.7}$$

155

where A is the normalized optical field distribution and E is the applied electric field. The total phase shift over the interaction length L is then

$$\Delta\beta L = -\pi n^3 r \Gamma \frac{V}{G} \frac{L}{\lambda} \quad . \tag{4.2.8}$$

The exact phase shift required to achieve complete intensity modulation or switching depends upon the modulator or switch type. The modulation condition can be written in general as

$$|\Delta\beta L| = p\pi \quad , \tag{4.2.9}$$

where p, which depends upon the modulator type, is on the order of one. The voltage length product required for modulation is thus

$$VL = \frac{p\lambda G}{n^3 r \Gamma} \quad . \tag{4.2.10}$$

The goal in switch/modulator design is low drive voltage, large modulation bandwidth and low optical insertion loss. We will show later that the achievable bandwidth also scales as $1/L$. Therefore, for a given electro-optic material and wavelength, one tries to minimize VL by optimizing the lateral geometric parameter G/Γ. Later we will find that modulation bandwidth and, more importantly, the insertion loss also depend upon the lateral geometric parameters, thus resulting in an interdependence that may require performance trade-offs that we will discuss.

The dependence of the induced Δn as well as of the electrical field/optical mode overlap parameter Γ upon the optical mode profile and electrode geometry for the structures of Fig. 4.5 have been calculated by several researchers [4.42–44]. Representative results from [4.42], where semi-infinite electrodes and no intermediate buffer layer were assumed, are shown in Fig. 4.6. The waveguide mode dimensions ($w_\parallel$ and $w_\perp$ are the mode 1/e intensity full width and depth, respectively) serve as the characteristic lengths

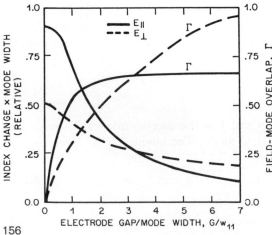

Fig. 4.6. Calculated electro-optically induced refractive index change and electrical field-optical mode overlap, Λ, versus electrode gap/mode width ratio

in the calculations. A mode profile typical of Ti : LiNbO$_3$ waveguides – Gaussian in width and Hermite-Gaussian in depth – was assumed. The results further assume that, for a utilization of $E_\perp$, the inner edge of the waveguide mode, and for $E_\parallel$, each electrode is placed symmetrically about the waveguide, as shown in inset of Fig. 4.5. Because it is convenient as well as instructive to measure the electrode gap in units of the optical mode width $w_\parallel$, the index change is calculated as the (index change) × (mode width) product. Numerical results of this analysis for both the relative index change mode width product and overlap parameter Γ versus normalized electrode gap $(G/w_\parallel)$ are summarized in Fig. 4.6.

Figure 4.6 shows that, although the induced index change continues to increase as the normalized gap $G/w_\parallel$ is decreased, the increase is relatively small for $G/w_\parallel < 0.5$. For modulators, a decrease in electrode gap beyond that which reduces the required drive voltage is not desirable because a small gap generally implies higher capacitance and ultimately lower speed. The parallel field $(E_\parallel)$ is more effective than the perpendicular component $(E_\perp)$ for a small gap, but the former also drops off more rapidly than the latter for a large gap. This effect has been qualitatively observed in phase modulators [4.45]. For devices with large mode size required for efficient fiber coupling, this effect becomes less important [4.46]. In fact, for $E_\parallel$, the (index change) × (mode size) product drops off as $1/G$ but, for $E_\perp$, this product depends upon $1/\sqrt{G}$. Thus, as shown in Fig. 4.6, Γ increases much more rapidly with increasing $G/w_\parallel$ for modulators that use $E_\parallel$ than for those that use $E_\perp$ but also maximizes to a lower value. The small value of Γ for small $G/w_\parallel$ implies less efficient utilization of the electric field for small values of G for which V/G is large. Because the calculated quantity (left ordinate in Fig. 4.6) is the (index change) × (mode size) product, if $w_\parallel$ and $w_\perp$ are reduced, then the e/o induced Δn increases as $1/G$, provided that fabrication techniques allow reducing G proportionately. Optimization of the electro-optic-induced index change thus requires both minimizing the waveguide mode size and fabricating electrodes with a gap $G \lesssim 0.5 w_\parallel$, although the latter requirement is not so critical for modulators that employ $E_\perp$. The effect of an intermediate insulating buffer layer is to decrease the applied field in the waveguide region. The increase in drive voltage depends upon buffer layer thickness and electrode geometry. For a typical SiO$_2$ buffer layer ($\sim 0.2\,\mu$m thick), the voltage increase is about 20–40 % [4.44].

Modulation Bandwidth

The potential bandwidth of waveguide modulators is, in practice, always limited by distributed-circuit effects. The electro-optic effect is an electronic phenomenon that has a subpicosecond response time. The achievable modulation speed depends upon several factors. Most important is the electrode type (lumped or traveling wave), shown schematically in Fig. 4.7.

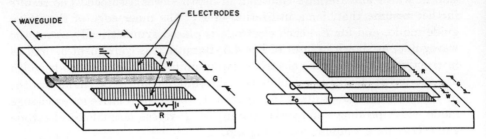

Fig. 4.7. Schematic of lumped and traveling-wave electrodes

The difference between these two electrode types is the electrical drive and termination. In both cases, an important parameter in the design and performance of Ti:LiNbO$_3$ modulators is the electrode capacitance/length, which depends upon the rf dielectric constant of lithium niobate and geometric parameters. To accommodate the different modulator structures, three principal electrode geometries have been employed for Ti:LiNbO$_3$ switch/modulators. They are the symmetric and asymmetric two electrode (also called coplanar strip), and the symmetric three electrode (or coplanar waveguide) (Fig. 4.8).

The capacitance/length for these three electrode structures, which can be calculated using conformal mapping techniques, can be written by a single expression [4.47, 48].

$$\frac{C}{L} = 2^{k-q} \varepsilon_{\text{eff}} \left(\frac{K'(u)}{K(u)} \right)^q \tag{4.2.11a}$$

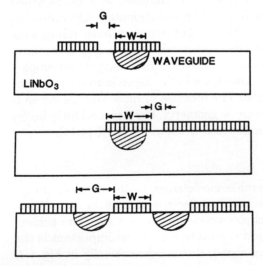

Fig. 4.8. Symmetric and asymmetric coplanar strip and coplanar waveguide electrodes (*top* to *bottom*)

where ε_{eff} is the effective microwave dielectric constant, given by $\varepsilon_{eff} = (\varepsilon_0/2)(1+\varepsilon_s/\varepsilon_0)$ where $\varepsilon_s = (\varepsilon_x\varepsilon_y)^{1/2} = 38.5$ for lithium niobate and ε_0 is the permittivity of the air cladding. Here W is the width of the non-semi-infinite electrode(s) and G is the electrode gap; u is the normalized geometric factor, $u = [1+(2/k)(W/G)^q]^{-1/k}$ where $k = 1$ for the symmetric coplanar strip and coplanar waveguide while $k = 2$ for the asymmetric coplanar strip; $q = -1$ for the three electrode geometry and $+1$ for the other two [4.49]. Finally, $K(u)$ is the complete elliptical integral of the first kind and $K'(u) = K(\sqrt{1-u^2})$. The calculated capacitance/length versus W/G for the three electrode structures is shown in Fig. 4.9. We now consider the modulation-frequency response for lumped and traveling-wave type modulators.

Lumped Electrode. The modulation bandwidth of the lumped electrode modulator (whose electrode length is small compared to the rf wavelength) is the smaller of the inverse of the optical or electrical transmit times, or the time constant of the lumped-circuit parameters. The latter is usually the more restrictive. The parallel resistor (Fig. 4.7) is used to allow broad-band matching to the impedance of the driving source, typically 50 Ω. The modulation bandwidth for the lumped modulator is then determined by the RC time constant, $\Delta f = 1/\pi RC$. For fixed termination resistance R and lateral electrode geometry, Δf decreases linearly with the electrode length L, (4.2.10). Therefore, the relevant parameter is the bandwidth-length product whose calculated value versus the normalized electrode width and gap parameter is also presented in Fig. 4.9 for each of the three electrode structures. A 50 Ω termination is assumed.

Although C/L increases and consequently Δf decreases as G/W is decreased (Fig. 4.9), both depend essentially only logarithmically on G/W. The electrical transit time cut-off frequency, where c is the speed of light,

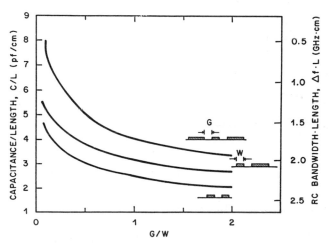

Fig. 4.9. Calculated capacitance per unit length and RC 3 dB bandwidth × length product versus lumped electrode gap/width ratio for the three electrode geometries

$$f_t = \frac{c}{\pi\sqrt{\varepsilon_{\text{eff}}}L} \qquad\qquad (4.2.11\text{b})$$

is $\sim 2.2\,\text{GHz}\cdot\text{cm}$ for lithium niobate, so there is no real advantage to design the capacitance/length below $\sim 2\,\text{pf/cm}$. Experimental results consistent with these expected values have been realized [4.50, 51].

As discussed earlier, an appropriately small electrode gap, comparable to the optical mode size, is important to reduce the required drive voltage. This requirement, together with that of small G/W for high bandwidth, necessitates a relatively narrow electrode width. However, W should not be designed much narrower than the waveguide width to insure good overlap of the electric and optical fields, yet it should be chosen sufficiently wide so that static electric resistance does not become the limiting factor in achievable bandwidth.

Traveling Wave. The goal of the traveling-wave electrode (Fig. 4.7) is to make the electrode appear as an extension of the driving transmission line. As such, it should have the same characteristic impedance as the driving source. In this case, the modulator speed is not limited by the electrode charging time but rather by the difference in transit time for the optical and modulating rf waves [4.52]. The characteristic impedance depends simply upon the capacitance/length

$$\frac{1}{Z} = \frac{c}{\sqrt{\varepsilon_{\text{eff}}}}\left(\frac{C}{L}\right) \quad . \qquad\qquad (4.2.11\text{c})$$

The calculated impedance results for the three electrode geometries are shown in Fig. 4.10. The coplanar waveguide has the lowest impedance, followed by the asymmetric coplanar-strip and symmetric-strip structures. The

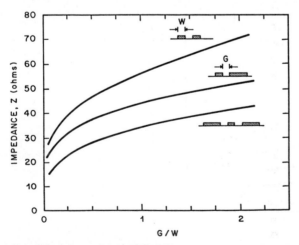

Fig. 4.10. Calculated electrode impedance versus gap/width ratio for three electrode geometries

required values of G/W to achieve an impedance of $50\,\Omega$ are 3.0, 1.7 and 0.6, respectively.

To examine the bandwidth limitation of traveling-wave modulators, we assume that the electrode impedance is matched to the driving source. We consider the phase modulation in response to a sinusoidal electrical drive voltage. The drive voltage along the electrode can be written as

$$V(z,t) = V_0 \sin\left(\frac{2\pi N_{\rm m}}{\lambda_{\rm m}}z - 2\pi f t\right) \ , \tag{4.2.12}$$

where $N_{\rm m} = \sqrt{\varepsilon_{\rm eff}/\varepsilon_0}$, $\lambda_{\rm m}$ and f are the microwave refractive index, free space wavelength and frequency, respectively, and z is the position along the electrode. For simplicity, we ignore microwave loss for the moment. The voltage seen at any point along the electrode by a photon that enters the waveguide at any time t_0 can then be written as

$$V(z,t_0) = V_0 \sin\left[\frac{2\pi N_{\rm m} f}{c}\left(1 - \frac{N_0}{N_{\rm m}}\right)z - 2\pi f t_0\right] \ , \tag{4.2.13}$$

where N_0 is the effective index of the guided optical mode. Then, because the electro-optically induced phase shift $\Delta\beta$ is proportional to V, the integrated $\Delta\beta$, is

$$\int_{z=0}^{z=L} \Delta\beta(f)dz = \overline{\Delta\beta_0}\frac{\sin(\pi f/f_0)}{(\pi f/f_0)}\sin(2\pi f t_0 - \pi f/f_0) \quad\text{where} \tag{4.2.14}$$

$$\overline{\Delta\beta_0} = \Delta\beta_0 L = \frac{\pi n^3 r V_0 \Gamma L}{\lambda G} \quad\text{and}\quad f_0 = (c/LN_{\rm m})[1 - N_0/N_{\rm m}]^{-1} \ .$$

The term $1 - N_0/N_{\rm m}$ is a measure of the velocity mismatch between the optical and microwave signals. For $N_0 = N_{\rm m}$, the optical wave travels down the waveguide at the same speed as the microwave drive signal moves along the electrode and "sees" the same voltage over the entire electrode length. In this case the integrated value of $\Delta\beta$ is proportional to $V_0 L$, and arbitrarily long electrodes can be used to reduce the required drive voltage with no frequency limitation. However, for $N_0 \neq N_{\rm m}$, as in lithium niobate, there is walk-off between the optical wave and microwave drive signal which results in a reduction and, for sufficiently large L or f, a complete cancellation of $\Delta\beta$. The result is the sinc function frequency response of (4.2.14). According to (4.2.14), the frequency for which the integrated value of $\Delta\beta$ is reduced by 50 % from its value for $f = 0$ (the 3 dB optical modulation bandwidth) is

$$\Delta f \cdot L \approx \tfrac{2}{\pi}f_0 L \ . \tag{4.2.15}$$

Therefore, for given electrode length, the achievable bandwidth is critically dependent upon the mismatch between the optical and microwave velocities.

The value of N_0/N_m for lithium niobate is approximately $\frac{1}{2}$. Therefore, the potential (velocity-mismatch limited) 3 dB optical bandwidth limitation for phase modulation is close to 9.6 GHz cm. This is the modulation frequency for which the degree of optical modulation decreases to 50 % of its value at $f = 0$. The detected electrical modulation is down by 6 dB.

Electrode loss reduces the effective $\Delta\beta$ seen along the modulator length. Frequency-independent loss can be compensated for by an increased driving voltage. However, if the loss depends upon f, as is generally the case, the frequency bandwidth is reduced.

For given electrode dimension, the high-frequency microwave loss is determined by the skin depth and one expects a loss of $a = a_0 f^{1/2}$ in dB/cm, where a_0 depends upon electrode conductivity and geometry. This frequency dependence has been verified experimentally [4.47, 53]. Assuming no velocity mismatch, the effect of loss on the integrated $\Delta\beta$ is

$$\int \Delta\beta(z)dz = \frac{\overline{\Delta\beta_0}}{\alpha_m}(1 - e^{-\alpha_m L}) \quad , \tag{4.2.16}$$

where $\alpha_m = a/4.3$ converts from loss in dB/cm to an exponential loss coefficient. Assuming a skin depth frequency dependent loss with $a_0 = 1\,\mathrm{dB/cm(GHz)}^{1/2}$, a value achieved experimentally, and a 1.5 cm long electrode, the frequency for which the integrated $\Delta\beta$ is reduced by 50 % relative to low frequency is about 10 GHz. This limit is somewhat less restrictive, though comparable to the ~ 7 GHz bandwidth limit due to velocity mismatch for lithium niobate. However, for larger electrode transmission loss coefficients, electrode loss can be the dominant frequency limiting factor.

Voltage/Bandwidth. The required modulation voltage scales inversely with length, as does, in the absence of electrode resistance, the modulation bandwidth. Thus, increased bandwidth can be achieved at the expense of drive voltage by simply decreasing the device length. The appropriate modulator figure-of-merit is thus the voltage/bandwidth ratio. This figure-of-merit allows for meaningful comparison of modulators of different lengths and types. The power/bandwidth ratio is the other frequently used figure-of-merit. However, it scales as $1/L$ and does not indicate the trade-off between bandwidth and drive voltage (or power) which can be achieved by simply using the same modulator design with a different interaction length. The voltage/bandwidth ratio for lumped modulators is

$$V/\Delta f = \pi R\left(\frac{\varepsilon_{\mathrm{eff}}}{n^3 r}\right)(p\lambda)\left(\frac{G}{\Gamma}\frac{K'(u)}{K(u)}\right) \quad , \tag{4.2.17}$$

where we have separated the material constants, modulator type, wavelength and geometric dependence.

With respect to geometric dependence, $V/\Delta f$ is minimized for $G/w_{\parallel} \approx G/w \approx 0.5$. Further reduction in the electrode gap increases the capacitance but does not appreciably reduce the required drive voltage. The minimum effective gap can then be determined from the minimum achievable mode size, which in turn depends upon the maximum Δn. That relation is $w_{\min}/\lambda \sim 0.5/\sqrt{n_s \Delta n}$ [4.54]. For lithium niobate, assuming $\Delta n \simeq 0.01$, we find $G_{\min} \simeq 1$ and $2\,\mu\mathrm{m}$ for $\lambda = 0.63$ and $1.3\,\mu\mathrm{m}$, respectively. If this value is used in (4.2.17) as the optimum value of $K'(u)/K(u)$ ($=2$ for $G/W = 0.5$) we find

$$(V/\Delta f)_{\min} \simeq \frac{\pi}{2} R \left(\frac{\varepsilon_{\mathrm{eff}}}{n^{7/2} r \sqrt{\Delta n}} \right) \frac{p\lambda^2}{\Gamma} \quad , \tag{4.2.18}$$

where Γ is 0.3 or 0.2 for modulators employing $E_\parallel$ or $E_\perp$, respectively. For lithium niobate, assuming $p = 1$ (π phase modulation) and that the electrode resistance can be ignored, the approximate minimum value of $V/\Delta f$ for lumped modulators is ~ 0.5 and $1.5\,\mathrm{V/GHz}$ for $\lambda = 0.6328$ and $1.32\,\mu\mathrm{m}$, respectively.

Equation (4.2.18) indicates that, for optimum design, $V/\Delta f$ scales as λ^2. Thus, there is a voltage penalty paid for operation at the long wavelengths (1.3–1.5 $\mu\mathrm{m}$) required for lightwave communication systems. Because the available drive voltage is generally limited, especially at very high frequencies, the trade-off at long λ is generally to use longer L to keep V acceptable and reduce the achievable bandwidth.

If we ignore loss, the voltage/bandwidth ratio for traveling-wave electrode modulators is

$$V/\Delta f = \frac{\pi}{2c} \left(\frac{N_{\mathrm{m}}[1 - N_0/N_{\mathrm{m}}]}{n^3 r} \right) (p\lambda) \frac{G}{\Gamma} \quad , \tag{4.2.19}$$

where, if we further assume the mode-size limited minimum gap as above, then

$$(V/\Delta f)_{\min} = \frac{\pi}{8c} \left(\frac{N_{\mathrm{m}}[1 - N_0/N_{\mathrm{m}}]}{n^{7/2} r \sqrt{\Delta n}} \right) \frac{p\lambda^2}{\Gamma} \quad . \tag{4.2.20}$$

Equation (4.2.20) is probably over-optimistic. To simultaneously achieve an electrode impedance of 50 Ω and a small electrode gap needed to minimize $V \cdot L$, a very narrow electrode width is required. For example, for $G = 5\,\mu\mathrm{m}$ an electrode width of only 3 $\mu\mathrm{m}$ is necessary for $Z = 50\,\Omega$ with an asymmetric electrode. As a result, researchers have used either a relatively narrow gap or 50 Ω electrodes, but not both. Assuming a 50 Ω impedance source in both cases, the first approach allows more efficient use of the applied voltage but suffers a reduction in effective applied voltage due to impedance mismatch reflection and due to the lower impedance. In the latter case, all the microwave power delivered by the source is coupled onto the electrode, but the large gap results in a less than optimum induced phase shift. A source with lower impedance can be used but because the important modulator drive parameter is voltage, for the same source power the applied voltage is decreased as $\sqrt{Z}$.

Velocity-Matching Techniques. The ultimate limit of the bandwidth for traveling-wave modulators in lithium niobate is set by the difference in velocity of the optical and electrical signals. Several techniques can be used to overcome this velocity mismatch limitation. Two separate techniques which can be generally described as real-velocity matching and artificial-velocity matching have been proposed. In the former technique one designs a structure to modify the effective propagation velocity of either the optical or, more typically, the electrical signal to equalize N_0 and N_m. Techniques include a high-index overlay and etching of lithium niobate to increase the electrical field into the lower-index air to speed up the electrical signal. In principle, such velocity-matching techniques allow one to increase the bandwidth for a given length device; however, they also tend to reduce the efficiency of interaction between the electrical and optical signals, resulting in higher drive voltage. Using, for example, an etched groove, an increase in bandwidth of as much as 20 % has been achieved [4.55].

In artificial-velocity-matching techniques, one seeks not necessarily to increase the bandwidth of the modulator (for fixed electrode length), but rather to shape and shift the frequency response. For example, artificial-velocity matching can be used to maximize the response at some arbitrarily high frequency [4.56, 57]. This can be done using techniques very similar to the techniques described later to achieve phase matching between two non-synchronous optical modes. An example electrode structure to achieve velocity matched interaction for a given design frequency is shown in Fig. 4.11.

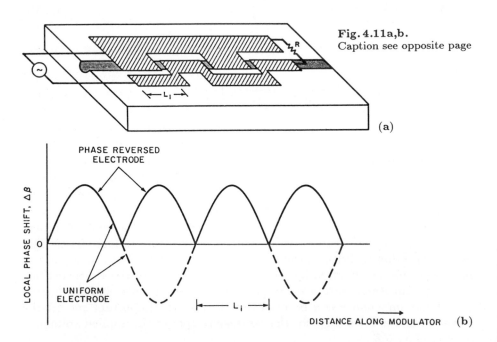

Fig. 4.11a,b.
Caption see opposite page

(a)

(b)

164

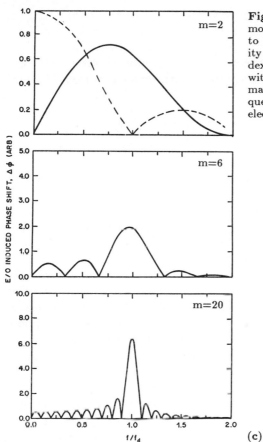

Fig. 4.11. (a) Traveling-wave phase modulator with periodic phase reversal to provide frequency dependent velocity matching. (b) Relative induced index change along the waveguide with and without phase reversal for the velocity-matched frequency. (c) Computed frequency response; m is the number of electrode sections

In this microwave phase shifted structure, the electrode is laterally shifted periodically along the length of the modulator to reverse the polarity of the effective applied field. At the design drive frequency, this physically induced phase reversal occurs exactly at the point where the electrical and optical signals are 180° out of phase due to velocity mismatch. Therefore the phase reversal effectively undoes the effect of velocity mismatch, as shown schematically in Fig. 4.11b. As a result, at the design frequency, ignoring microwave propagation losses, the electro-optic interactions of each of the electrode sections add up in phase. The electrode structure can be made arbitrarily long and still achieve modulation efficiently at the design frequency approximately equivalent to that achievable at dc in a standard modulator of the same length, again in the absence of electrode loss.

The relationship between the velocity-matched design frequency f_d and the phase reversal length can be found by setting the position dependent phase term of the effective voltage expression (4.2.13) equal to π for the phase reversal length $z = L_i$. The result is

165

$$f_\mathrm{d} = \frac{c}{2N_\mathrm{m}L_i}(1 - N_0/N_\mathrm{m})^{-1} \ . \tag{4.2.21}$$

Following the earlier procedure, the integrated e/o induced phase shift along the waveguide using a phase-reversed drive with m sections each of length L_i is [4.57]

$$\int_{z=0}^{L} \Delta\beta(f)dz = \left[\frac{\Delta\beta_0 L_i}{(\pi f/f_\mathrm{d})}\sin\left(\frac{\pi f}{f_\mathrm{d}}\right)\right] \times \left[\left(\frac{\sin\left[\frac{m}{2}(\frac{\pi f}{f_\mathrm{d}}) + \pi\right]}{\cos(\pi f/2f_\mathrm{d})}\right)\right]$$

$$\times \sin\left[\left(\frac{m-1}{2}\right)\left(\frac{\pi f}{f_\mathrm{d}} + \pi\right) + \frac{\theta_i}{2}\right] \tag{4.2.22}$$

where $\Delta\beta_0$ and f_d are previously defined. The frequency response of (4.2.22), shown in Fig. 4.11c for $m = 2$, 6 and $m = 20$, is characterized by the same $\sin(x)/x$ envelope functional form as the non-velocity-matched case, except in this case the walk-off over only one section length L_i is important. The second term on the right hand side in (4.2.22) describes the additive effect of the m sections and for $f = f_\mathrm{d}$ has a value of m. Therefore, at the design frequency $f = f_\mathrm{d}$ the required drive voltage, ignoring electrode loss, is that set by the total device length $L = mL_i$. The bandwidth that one achieves at the design frequency is still limited by the total length of the device. The advantage is that one can place that bandwidth at the desired frequency interval or, in general, synthesize a desired response. Artificial-velocity-matched devices are potentially important for applications such as time division multiplexing [4.58, 59] or overdriven switch/modulators for proposed ultra-short optical gate devices [4.60, 61] where the optical switch need be driven only in a strictly periodic fashion. In these applications, broad bandwidth is not necessary while high-frequency operation is essential.

In addition to providing the possibility of modulation at very high frequencies, phase-reversed electrode techniques can provide compensation or equalization for finite electrode losses. For example, by designing a device with just two sections or a section and one-half designed for velocity match at a frequency slightly above the 3 dB cutoff point of a uniform electrode structure, one can increase the response at the higher frequency where normally the response would be decreasing because of higher microwave losses [4.62]. This is achieved at the expense of a higher drive voltage. In this way, artificial-velocity matching techniques can be made to level or equalize the response of modulators that are degraded by microwave loss.

The modulator frequency response can also be altered by using a structure which includes both phase reversal and intermittent interaction in which the electrode is, in fact, periodically removed from the waveguide region. Using this technique, one can get efficient response not only at the design frequency but also at arbitrarily many harmonics of the design frequency. Furthermore by aperiodic phase reversal [4.63], that is by varying L_i along the electrode length, and by pseudo-random phase reversal [4.64], that is by reversing or not reversing the phase over each section, based upon a pseudo-random code, added flexibility in shaping of the amplitude response can be achieved.

4.2.3 Insertion Loss

The source of losses in waveguide devices include: coupling loss, propagation loss — both scattering and absorption — electrode loading loss and possible waveguide bend loss. For most practical applications, light is coupled in and out of the device with a single-mode fiber. Therefore we examine the requirements for efficient fiber-waveguide coupling.

Fiber-Waveguide Coupling

Efficient fiber-waveguide coupling, which is essential for most applications of these devices, has been studied by many groups [4.65–72]. Assuming perfect alignment between the fiber and Ti:LiNbO$_3$ waveguide, the two contributions to fiber-waveguide coupling loss are reflection (or Fresnel loss) and the loss caused by mismatch between the fiber and waveguide modes. The former can be greatly reduced by using either index-matching fluid or antireflection coatings on the lithium niobate end faces [4.38]. The latter arrangement is preferred when permanently attaching fibers. The loss resulting from mismatch between the fiber and waveguide modes is then the principal source of fiber-waveguide coupling loss. Indeed, the total device insertion loss in early devices was dominated by fiber-waveguide coupling loss due to mode mismatch.

Mode Matching. To achieve efficient coupling, it is necessary to fabricate titanium-diffused waveguides with a spot size which is well matched to that of the fiber. Because of the large index difference between lithium niobate and air compared to the small index difference of the waveguide region relative to the substrate, the waveguide mode shape in depth is typically quite asymmetric, while the fiber is circularly symmetric. Furthermore, for operation at $\lambda = 1.3$–$1.5\,\mu$m and for fiber core sizes typical for lightwave communciation applications, the fiber mode is relatively large. Thus, relatively deep diffusion of the titanium into the lithium niobate is required. Fortunately, a deep diffusion also helps to reduce propagation loss that may result from surface scattering.

As discussed earlier, the diffusion depth is determined by the diffusion time t, temperature T and activation temperature T_A. The titanium-metal thickness τ should satisfy two requirements. First, for the diffusion parameters required to achieve the desired depth, the titanium metal must be thin enough to allow complete diffusion. However, it must be sufficiently thick to produce a waveguide/substrate index difference, Δn, to provide strong guiding. Within these general constraints, one should choose t, T, τ and the waveguide strip width w to produce a waveguide mode that matches the fiber mode as closely as possible. While initial values of these parameters can be obtained from waveguide models, the final selection of diffusion parameters to ensure the waveguides are optimized must, of course, be determined empirically.

The waveguide and fiber modes can be measured by imaging the near-field patterns onto an infrared vidicon. The output from the vidicon is displayed on a signal-averaging oscilloscope and waveguide-mode intensity profiles recorded in the width and depth directions. The mode size depends upon wavelength and polarization.

An example of the measured width and depth waveguide mode profile with a 9 μm metal-strip width and diffusion conditions (given in caption) is depicted in Fig. 4.12. Also shown is the measured mode profile of a standard single-mode fiber. The waveguide mode in the crystal plane is essentially Gaussian and is an excellent match to the fiber. The mode profile in depth is asymmetric, as expected, because of the large waveguide-cladding index difference. The mode shape in depth is approximated by the Hermite-Gaussian function over most of the profile. The intensity roll off both near the surface and into the substrate is slower, however, than the Hermite-Gaussian function.

The longer tail on the air side of the waveguide is especially significant because of its better compatibility with the symmetric fiber mode. The measured 1/e intensity full width w_x and w_y, for these waveguides, all of

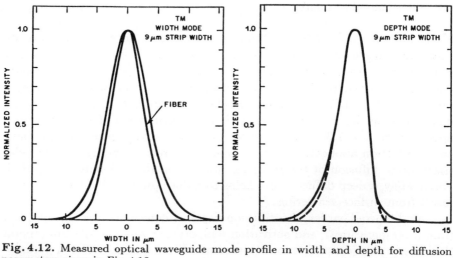

Fig. 4.12. Measured optical waveguide mode profile in width and depth for diffusion parameters given in Fig. 4.13

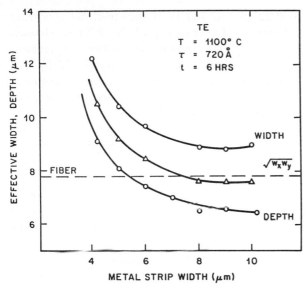

Fig. 4.13. Measured optical mode width, depth and geometric mean versus titanium metal strip width for waveguides fabricated with the diffusion parameters given. Results are for TE polarization

which were single mode, is shown versus metal-strip width in Fig. 4.13 for the TE polarizations. Qualitatively similar results hold for the TM mode. As discussed below, it is convenient to describe the waveguide spot size by the geometric-mean diameter $w = \sqrt{w_x w_y}$, and the mode eccentricity, $\varepsilon = w_x/w_y$. The geometric mean mode diameter is also displayed in Fig. 4.13. The mode dimensions are relatively constant over the range of strip widths between ~ 6 and $10\,\mu$m, which correspond to well-guided modes. The rapid increase in mode size for small strip width results from poor confinement as these waveguides approach cutoff. The mode eccentricity is approximately 1.5 for the well-confined modes and approaches unity for waveguides near cutoff. As discussed below, this can be altered by burying the waveguide. For these diffusion conditions, there is a very good match between the fiber mode diameter and the waveguide mean mode diameter for strip widths from approximately 7 to $10\,\mu$m.

The expected coupling efficiency can be determined by calculating the overlap between the fiber and waveguide modes. The power coupling coefficient can be written conveniently as [4.70, 73]

$$\kappa = 0.93 \left(\frac{4(w/a)^2}{[(w/a)^2 + \varepsilon][(w/a)^2 + 1/\varepsilon]} \right) \quad , \tag{4.2.23}$$

where a is the fiber-mode $1/e$ intensity diameter, w is the geometric-mean waveguide diameter, and ε is the ratio of the waveguide mode width and depth, as defined above.

169

The constant factor of 0.93 in (4.2.23) is determined by modeling the mismatch between the symmetric fiber mode and the diffused channel waveguide mode assuming identical 1/e dimensions for the two. Experimentally, this coefficient is found to be somewhat closer to unity [4.67, 70]. The second term includes the further reduction in coupling efficiency due to any difference in the dimensions of the fiber and waveguide modes, but assuming perfect alignment and ignoring interface reflections.

The dimensional dependence of the expected coupling loss (4.2.23) indicates that, regardless of the mode eccentricity, the optimum coupling is achieved for $w = a$. The dependence of the coupling loss on mode mismatch is weak for w near a. A mismatch in the mean waveguide mode size of 10 % causes an increase in coupling loss of only ~ 0.04 dB relative to its value for $w/a = 1$ for ε as large as 2. The coupling loss is less sensitive to non-unity eccentricity. A 10 % error from the optimum condition $\varepsilon = 1$ increases the coupling loss by only ~ 0.01 dB for $w/a = 1$.

Examples of measured loss results including separate propagation and coupling contribution are shown in Fig. 4.14 [4.70]. Results are shown for the TE mode, but, for these diffusion conditions, comparable results apply for the TM mode. The results are for z cut, y propagating crystals. For the waveguides with minimum total insertion loss, the coupling loss per face is ~ 0.35 dB and the propagation loss is ~ 0.3 dB/cm. This measured coupling loss includes a calculated residual Fresnel loss of 0.09 dB per face for TE for a matching index of 1.65. The rapid increase in total insertion

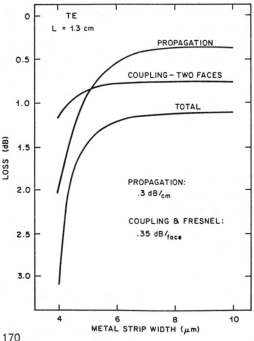

Fig. 4.14. Measured total fiber-waveguide-fiber loss and approximate fiber-waveguide coupling and propagation loss contributions versus titanium metal width for waveguides fabricated using parameters of Fig. 4.13

loss for small metal-strip widths is principally due to large propagation loss resulting from poor mode confinement as waveguide cutoff is approached. Based on resonators, propagation losses for Ti : LiNbO$_3$ waveguides as low as 0.1 dB/cm have been inferred [4.74].

Calculated coupling losses using (4.2.23) and the measured mode sizes correlate accurately with the measured coupling loss displayed in Fig. 4.14. The insensitivity of the mode overlap to small changes in w/a, together with the fact that (for well-confined modes) the mode width and depth are relatively constant over a large range of channel waveguide strip widths, results in a coupling loss that is very insensitive to channel width. Very good coupling efficiency can be achieved for both TE and TM modes if one uses diffusion conditions for which Δn is comparable for the ordinary and extraordinary indices (see Fig. 4.2). However, for large values of titanium concentration, the index changes for the ordinary and extraordinary indices may be very different and simultaneous mode matching of the TE and TM modes is more difficult. Fortunately, for the typically large fiber mode sizes, only modest values of Δn are necessary. It has also been observed that the vertical position of the mode may be slightly different for the two polarizations [4.71]. While results described here are for Z cut crystals, similar results have been achieved for the X cut orientation in spite of a slightly larger asymmetry factor [4.107].

Alignment Sensitivity. The sensitivity of the fiber-waveguide coupling efficiency to lateral and vertical (depth) translational misalignment is an important issue in considering manufacturable fiber attachment techniques. Example results for 1.3 μm are shown in Fig. 4.15 [4.75]. A slight asymmetry with respect to the vertical offset, which originates from the verti-

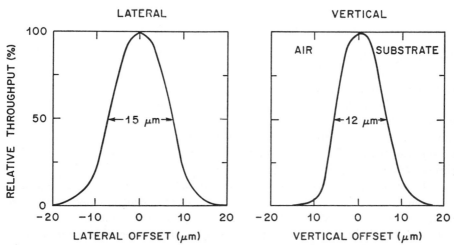

Fig. 4.15. Measured relative fiber-waveguide-fiber throughput versus fiber-waveguide misalignment for vertical and lateral offset

cal waveguide mode shape, is evident. Because the mode size is relatively large to match typical fibers, the coupling efficiency is not very sensitive to translational-alignment errors. For both vertical and lateral translations, an offset of $\pm 2\,\mu$m increases the coupling loss by only ~ 0.25 dB relative to the best alignment.

In addition to the above contributions to the coupling loss, one must also consider the consequences of misalignments of the waveguide and fiber axes. In general, angular misalignment must be maintained below 0.5° in order to attain a misalignment loss below 0.25 dB [4.77]. Also, because the sensitivity to angular misalignment is greater for larger mode sizes, the angular tolerances for Ti : LiNbO$_3$ waveguide-to-fiber coupling are slightly more stringent for angles in the waveguide plane than out of the plane. This is a consequence of the slight asymmetry of the Ti : LiNbO$_3$ waveguide mode.

Permanent Fiber Attachment. For practical application, permanent fiber attachment is essential. Typically one needs to attach several fibers for which the interfiber spacing must match that of the waveguides. This can be best accomplished using photolithographically defined V-grooves in silicon [4.67, 76]. The etching can be precisely controlled to provide desired fiber-to-fiber spacing and to insure that the vertical positions of the cores of fiber placed in the V grooves are aligned. The silicon also provides a massive support to minimize movement and provides a larger area of contacting for permanent bonding. Bonding between the lithium-niobate-chip and V-groove fiber assembly has been achieved with uv curable epoxies. Using this technique arrays of up to 12 fibers have been permanently attached with average excess loss (due to array and attachment) of 0.3 dB per interface [4.76].

4.2.4 Voltage/Loss Tradeoffs: Waveguide Tailoring

To minimize the switching voltage, a small optical mode is important; however, to achieve good fiber-waveguide coupling to standard telecommunication fibers, a relatively large mode is necessary. As a result, there is typically a trade-off between insertion loss and drive voltage. This trade-off is summarized by the measured results [4.40, 72, 75] shown in Fig. 4.16 where the mode size has been varied primarily through the diffusion temperature. A reasonable trade-off between low loss and low voltage can be achieved although at values higher than that separately obtainable.

To avoid the compromise or trade-off between low drive voltage and low fiber-coupled insertion loss, it is necessary to fabricate waveguides whose mode characteristics at the edge of the crystal coupled to a single-mode fiber match those of the fiber, but in the interaction region provide a small mode necessary to achieve efficient electro-optic device operation. To obtain re-

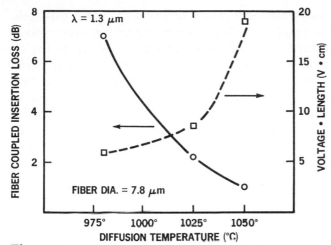

Fig. 4.16.
Trade-offs between efficient fiber-waveguide coupling and low voltage × length parameter versus diffusion temperature

sults appreciably better than those achievable by simply using trade-off diffusion conditions as above, extremely good waveguide tapers between these two regions are required. To achieve optimum fiber-waveguide coupling, a symmetric-waveguide optical mode is desirable. This can be achieved by burying the waveguide by, for example, a second shallow diffusion of a material that lowers the refractive index compared to the titanium diffused waveguide. Using magnesium oxide for the second diffusion, essentially circular modes have been fabricated and waveguides of a total fiber-device-fiber insertion loss as low as 0.5 dB have been achieved [4.78].

However, for optimum interaction with the applied electric field in the device interaction region the waveguide mode should not be buried. Therefore the magnesium oxide thickness is tapered to zero at the active device position via shadow masking during deposition. The fabrication steps to

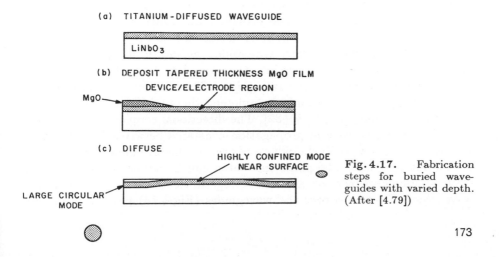

Fig. 4.17. Fabrication steps for buried waveguides with varied depth. (After [4.79])

173

form tapered, buried waveguides are shown in Fig. 4.17. Using this technique very low total insertion loss ($\sim 1\,\mathrm{dB}$) and relatively low drive voltage have been achieved. To further reduce the drive voltage, an increasing taper of the titanium thickness in the device region is also required [4.79]. Techniques to customize the waveguides over different regions of the optical circuit are just beginning to be explored. These techniques will be especially important for many device circuits, such as the optical switch arrays discussed below, where minimizing the drive voltage of each switch is essential to keep the total drive power required for the array at an acceptable level.

4.3 Switch/Modulator

Optical switches are required for signal routing. Applications include switching of wide-band services, protection and facility switching in lightwave networks and bypass switching in local area networks. Signal-processing application includes programmable delay lines. In addition, high-speed switches may be used for time-division multiplexing to allow several lower bit rate channels to cooperatively utilize the high bandwidth of a common single-mode fiber and as external modulators for signal encoding. The most basic optical switch must have at least two usable output ports and, for symmetry, two input ports are also desirable. Important switch parameters include the required switching voltage, crosstalk of the two switch states and optical insertion loss. For time-division multiplexing and signal encoding, the switching speed is also important. Several switch types — the directional coupler, the balanced bridge interferometer and the intersecting waveguide switch — have demonstrated two low crosstalk switching states. In addition, branching-type switches have been covered in Chap. 3.

4.3.1 Directional Coupler

The directional coupler is a very versatile device geometry. This device, in several forms, has been used to provide many functions like wavelength filtering and polarization selection, as discussed later. It consists of a pair of closely spaced identical strip waveguides, as shown in Fig. 4.18a. Light input to one waveguide couples into the second as a result of the overlap in the evanescent fields of the two guides. The coupling per unit length, κ, depends upon the waveguide parameters, the guided wavelength λ and the interwaveguide gap g [4.80–82]. The directional coupler is also characterized by the difference in propagation constants between the two waveguides, $\Delta\beta = 2\pi(N_2 - N_1)/\lambda$, where N are the effective indices, and by the interaction length L. The phase mismatch can be adjusted via the linear electro-optic effect by application of voltage to electrodes over or along the waveguides. Electrodes over the waveguides (Fig. 4.18) provide push/pull

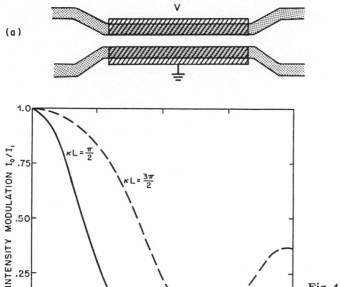

(a)

Fig. 4.18. (a) Directional coupler switch. (b) Directional coupler switch response

refractive index change between the two guides. For z-cut lithium niobate, this electrode placement utilizes the strongest electro-optic coefficient. For the x-cut orientation, a three electrode configuration is required to use the r_{33} coefficient with the push/pull effect. To first order, the applied field produces only small changes in κ [4.83]. However, κ modulation has been achieved using a three electrode structure [4.84].

The operation of the directional coupler, as well as several other important Ti : LiNbO$_3$ waveguide devices, can be described by the well-known coupled-mode equations. Because of its central importance to several of the interesting Ti : LiNbO$_3$ devices, a good understanding of coupled mode theory, including the effects of weighted coupling, periodic coupling and periodic phase reversal, is essential. We therefore momentarily divert to examine the coupled-mode equations which can be written as [4.85, 86],

$$R' - j\delta R = -j\kappa S \quad , \quad \text{and} \tag{4.3.1}$$

$$S' + j\delta S = -j\kappa R \quad , \tag{4.3.2}$$

where $\delta = \Delta\beta/2$, R and S represent the complex amplitudes in the two guides (or, in general, of the two modes) and the primes represent differentiation with respect to the propagation direction. For unit input to one waveguide (or mode), the coupled intensity (efficiency) into the second waveguide (or mode) is found by solving (4.3.1, 2), which yields

$$\eta = \frac{1}{1 + (\delta/\kappa)^2} \sin^2 \kappa L [1 + (\delta/\kappa)^2]^{1/2} \quad . \tag{4.3.3}$$

For identical waveguides and no applied voltage, $\delta = 0$, $\eta = \sin^2 \kappa L$ and complete crossover occurs for $\kappa L = n\pi/2$, with n an odd integer. The length for complete coupling is $l = \pi/2\kappa$, which is called the coupling or transfer length. For $\delta \neq 0$, complete crossover is impossible regardless of the value of κL. For Ti-diffused lithium niobate waveguides, coupling lengths from $\sim 200\,\mu$m to ~ 1 cm have been demonstrated [4.81]. For the appropriate device length, the complete crossover or cross state (denoted by $\otimes$) can be achieved, in principle, without any applied voltage.

The straight through or bar state (denoted by $\ominus$) is achieved electro-optically by inducing a mismatch $\Delta\beta$ so that $\eta = 0$. To minimize the necessary $\Delta\beta L$ for 100 % modulation, a single coupling length, $l = \pi/2\kappa$, is desirable. For $\kappa L = \pi/2$, a value of $\Delta\beta L = \sqrt{3}\pi$ is required to reduce the efficiency from 100 % to 0. Multiple-coupling-length devices result in a significantly larger required $\Delta\beta L$. The switch response is shown in Fig. 4.18b; for $L \gg l$ a value of $\Delta\beta L \sim 2\kappa L$ is required for complete switching.

Directional Couplers with Weighted Coupling

The response of directional couplers and other coupled-mode devices can be altered by varying either the phase mismatch or the coupling coefficient along the interaction length [4.87a–90]. A schematic drawing of a directional coupler with weighted coupling coefficient is shown in Fig. 4.19a.

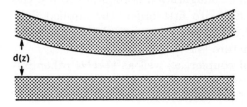

d(z)

z = −L/2   z = L/2

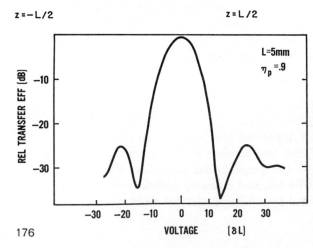

Fig. 4.19. (a) Directional coupler with weighted coupling. (b) Measured switch response for directional coupler with Hamming function weighting of the coupling coefficient

The differential equations that describe codirectional coupling when either κ or $\Delta\beta$ vary as a function of position along the interaction length can be written as [4.88]

$$R' - j\delta R = -j\kappa(z)Se^{-j\phi(z)} \quad , \quad \text{and} \tag{4.3.4}$$

$$S' + j\delta S = -j\kappa(z)Re^{j\phi(z)} \quad , \tag{4.3.5}$$

where, in addition to the previously defined parameters, we have

$$\phi(z) = \int\limits_0^z [\Delta\beta(z') - \Delta\beta(0)]dz' \quad . \tag{4.3.6}$$

Thus $\phi'(z)$ is the position dependent variation in phase mismatch along the coupled mode device which can be continuous, as implemented in, for example, tapered velocity couplers [4.90] or discrete as used in the reversed $\Delta\beta$ coupler. Equations (4.3.4, 5) can be converted to a single nonlinear Riccati equation [4.87a, 88]

$$\varrho' = -j(2\delta + \phi') + j\kappa(\varrho^2 - 1) \quad , \quad \text{where} \tag{4.3.7}$$

$$\varrho = \frac{S}{R}e^{-j\phi} \quad . \tag{4.3.8}$$

By application of boundary conditions and energy conservation, and for unity input to R, the coupling efficiency is

$$\eta = \frac{|\varrho|^2}{1 + |\varrho|^2} \quad . \tag{4.3.9}$$

For given spatial variation of κ or ϕ, ϱ can be found by solving (4.3.7) numerically. As an alternate approach, the coupled-mode solutions for constant κ and $\Delta\beta$ (Sect. 2.6) can be used to form the transfer matrices over small propagation intervals. The net coupling for arbitrary spatial (longitudinal) profile can then be determined by matrix multiplication of these step-wise constant transfer matrices.

Several important and useful closed-form results can be derived from (4.3.7). First, for coupled-mode devices in which only κ is spatially weighted ($\phi' \equiv 0$), the coupling efficiency for no mismatch ($\delta = 0$) is simply

$$\eta = \sin^2\left[\int\limits_{-L/2}^{L/2} \kappa(z)dz\right] \quad . \tag{4.3.10}$$

Thus, only the integrated coupling strength is important and 100 % efficiency is possible. A qualitative understanding of the transfer response (η versus δ) of generalized coupled-mode devices can also be obtained from

(4.3.7). For small depletion of the initially excited mode ($|R| \sim 1 |S| \ll 1$), the coupling efficiency can be written as

$$\eta(\delta) \alpha \int_{-L/2}^{L/2} \kappa(z) e^{-j[\phi(z)+\delta]} dz \quad . \tag{4.3.11}$$

Because $\kappa(z) = 0$ for $|z| > L/2$, the limit can be extended to $\pm\infty$. Therefore, for low values of efficiency, the filter or transfer response is functionally equivalent to the Fourier transform of the spatially dependent coupling coefficient. As an example, for uniform coupling the Fourier transform is the sinc $\equiv (\sin x)/x$ function, which has the same functional features as the real solution (Fig. 4.18b) [4.88].

For several applications, including wavelength filters and polarization-independent switches, a step-like transfer response with negligible sidelobes is required. Such a response can be approximated by choosing as $\kappa(z)$ a weighting or apodizing function that has low Fourier-transform sidelobes. Many possible functions have been examined for other applications. Most require that κ slowly approach zero at the coupler extremes. For waveguide couplers, this is readily achieved due to the exponential falloff in κ with interwaveguide separation. The measured filter response for one such apodizing function, Hamming-function weighting of κ, is shown in Fig. 4.19b [4.91].

Another important technique to shape or synthesize the response of a coupled-mode device is to periodically vary κ to achieve efficient coupling between strongly mismatched ($\delta \gg 1$) modes. A specific case, the TE $\leftrightarrow$ TM mode coverter will be considered in detail later. The periodic coupling coefficient can be written as

$$\kappa(z) = \frac{\kappa_0}{2} \left[1 + m \, \sin\left(\frac{2\pi}{\Lambda} z\right) \right] \quad (0 \le m \le 1) \quad . \tag{4.3.12}$$

On the basis of the Fourier-transform relationship of (4.3.11), one can expect a finite net coupling efficiency (despite the very large phase mismatch) for values of $\Delta\beta$ that satisfy the phasematch condition [4.87]

$$\frac{2\pi}{\lambda} |N_2 - N_1| = \frac{2\pi}{\Lambda} \quad . \tag{4.3.13}$$

In fact, it can be shown that, when this phase-match condition is satisfied, 100 % coupling efficiency is possible, and all the above coupled-mode results remain valid by simply translating $\Delta\beta' = \Delta\beta - 2\pi/\Lambda$ and $\kappa' = m\kappa/2$. Because the phase-match condition depends explicitly on λ, periodic-coupled devices are strongly wavelength selective, as will be discussed below.

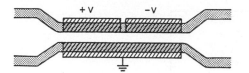

Fig. 4.20. Directional coupler with reversed $\Delta\beta$ control electrodes

Reversed $\Delta\beta$ Directional Coupler

The response of the directional-coupler switch can also be modified by spatially changing $\Delta\beta$. Most important is the stepped or reversed $\Delta\beta$ coupler in which the polarity of the applied $\Delta\beta$ is periodically reversed along the interaction length [4.92, 93]. This device (Fig. 4.20) overcomes two severe limitations of the standard switched directional coupler. The first is the fabrication difficulty associated with making L exactly an integer number of coupling lengths required to assure 100 % crossover in the absence of voltage. Note that this would not be a problem if κ could be efficiently changed electro-optically. The second disadvantage is the fact that a larger value of $\Delta\beta L$ is required if L corresponds to multiple-coupling lengths. The former is more critical for switching applications where low crosstalk for both switch states is typically required. In modulators $L \neq nl$ simply results in loss on the "on" state. The second problem is more critical for high-speed modulation where low voltage is essential. An appropriately small electrode gap is desirable to reduce the drive voltage. However, for typical values of G, the transfer length l is small compared to the value of L required for acceptably low drive voltage. In this predicament, the full advantage in low values of V for longer L is not realized. However, for the reversed $\Delta\beta$ modulator as shown below, the required swing in $\Delta\beta L$ is $\sim \sqrt{3}\pi$ even for $L/l > 1$ provided that the number of electrode sections is equal to the approximate number of coupling lengths. The voltage response for the reversed $\Delta\beta$ coupler can be calculated most conveniently by multiplication of the transfer matrices for each electrode section which are identical except for the periodic sign change of $\Delta\beta$ [4.92]. The voltage response for an N (N even) sections, $\Delta\beta$-reversed modulator can be compactly written as [4.92]

$$\eta = \sin^2 \kappa_{\text{eff}} L \quad , \quad \text{where} \tag{4.3.14}$$

$$\kappa_{\text{eff}} = \frac{N}{L} \sin^{-1} \sqrt{\eta_{\text{s}}} \quad . \tag{4.3.15}$$

Here η_{s} is the crossover efficiency of one section of length L/N given by (4.3.3). Note that perfect $\otimes$ and $\ominus$ states can be achieved for electro-optically adjustable values of $\Delta\beta$. The values of $\Delta\beta L$ required to achieve $\otimes$ and $\ominus$ states as a function of κL can be conveniently displayed in the switching curve shown in Fig. 4.21. The ordinate is the fabrication determined value of L/l while the abscissa is the (generally) electro-optically induced value of $\Delta\beta L$. The curves show the required combinations to achieve the $\otimes$

179

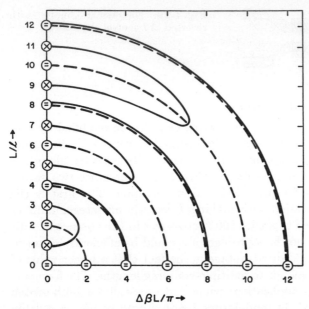

Fig. 4.21. Reversed $\Delta\beta$ switching diagram. Curves show combination of L/l and $\Delta\beta L$ values required for $\otimes$ and $\ominus$ states. Dashed curves are switching curves for $\ominus$ state of a directional coupler with uniform control electrodes. (After [4.92])

or $\bigcirc$ state as indicated. It is important to note again that for the uniform $\Delta\beta$ coupler $\otimes$ states can be achieved only for isolated points $L/l = 1, 3, \ldots$ on the ordinate, which places generally unachievable demands upon the fabrication process. Measurement-limited crosstalk values (~ 43 dB) have been achieved with reversed $\Delta\beta$ electrodes [4.94]. The flexibility achieved with the reversed-$\Delta\beta$ technique is essential for a number of Ti : LiNbO3 coupled-mode devices, in addition to the directional-coupler switch, as will be discussed below. The traveling-wave principle can be combined with that of $\Delta\beta$ reversal to achieve high-speed, low-crosstalk switching [4.95, 96].

4.3.2 Balanced-Bridge Interferometer

The balanced-bridge interferometric switch [4.97, 98], shown schematically in Fig. 4.22a, is an analog of the bulk Mach-Zehnder interferometer. Light entering either of two input waveguides is divided by a 3 dB (50/50 splitter) directional coupler; the waveguides are then sufficiently separated so that they are not coupled, then combined again by a 3 dB coupler. While separated, the relative phase between the two arms can be changed via the electro-optic effect. The output crossover efficiency versus the induced phase shift can be calculated with the coupled-mode equations by cascading the results for the 3 dB coupler, the phase-shifter section (with $\kappa = 0$), and the final 3 dB coupler. For unit input to one waveguide, the crossover efficiency

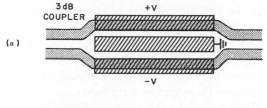

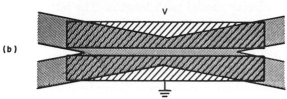

Fig. 4.22. (a) Balanced bridge interferometric switch. (b) Intersecting waveguide switch

into the second output waveguide is

$$\eta = \cos^2 \frac{\Delta\beta L}{2} \quad , \tag{4.3.16}$$

where L is the phase-shift electrode length. This switch has a periodic response with respect to $\Delta\beta$ with multiple $\otimes$ and $\ominus$ states, in contrast to that of the switched-directional coupler but similar to the multiple-section reversed $\Delta\beta$ coupler. The required phase shift to switch from the cross to bar states is $\Delta\beta L = \pi$, which is $\sqrt{3}$ times smaller than for the directional-coupler switch. Because the efficiency depends only upon a single, voltage-controlled parameter $\Delta\beta$, both switch states can be achieved electrically.

Perfect switch states (i.e., low crosstalk) require that the splitter and combiner be perfect 3 dB couplers and that the losses in each arm be identical. For small coupler errors, $\eta = 1$ cannot be achieved. Fortunately, 3 dB couplers can be made either by fabricating such that $\kappa L = \pi/4$ and $\Delta\beta = 0$ (identical waveguides) or by making $\kappa L > \pi/4$ and electro optically adjusting $\Delta\beta$, (4.3.3), to achieve 3 dB coupling.

Two variations of this device that use closely spaced waveguides have been reported [4.99, 100]. To eliminate coupling in the phase shift section, one uses an etched gap, the other uses mismatched waveguides. However, generally only poor crosstalk results have been achieved for these configurations. These structures do have the advantage that no bends are required, which can potentially reduce optical loss.

4.3.3 Intersecting-Waveguide Switch

Several types of intersecting-waveguide switches, one of which is shown schematically in Fig. 4.22b, have been demonstrated with Ti : LiNbO$_3$ waveguides [4.101–105]. Although an early device that employed multimode waveguides apparently operated by the principle of total internal reflection [4.104], the single-mode version can be viewed as a zero-gap directional coupler or a modal interferometer. It is convenient to study the symmetric

181

directional coupler by considering the two eigenmodes of (Chap. 3) the total structure rather than the spatial modes of the individual waveguide as in the coupled-mode formalism [4.105]. Eigenmodes for a totally symmetric structure do not couple but propagate with different phase velocities. Coupling light into either well-separated input waveguide equally excites the two normal modes, which propagate through the structure unperturbed but interfere based upon their different modal path lengths. The latter can be altered with a symmetric transverse electro-optic field to change the relative output amplitudes in the two output waveguides. Because it is an interferometer, ideally both switch states can be achieved with low crosstalk. However, to achieve that, the structure both with and without applied fields must be symmetric to avoid coupling between the local normal modes to insure good modal interference and resulting low switch crosstalk. By careful fabrication, electrode design and alignment, crosstalk approaching $-30\,\mathrm{dB}$ has been achieved. These switches exhibit voltage times length products comparable to finite gap directional coupler switches [4.106].

4.4 On/Off Modulators

High-speed waveguide intensity modulators may be essential components for very-high-bit-rate lightwave systems. Direct current modulation of semiconductor lasers is the most convenient and presently used method for data encoding for data rates to ~ 2 Gigabit/s. However, in most semiconductor lasers fast current modulation also results in an undesirable wavelength modulation or chirp which limits bandwidth for propagation at the lowest-loss region of the standard silica fiber $\lambda = 1.55\,\mu\mathrm{m}$, where finite dispersion is encountered. External modulation eliminates chirp to allow longer-distance transmission. We will examine this application in more detail later.

An on/off modulator is inherently more simple than a 2×2 switch because only a single usable output port is required. In addition to the switches mentioned above, electro-optic $\mathrm{Ti:LiNbO_3}$ on/off waveguide modulators based upon Y branch interferometers [4.51, 97, 108–110], active Y branching [4.111], induced waveguide cutoff [4.112], and $\mathrm{TE} \leftrightarrow \mathrm{TM}$ mode conversion [4.113, 114] have been demonstrated. Important modulator characteristics are the modulation depth or on/off extinction ratio, required drive voltage and the modulator speed or bandwidth. An extinction ratio of $\sim 15\,\mathrm{dB}$ is generally considered sufficient.

4.4.1 Y-Branch Interferometer

The Y-branch interferometer, which operates by the same principle as the balanced-bridge modulator, is shown in Fig. 4.23. However, unlike the balanced-bridge switch, it has only a single accessible input and output port. It

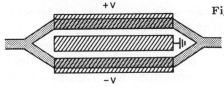

Fig. 4.23. Y-branch interferometric modulator

has been an especially popular structure for on/off modulation because of its simplicity. The 3 dB splitter and combiner are power splitting Y branches. The input wave is split into equal components. Each propagates over one arm of the interferometer. The interferometer waveguide arms are sufficiently separated to prohibit evanescent coupling between them and thus are not coupled. The optical paths of the two arms are typically equal. Therefore if no phase shift is introduced between the interferometer arms, the two components combine in phase at the output Y-branch 3 dB coupler and continue to propagate undiminished in the output waveguide. For $\Delta\beta L = \pi$, however, the combined mode distribution is double-mode-like; light is radiated into the substrate and the transmitted light is a minimum. The transmission-efficiency response is the same as for the balanced-bridge switch/modulator. A three-electrode configuration can be used to achieve a push-pull phase change; however the capacitance is then also increased, which is disadvantageous when high-speed operation is desired. For modulator applications the interferometer is typically dc biased for 50 % transmission and an ac modulation signal added. Bias can be achieved either electro-optically by an applied bias voltage or optically by using interferometer arms of slightly different length [4.115].

A variation of this device is to use an asymmetric branch on the output side of the interferometer to provide two usable output ports thus converting it to a one by two switch [4.115a].

Other types of waveguide intensity modulators include the cutoff, and polarization modulators. The cutoff modulator is simply a phase modulator in which the induced index change reduces the waveguide-substrate index so that guiding is eliminated and light radiates into the substrate. Its principal advantage is its simplicity. However, for a low modulation voltage, the waveguide must be near cutoff which usually implies significant optical loss for the on ($V = 0$) state. Furthermore, this condition requires careful control of waveguide parameters. The polarization modulator will be discussed in a later section.

4.4.2 Voltage and Bandwidth Consideration for Switch/Modulators

The expected voltage $\times$ length product for directional coupler and interferometric type switch/modulators can be determined from (4.2.10) with $p = \sqrt{3}$ and 1, respectively. Experimentally, VL for directional coupler switches using z-cut $\text{Ti} : \text{LiNbO}_3$ waveguides scales as λ^2 and can be written as $VL \sim 3.2\,\text{V cm}\,\lambda^2$ with λ in units of μm [4.116].

The electrical frequency response results from phase modulators, for both the lumped and traveling-wave (TW) cases, translate directly to those for interferometric modulators because the amplitude modulation depends simply upon the integrated phase change. The same is true for the lumped directional-coupler device. However, the situation is quite different for the TW directional coupler where the distribution of $\Delta\beta$ along the coupler, not just its integrated value, is important. For example, for a directional coupler modulator with $L = l$, the phase mismatch at either end of the coupler, where most of the light is in only one of the waveguides, is relatively unimportant. Physically the effect of velocity mismatch is weighted along the coupler and is most critical at the center of the coupler and least sensitive at the ends. As a result, with respect to velocity mismatch the effective coupler length is shorter, $\sim L/\sqrt{3}$, than the actual length [4.117, 118]. The result is that for equal interaction lengths, directional-coupler modulators provide greater velocity-mismatch limited bandwidth than interferometric ones. Indeed, this increase in bandwidth is approximately equal to the increase in drive voltage required for the directional coupler compared to the interferometer. Calculated frequency responses for the two modulator types are shown in Fig. 4.24. The effective weighting between the optical and electrical signals in the case of the directional coupler results in low sidelobes of the frequency response.

Numerous demonstrations of very broadband interferometric and directional coupler modulators have now been reported [4.39, 40, 119–122]. Applications will be discussed later.

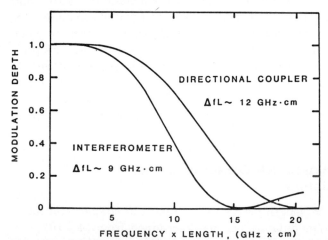

Fig. 4.24. Calculated optical modulation frequency response for traveling-wave interferometric and directional coupler modulators. (After [4.117])

4.5 Polarization Devices

Devices to modulate, to separate or to control polarization may be required in lightwave systems, especially for future coherent systems. Devices to peform all these functions have been demonstrated with Ti : LiNbO$_3$ waveguides.

While the optical polarization state can be defined in a number of bases, for our purposes it is convenient to define the polarization in terms of the polarization angle, θ and the phase angle ϕ. The normalized TE and TM complex amplitudes can be written as

$$
\begin{pmatrix} A_{\text{TE}} \\ A_{\text{TM}} \end{pmatrix} = \begin{pmatrix} \cos\theta \\ \sin\theta\, e^{j\phi} \end{pmatrix} \quad , \tag{4.5.1}
$$

where θ specifies the relative TE/TM amplitudes while ϕ is the phase difference between the TE and TM components. Light is linearly polarized at an angle θ if $\phi = 0$; $\theta = 0$ represents purely TE polarized, while $\theta = \frac{\pi}{2}$ is purely TM. Right circularly polarized light, for example, is given by $\theta = \frac{\pi}{4}$ and $\phi = \frac{\pi}{2}$. In passive Ti : LiNbO$_3$ waveguides, light linearly polarized along a principal axis has its polarization maintained for propagation along a principal axis. Thus, for example, light incident as TE (TM) polarization to waveguides in z (or x, or y)-cut lithium niobate exits in the same state. Further, for elliptical polarization input, the relative TE/TM amplitude of the output wave is unchanged although the relative phase is changed if the waveguides exhibit birefringence as in the case of Ti : LiNbO$_3$ waveguides.

4.5.1 TE $\leftrightarrow$ TM Conversion

A device for electro-optically producing TE to TM polarization conversion that utilizes the strongest coefficient in lithium niobate is shown in Fig. 4.25. This device uses an off-diagonal element, r_{51}, of the electro-optic tensor to convert incident TE (TM) to TM (TE) polarized light. A periodic electrode is required because lithium niobate is birefringent and the TE and TM modes have different effective indices. As discussed in Sect. 4.3, efficient coupling can be achieved for Λ given by

$$
\frac{2\pi}{\lambda}|N_{\text{TM}} - N_{\text{TE}}| = \frac{2\pi}{\Lambda} \quad . \tag{4.5.2}
$$

For lithium niobate, the required period is $\Lambda = 7$ and $18\,\mu\text{m}$ for $\lambda = 0.6$ and $1.3\,\mu\text{m}$, respectively [4.113]. As discussed, this coupling process is described by the coupled-mode equations with $\Delta\beta = 0$, so that

$$
\eta = \sin^2 \kappa L \quad , \quad \text{where} \tag{4.5.3}
$$

$$
\kappa = \frac{\Gamma\pi}{\lambda} n^3 r_{51} \frac{V}{G} \quad . \tag{4.5.4}
$$

185

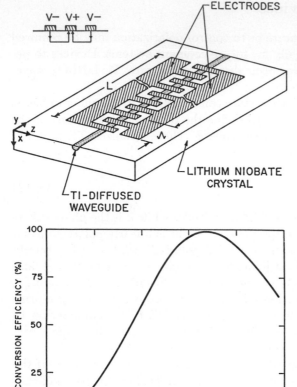

Fig. 4.25. (a) Waveguide electro-optic TE ↔ TM converter with phase matching periodic electrode. (b) Measured phase-matched TE ↔ TM conversion efficiency versus applied voltage. $L = 6\,\mathrm{mm}$ and $\lambda = 1.32\,\mu\mathrm{m}$

According to the electro-optic tensor for LiNbO$_3$, (4.2.3), an x-directed electric field can couple the x- and z-polarization components through the r_{51} coefficient, for example. For 100 %-polarization modulation $\kappa L = \pi/2$ or $p = \frac{1}{2}$, (4.2.9), which is the lowest modulation condition for any modulator device considered. Experimental results for a TE ↔ TM mode converter phase matched for $\lambda = 1.32\,\mu\mathrm{m}$ are shown in Fig. 4.25b [4.123].

4.5.2 Polarization Controller

For some applications, it is important to convert an input signal of arbitrary polarization (θ_i, ϕ_i) into an arbitrary output polarization. Polarization control is important because a typical single-mode fiber does not maintain polarization; while, e.g., coherent communication systems require a known signal polarization. In many cases it is sufficient that the output signal be either TE or TM. An unknown and typically elliptically polarized input can be expected from a standard (non-birefringent) single-mode fiber. To effect

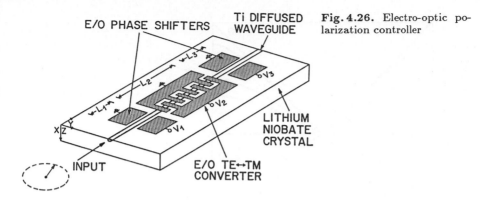

E/O PHASE SHIFTERS Ti DIFFUSED WAVEGUIDE

Fig. 4.26. Electro-optic polarization controller

LITHIUM NIOBATE CRYSTAL

INPUT

E/O TE↔TM CONVERTER

general polarization conversion, a more complicated circuit than the simple TE $\leftrightarrow$ TM converter is required. A generic structure for general polarization conversion is shown in Fig. 4.26 which consists of a phase-matched TE $\leftrightarrow$ TM mode converter between two phase shifters that can provide electrical control over the relative TE/TM phase angle, ϕ [4.124, 125]. For generalized input (θ_i, ϕ_i) to the mode converter, we can use multiplication of the three transfer matrices for the electrode sections to find the output's TE and TM components (note that for the moment we ignore all static TE/TM phase shifts due to birefringence outside the electrode regions):

$$
\begin{pmatrix} A_{TE} \\ A_{TM} \end{pmatrix}_o = \begin{pmatrix} \cos \kappa L_2 \cos \theta_i - \mathrm{j} \, \mathrm{e}^{\mathrm{j}(\phi_i + \Delta\phi_1)} \sin \kappa L_2 \sin \theta_i \\ \mathrm{e}^{\mathrm{j}(\Delta\phi_3)} (\mathrm{e}^{\mathrm{j}(\phi_i + \Delta\phi_1)} \cos \kappa L_2 \sin \theta_i - \mathrm{j} \sin \kappa L_2 \cos \theta_i) \end{pmatrix} .
$$

$$(4.5.5)$$

Reducing this to the form of (4.5.1) one can write for the output polarization angle

$$
\theta_o = \tfrac{1}{2} \cos^{-1}(\cos 2\theta_i \cos 2\kappa L_2 + \sin 2\kappa L_2 \sin 2\theta_i \sin \phi_i')
\tag{4.5.6}
$$

and the relative TE/TM phase

$$
\phi_o = \phi_i' + \Delta\phi_3 + \tan^{-1}\left(\frac{\cos \kappa L_2 \sin \theta_i \sin \phi_i' - \sin \kappa L_2 \cos \theta_i}{\cos \kappa L_2 \sin \theta_i \cos \phi_i'} \right)
$$

$$
+ \tan^{-1}\left(\frac{\sin \kappa L_2 \sin \theta_i \cos \phi_i'}{\cos \kappa L_2 \cos \theta_i + \sin \kappa L_2 \sin \theta_i \sin \phi_i'} \right) , \tag{4.5.7}
$$

where $\phi_i' = \phi_i + \Delta\phi_1$ is the relative phase input to the mode converter section and

$$
\Delta\phi_{1,3} = \frac{2\pi}{\lambda} \frac{V_{1,3}}{G} L_{1,3} [n_{TE}^3 \Gamma_{TE} r_{TE} - n_{TM}^3 \Gamma_{TM} r_{TM}]
$$

are the induced TE/TM phase shifts.

187

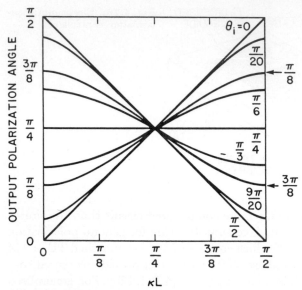

Fig. 4.27. Calculated rotated output polarization angle versus mode converter coupling strength, κL, with input polarization angle θ_i as a parameter. The TE/TM phase relative to the mode converter is assumed to be 0

The importance of the first phase shifter for achieving general polarization transformation is illustrated in Fig. 4.27. In that figure, we plot the calculated, from (4.5.6), output polarization angle θ_o versus the mode converter coupling strength for several values of θ_i. The results are calculated for $\phi_i' = 0$. As expected and shown in Fig. 4.27 for an incident wave that is either pure TE or TM ($\theta_i = 0$ or $\frac{\pi}{2}$, respectively), all possible output values of θ_o are achievable with the appropriate mode-converter voltage (κL_2). This is, of course, true for arbitrary ϕ_i'. However, as θ_i increases from 0 or decreases from $\frac{\pi}{2}$, the range of achievable values of θ_o becomes increasingly limited. Indeed, for $\theta_i = \frac{\pi}{4}$, the output polarization angle θ_o remains unchanged regardless of the mode-converter coupling strength. In general, for other values of ϕ_i', the range of achievable θ_o is limited, too. Thus, while for the special case of TE or TM input, arbitrary output polarization angles can be achieved by the mode converter, for general θ_i and ϕ_i', arbitrary transformations of θ do not exist.

The key to achieving an arbitrary transformation in θ is to use the first phase shifter to adjust the relative phase between the TE and TM components input to the mode converter to $\phi_i' = \pm \frac{\pi}{2}$. Then, (4.5.6) simply becomes

$$\theta_o - \theta_i = \pm \kappa L_2 \quad , \quad \phi_i' = \pm \frac{\pi}{2} \quad . \tag{4.5.8}$$

In these cases, an arbitrary $\theta_i \to \theta_o$ transformation is possible and, indeed, the mode converter acts as a *linear* rotator with respect to θ. In fact, only for those special cases is a general θ transformation possible. Thus, the first phase-shifter and the mode-converter combination are required to provide complete control over θ.

The output's relative phase ϕ_o from the device is then determined by the second phase shifter. For the preferred special case of $\phi_i' = \pm\frac{\pi}{2}$, (4.5.7) becomes

$$\phi_o = \Delta\phi_3 \pm \frac{\pi}{2} \ . \tag{4.5.9}$$

Equation (4.5.9) shows that the relative-phase output from the mode converter is identical to the input to it. This is in contrast to the general case (arbitrary ϕ_i'), where the output phase from the mode converter depends in a complicated way upon θ_i, ϕ_i', and κL_2. Therefore, for $\phi_i' = \pm\frac{\pi}{2}$, the desired output phase ϕ_o is achieved by applying a voltage V_3 to the second phase shifter so that

$$\phi_o = \phi_i + \Delta\phi_f + \Delta\phi_1 + \Delta\phi_3 \ , \quad \text{where} \tag{4.5.10}$$

$$\Delta\phi_f = \frac{2\pi}{\lambda} L_T (N_{\mathrm{TM}} - N_{\mathrm{TE}}) \tag{4.5.11}$$

is the now added phase shift due to material and modal birefringence through the entire crystal length L_T. The phase shift $\Delta\phi_1$ is fixed by the condition that $\phi_i' = \pm\frac{\pi}{2}$. Note that θ_o is unaffected by the second phase shifter, see (4.5.8).

Several special modes of operation of this device are worth noting. As a linear rotator, the first phase shifter is adjusted so that $\phi_i' = -\frac{\pi}{2}$ and the second one is adjusted so that the output phase is zero. The rotation of the polarization angle is proportional to the mode-converter voltage. For arbitrary input polarization, if the desired output is TE or TM, then ϕ_o is meaningless and the second phase shifter is not required. For input light that is pure TE or TM, only the mode converter and the second phase shifter are required to achieve output light of arbitrary polarization.

The electro-optic polarization controller described above has an important feature for active feedback control, namely, that the values of the phase shift and mode converter voltages can be independently optimized. Suppose an input signal must be converted to TE polarization. The state of polarization can be sensed with the polarization splitter described below. As the input polarization to the polarization controller changes, the V_1 is varied to find the value that minimizes the error signal. Voltage V_1 is then fixed at that value and V_2 swept to further reduce the error signal. This done, there is no need to further optimize V_1 until the input polarization changes.

To avoid the wavelength dependence of this device, z-propagation devices for which there is little birefringence are preferred [4.126–128]. The electro-optic coefficient for mode conversion in this case is reduced by a factor of about ten. The theory of operation is the same. This z-propagating orientation also reduces the effects of the photorefractive phenomenon. To reduce, though not eliminate, the limitations of wavelength sensitivity while utilizing the strongest electro-optic coefficient, another polarization controller design which uses a single voltage to provide wavelength tunability and phase control has been demonstrated [4.129].

4.5.3 Polarization-Selective Devices

Polarization-selective devices may be required for several applications including polarization sensing mentioned above for the feedback loop of a polarization controller and for polarization multiplexing or polarization diversity. These devices can be grouped as polarizing or polarization splitting components which are resistive and reactive, respectively.

The simplest linear polarizer is a metal-clad waveguide for which induced currents in the metallic overlay result in as much as a 10 dB/cm loss to the TM-polarized component while producing negligible loss increase to the TE mode [4.32, 130]. Their effect can be enhanced by placing a thin dielectric layer between waveguide and electrode to resonantly enhance coupling of the TM mode to the metallic overlay to increase its loss [4.131]. The differential TM/TE loss coefficient in this case can be as high as 35 dB/cm [4.131].

In another approach, polarizers take advantage of polarization-dependent index changes in waveguide fabrication processes such as proton exchange or out diffusion in $LiNbO_3$. The proton exchange process increases the extraordinary index but actually decreases the ordinary index of $LiNbO_3$. Therefore, proton-exchange waveguides fabricated in z-cut $LiNbO_3$, for example, guide only the TM mode resulting in a linear polarizer. By combining titanium diffusion and proton-exchange methods a short, high-extinction-ratio proton-exchange polarizing section can be fabricated on a substrate that otherwise guides both polarizations [4.132–134]. Polarizers with a 40 dB extinction and very low excess loss have been demonstrated by this technique.

Perhaps a more versatile device is the linear polarization splitter which is capable of spatially separating TE and TM components of a guided mode. This function has been achieved with specially designed Y-branch splitters [4.135], intersecting waveguides [4.136] and directional couplers [4.137–139]. As an example, we consider the polarization-selective directional coupler, shown schematically in Fig. 4.28. Polarization-selective inter-waveguide coupling can be achieved by making either κ or $\Delta\beta$ strongly polarization dependent. The substrate-waveguide refractive-index difference Δn, as discussed in Sect. 4.1, is generally different for the TE and TM modes of $Ti : LiNbO_3$ waveguides. Therefore, the lateral waveguide mode width and κ also depend upon polarization. This difference can be enhanced or reduced by choice of diffusion parameters. Therefore, by appropriate choice of diffusion parame-

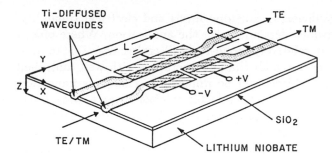

Fig. 4.28. Polarization-selective coupler

Ti-DIFFUSED WAVEGUIDES

TE
TM
G
L
Y
Z
X
+V
−V
SiO₂
TE/TM
LITHIUM NIOBATE

ters and interaction length L, one can approximately make $\kappa_{\text{TE}}L = \pi$ and $\kappa_{\text{TM}}L = \frac{\pi}{2}$. Generally, the limited value of $|\kappa_{\text{TE}} - \kappa_{\text{TM}}|$ demands a relatively long device, of about ~ 1 cm. In this case, for light incident in waveguide 1, the TE component exits waveguide 1 while the TM exits waveguide 2. In principle, being limited by fabrication tolerances, a polarization splitter of this type is purely passive and does not require the electrodes of Fig. 4.28.

Another approach is to utilize the polarization dependence of $\Delta\beta$. In one approach, this polarization dependence was achieved by a direct metallic electrode on one waveguide and a buffer layer and electrode on the other [4.137]. The metallic overlay loads the TM mode and changes its propagation constant but has little effect on the TE mode. Therefore, $\Delta\beta_{\text{TM}}$ can be made finite while $\Delta\beta_{\text{TE}} \approx 0$. By choosing L such that $\kappa_{\text{TE}}L = \frac{\pi}{2}$, and $\delta\beta_{\text{TM}}L \simeq \sqrt{3}\pi$ for light incident in the unloaded waveguide (this is necessary to avoid loss to the TM component) the TE component will couple to the second waveguide while TM component stays in the incident waveguide. In practice, it can be somewhat difficult to satisfy both conditions. However, a polarization splitter of this type in TiLiNbO₃ waveguides has achieved polarization crosstalk of ~ -17 dB [4.137].

To reduce tolerances and to allow input in either waveguide, one can use a weak polarization dependence of κ and the strong polarization dependence of the electro-optic induced $\Delta\beta$ in lithium niobate. These effects and the use of the reversed $\Delta\beta$ electrode (Fig. 4.28) to achieve tuning has resulted in a polarization splitter with simultaneous crosstalk below 20 dB for both polarizations and values as low as -27 dB when optimized for one polarization [4.139].

4.6 Wavelength Filters

Wavelength multiplexing offers enormous potential for utilizing the information capacity of single-mode fibers. The potential is especially impressive in view of the demonstration of fiber loss below 0.5 dB/km extending from $\lambda = 1.3\,\mu$m to $\lambda = 1.6\,\mu$m. Single-mode waveguide filters for multiplexing/demultiplexing have long been recognized as important communications devices. The important filter parameters are the filter center wavelength and bandwidth, peak filter efficiency, side-lobe levels and electrical tunability. The desired filter bandwidth depends upon the application. Where only a few channels are required, relatively broad-band filters are desirable to minimize tolerances. However, for multichannel operation with well controlled sources, narrow-bandwidth filters will be important. Thus it is important for the system designer to have available filters that cover a wide range of bandwidths. To fulfill this requirement, several filter types, each with its own characteristic bandwidth range, have been demonstrated. As with switch/modulators, they can be generically classed as coupled-mode or interferometric devices.

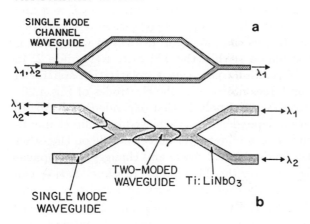

Fig. 4.29. (a) Differential path length interferometric filter. (b) Modal interferometric wavelength multiplexer/demultiplexer. (After [4.140])

4.6.1 Interferometric Filters

Two implementations of interferometric filters are shown in Fig. 4.29. For purposes of explanation, the Y-branch interferometer with unequal arms indicated in Fig. 4.29a is a good example. The Y-branch splitter and combiner perform the same function as in the interferometric modulator. Alternately these functions can be provided by a 3 dB directional coupler. However unlike the equal arm interferometer, as a result of the physical path length difference ΔL, the optical phase difference at the combiner is strongly wavelength dependent, as given by

$$\Delta\phi = \frac{2\pi}{\lambda} N \Delta L \quad . \tag{4.6.1}$$

As a result, the intensity transmittance (4.3.15) is periodic in wavelength

$$I_{\text{out}} = I_{\text{in}} \cos^2\left(\frac{\pi N \Delta L}{\lambda}\right), \tag{4.6.2}$$

with a null-to-null bandwidth

$$\Delta\lambda = \frac{\lambda^2}{N \Delta L} \quad . \tag{4.6.3}$$

As an example, for a path length difference of $100\lambda/N$ ($\sim 70\,\mu$m for $\lambda_0 = 1.5\,\mu$m, for example) the periodic response peaks at $150\,\text{Å}$ wavelength intervals.

Building an integrated-optic interferometer with different arm lengths is limited by the ability to make sharp bends. An alternate approach is the interferometer whose arms have equal physical length but unequal optical mode path lengths. One example is shown in Fig. 4.29b. Light from a single-mode waveguide is injected into a double-mode waveguide. After a length L, the light is split into two separate single-mode waveguides. Assuming a 3 dB modal splitter and combiner and no inter-mode coupling in the straight waveguide section, for input in the lower guide, the efficiency of light exiting in the upper waveguide is identical to (4.3.15) with

$$\frac{\Delta\phi}{2} = \frac{\pi}{\lambda}(N_2 - N_1)L = \frac{\pi}{\lambda}\Delta N L \quad , \tag{4.6.4}$$

where N_i are the effective refractive indices of each mode. Due to the different mode distributions, application of an electric field changes the difference in effective indices of the two modes resulting in tuning of the filter center wavelength. Using 1.2 cm long Ti : LiNbO$_3$ waveguides, a periodic filter response about $\lambda_0 = 1.5\,\mu$m with $350\,\text{Å}$ separation between wavelength peaks was demonstrated. The achieved tuning rate was $25\,\text{Å/V}$ [4.140].

The periodic filter response of a single interferometric device is appropriate for tree-type cascaded multi-channel multiplex/demultiplex architectures. It is also well suited to separate two appropriately designed wavelength channels.

4.6.2 Coupled-Mode Filters

For some applications greater flexibility for wavelength division multiplexing can be obtained with filters that have a passband response. Such a response is achieved with coupled-mode filters. In these devices, wavelength dependence results from the distributed coupling between two waveguide modes that generally have different propagation indices. Finite coupling at and about a single design wavelength results from achieving effective phasematching at that wavelength. Two general techniques have been used. The first is to use periodic coupling to achieve phasematching at the desired wavelength. The second is to design the two mode indices versus wavelength functions to be generally different but such that they intersect at the desired filter center wavelength. We now consider these in more detail.

Tunable Mode Converter Filter

When performed in a birefringent substrate like lithium niobate, electro-optic TE $\leftrightarrow$ TM conversion is wavelength dependent. Efficient coupling between the nonsynchronous TE and TM modes can be achieved by periodic coupling with an electrode period Λ, as discussed in Sect. 4.3, that satisfies

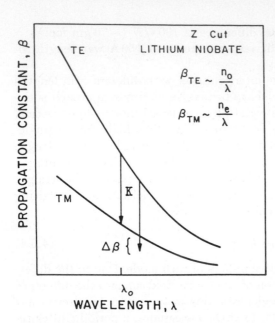

Fig. 4.30. TE/TM mode phase-matching via periodic ($\mathbf{K} = 2\pi/\Lambda$) coupling and wavelength dependent mismatch ($\Delta\beta$)

the phase-match condition (4.3.12). However, the phase-match condition depends explicitly on wavelength and is strictly satisfied only for wavelength λ_0. For $\lambda = \lambda_0 + \Delta\lambda$ the mismatch is (Fig. 4.30) [4.113]

$$\Delta\beta = -\frac{2\pi}{\Lambda}\frac{\Delta\lambda}{\lambda} \quad . \tag{4.6.5}$$

The conversion efficiency versus λ or filter response is given by the standard coupled-mode filter function (4.3.3) with κ given by (4.5.3). (The factor of two increase from the push/pull effect compensates for the fact that voltage is applied only along one half the interaction length.) The bandwidth (FWHM) is $\Delta\lambda/\lambda \approx \Lambda/L$. Because of its large birefringence, lithium niobate can be used to achieve rather narrow bandwidth. To realize demultiplexing, a polarization splitter is required to select the converted component, or mode conversion can be simultaneously combined with a mismatched directional coupler [4.141].

Phase-matched TE $\leftrightarrow$ TM conversion has been achieved in both x- and z-cut lithium niobate by using an interdigital or finger electrode, respectively. In either case, the x electric field component which couples to the strong off-diagonal r_{51} ($=28 \times 10^{-10}$cm/V) coefficient is utilized. Using electrode lengths from 0.5 to 6 mm, filter bandwidths between ~ 50 Å and 5 Å have been demonstrated in the visible wavelengths.

Mode-converter filters can be tuned by electro-optically changing the birefringence using, for example, a three electrode structure shown in Fig. 4.31a [4.142]. Because of the different electro-optic coefficients for the TE and TM modes, application of voltage V_2 changes the birefringence (and relative phase as in Sect. 4.5.2) as

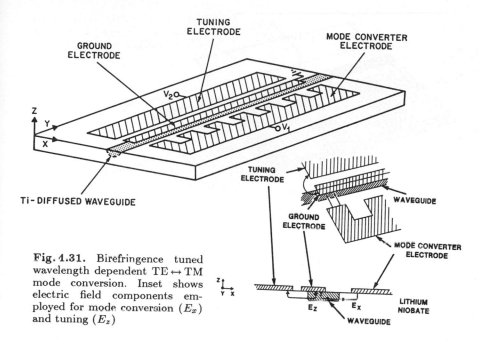

Fig. 4.31. Birefringence tuned wavelength dependent TE ↔ TM mode conversion. Inset shows electric field components employed for mode conversion (E_x) and tuning (E_z)

$$\Delta(\Delta N) = \Delta(N_{\text{TM}} - N_{\text{TE}}) = \frac{-V_2}{2C}(n_e^3 r_{33} - n_0^3 r_{13}) \qquad (4.0.0)$$

where for simplicity we assume equal overlap Γ for the TE and TM modes and here $\Delta N = N_{\text{TM}} - N_{\text{TE}}$. As a result, the phase-match condition is satisfied by a new wavelength $\lambda_T = \lambda_0 + \Delta\lambda_T$, where

$$\Delta\lambda_T \approx \lambda_0 \frac{\Delta(\Delta N)}{\Delta N(V_2 = 0)} \qquad (4.6.7)$$

In fact, the tuning can be also efficiently achieved by spatially alternating the mode conversion and tuning region using the device shown in Fig. 4.32 [4.129, 143] provided the individual mode converter sections are sufficiently short. The tuned filter response for this device, which can also be used as a polarization controller, is also shown [4.144].

Tunable Directional Coupler Filter

Broad-band filtering can be achieved with the specially designed wavelength selective directional coupler shown schematically in Fig. 4.33a [4.145–147]. The coupler consists of two-strip waveguides of different widths and refractive indices. The wider (narrower) guide has the lower (higher) refractive index. Because of the different dimensions and refractive indices, the two guides have distinct modal dispersion characteristics $N(\lambda)$. By proper design, the effective indices for the two guides can be made equal at the desired

195

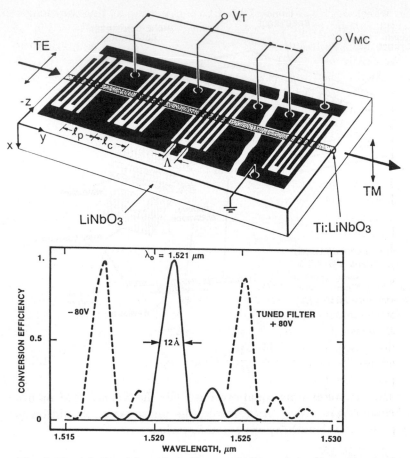

Fig. 4.32. (a) Tunable polarization controller using alternating tuning and mode converter sections. (b) Measured tuned TE $\leftrightarrow$ TM filter response [4.144]

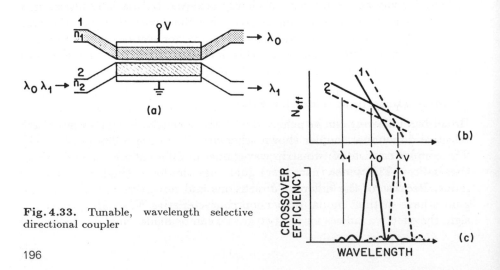

Fig. 4.33. Tunable, wavelength selective directional coupler

filter center wavelength λ_0 (Fig. 4.33b). Phase-matching at this wavelength makes possible complete light transfer between the guides. However, no transfer occurs for λ sufficiently different from λ_0. Thus, the demultiplexed light is conveniently separated spatially from the other input wavelength components. Again, the filter response is given by (4.3.3). In analogy with the previous filters, the mismatch for $\lambda = \lambda_0 + \Delta\lambda$ is $(2\pi/\Lambda_{\text{eff}})(\Delta\lambda/\lambda)$, and the fractional bandwidth is $\Delta\lambda/\lambda_0 \approx \Lambda_{\text{eff}}/L$, where

$$\Lambda_{\text{eff}} = \left[\frac{d}{d\lambda}(N_2 - N_1)\right]_{\lambda=\lambda_0} . \tag{4.6.8}$$

By applying voltage to electrodes over the waveguides, the dispersion curves are altered which results in a new filter center wavelength (Fig. 4.33b and c). Reversed $\Delta\beta$ can be used to insure 100 % crossover at the tuned center wavelength.

A tunable directional coupler filter has been demonstrated with Ti : LiNbO$_3$ waveguides. For an interaction length of 1.5 cm, the measured 3 dB bandwidth was 200 Å for $\lambda_0 = 0.6\,\mu$m and 700 Å for $\lambda_0 = 1.5\,\mu$m with a peak crossover efficiency of $\sim$100 %. Tuning over 1600 Å at a tuning rate of 85 Å/V for the latter device was demonstrated [4.146, 147]. Because this device is really a switch, which is also tunably wavelength selective, it can be used to provide a rather unique function in wavelength division multiplexed switching systems.

The expected electrical tunability of these filters can be calculated as follows. For the periodically coupled-mode converter filters, the filter center wavelength change $\Delta\lambda_V$ with respect to the electro-optically induced index change ΔN_V is

$$\Delta\lambda_V = \Lambda\Delta N_V , \tag{4.6.9}$$

where, for the mode-converter filter, ΔN_V corresponds to the change in the index difference of the two modes. Normalized by the 3 dB filter bandwidth, the result is

$$\frac{\Delta\lambda_V}{\Delta\lambda_{BW}} = \frac{\Delta N_V L}{\lambda} . \tag{4.6.10}$$

An identical result applies to the directional coupler filter using Λ_{eff} for Λ. Thus, for the same length and induced index change, the change in the filter center wavelength relative to its filter bandwidth is the same for each of these filters.

To minimize interchannel separation without introducing unacceptable crosstalk, it is important to reduce filter response sidelobes which can be achieved with weighted coupling.

4.7 Polarization-Insensitive Devices

The devices, except for the polarization controller, described thus far operate effectively for a single linear polarization, TE or TM. Such devices are not compatible with available low-loss single-mode fibers which do not preserve linear polarization [4.148]. Specially fabricated fibers with stress-induced birefringence that maintain linear polarization have been demonstrated but typically the losses are higher than for standard fiber. Furthermore, if single-polarization fibers become practical, it may be desirable to multiplex both polarization components. In either case, polarization insensitive devices are important.

The polarization dependence of switches and modulators results primarily from the othogonal TE and TM modes seeing unequal electro-optic coefficients. For example, for Z-cut lithium niobate with the field applied in the Z direction, the TE mode sees the r_{13} coefficient while the TM mode sees r_{33} and $r_{33}/r_{13} \simeq 3$. As a result, for the same applied voltage, the induced shift $\Delta\beta$ is different for the two polarizations. In addition, if the guide-substrate refractive-index difference Δn is unequal for the TE and TM modes (as it typically is for Ti : LiNbO$_3$), the mode confinement and consequently the coupling strength (κ) for directional couplers depends upon polarization. The polarization dependence of $\Delta\beta L$ and, for directional coupler devices, κL translates into the polarization dependence of the switching or on/off states through the switch response. The two basic approaches to achieving polarization-independent operation have been either to make the parameters $\Delta\beta$ and κ relatively insensitive to polarization or to design a device whose switch response is insensitive or self-compensating for polarization-dependent differences in these parameters. The first approach has been use for the Y-branch on/off modulator and the latter for a specially designed directional coupler switch.

The polarization-independent Y-branch interferometric modulator which uses a dual set of electrodes is shown in Fig. 4.34 [4.149]. The device is fabricated on a Z-cut X-propagating lithium niobate crystal. Voltage applied to electrodes alongside the two waveguides (Fig. 4.34) results in a predominantly y-directed field within the waveguides while that applied to electrodes on top of the waveguides produces primarily a z-directed field. Both fields contribute to the electro-optically induced TE mode (polarized along y) phase change through r_{13} and r_{22}. However, to first order, the TM-mode phase shift results only from E_Z. Thus, in spite of the different electro-optic coefficients, there is a pair of voltages for the two electrodes such that a π phase shift and thus output minimum can be achieved for both TE and TM simultaneously. For no applied voltage, there is an output maximum for both polarizations.

A variation on this approach is to use a single-electrode structure but to laterally offset it over the waveguide. One finds that the electrical/optical

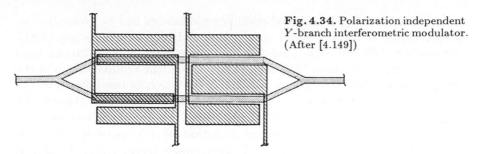

Fig. 4.34. Polarization independent
Y-branch interferometric modulator.
(After [4.149])

overlap parameters have different dependence for the TE and TM modes as a function of offset. For the appropriate offset $\Delta\beta_{TE}$ can be made equal to $\Delta\beta_{TM}$; however, at a voltage penalty due to less than optimum overlap [4.150].

Polarization-independent switching has been achieved with the Ti: LiNbO$_3$ weighted directional coupler 2×2 switch with reversed $\Delta\beta$ electrodes shown in Fig. 4.35a [4.151]. The polarization limitations outlined above have been overcome by several design features. In spite of different values of the guide-substrate index difference for the two modes, the rel-

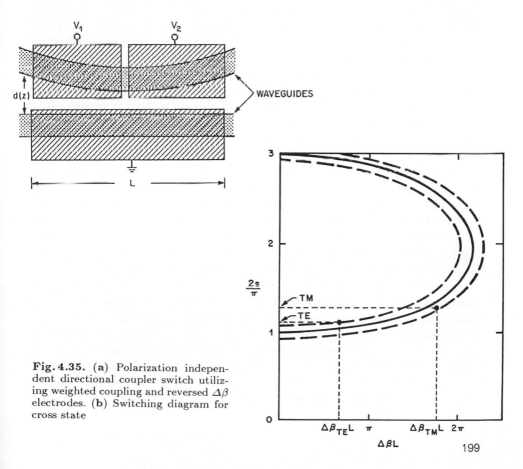

Fig. 4.35. (a) Polarization independent directional coupler switch utilizing weighted coupling and reversed $\Delta\beta$ electrodes. (b) Switching diagram for cross state

199

ative values of the TE and TM coupling coefficients can be controlled by proper design of the waveguide dimensions and interwaveguide gap [4.81]. Alternately, diffusion conditions that give approximately equal Δn for both polarizations can be used (Fig. 4.2). Furthermore, the relative values of the integrated coupling strength, $s = \int_0^L \kappa(z)dz$, which is the important parameter for weighted couplers, can similarly be controlled by design. For a Z-cut crystal in which the TM mode sees the r_{33} coefficient and the TE sees r_{13}, the coupler is designed such that $s_{TM} \gtrsim s_{TE} \approx \frac{\pi}{2}$. For these values of s and for the reversed $\Delta\beta$ electrodes, a larger value of $\Delta\beta_{TM}L$ than $\Delta\beta_{TE}L$ is required to achieve the cross state as shown in the switching diagram of Fig. 4.35b, which indicates the $\pm\Delta\beta L$ values required to achieve a perfect cross state as a function of s. This requirement of a larger $\Delta\beta$ for TM than for TE compensates for the fact that, for given voltage, $\Delta\beta_{TM} \approx 3\Delta\beta_{TE}$. Precise values of s_{TE} and s_{TM} are not required because a good cross state (< -20 dB crosstalk) can be achieved for a reasonable range of $\Delta\beta L$ (Fig. 4.35b). Fabrication tolerances are thus within achievable limits.

The bar state is achieved with uniform mismatch ($V_1 = V_2$ in Fig. 4.35a). Polarization-independent operation results directly from sidelobe reduction of the switch response due to weighted coupling. By reducing the sidelobes below an acceptable crosstalk level (for example, -25 dB), a switch voltage which yields the first crossover null (bar state) for the TE mode that sees the weaker coefficient also provides a good TM mode bar state. For simple on/off modulation applications, uniform mismatch voltage alone can be used. Essentially the switch response is clipped so that it remains in a good bar state for all values of $(\Delta\beta)$ greater than a fixed value. Because the weighted coupling reduces the effective electrode length, the switching voltage is larger than for a uniform coupler of the same length.

Crosstalk levels below -23 dB for both switch states and for arbitrary input polarization have been demonstrated with this device using a Hamming function weighting of κ [4.151]. A 1×16 switch array of polarization independent switches for operation at $\lambda = 1.3\,\mu$m has been reported [4.152].

Recently an asymmetric branching switch which gives a saturating or digital switch response for both the $\otimes$ and $\ominus$ states has been used to demonstrate polarization and wavelength independent switching [4.153].

The polarization dependence of wavelength filters is due primarily to birefringence, both material and modal. For example, the curves of Fig. 4.33b depend upon polarization because of the different Δn for the two polarizations. Therefore, the phase-match wavelength depends upon polarization. κ and therefore the peak coupling efficiency also depends upon polarization. While the TE $\leftrightarrow$ TM interaction is symmetric and therefore insensitive to polarization, the need to select the converted polarization dictates that this filter can be used effectively for a single linear polarization.

However, the symmetry of the TE ↔ TM interaction has been employed with a pair of polarization splitter/combiners (4.5.3) to demonstrate a tunable, narrowband (12 Å for $\lambda_0 \approx 1.5\,\mu\text{m}$) filter whose filter centerwavelength and tuning are both polarization independent [4.154].

An alternate approach to polarization-independent devices is to align the waveguide along the optical axis of the crystal. In this case the TE and TM modes both see the ordinary index and will be affected similarly by an applied field. The e/o coefficient used for switching or modulation in this case is r_{22} (4.2.3) which is nearly ten times smaller than r_{33}, resulting in a significant voltage penalty.

4.8 Some Ti : LiNbO$_3$ Integrated-Optic Circuits

4.8.1 Coherent Lightwave Receiver

Lightwave systems that employ coherent detection techniques have need of advanced optical control devices including polarization controllers, frequency shifters [4.155, 156], directional couplers and perhaps tunable filters. A significant advantage could be achieved if several of these functions could be integrated onto a single chip to help produce a coherent receiver. An example of one such circuit is shown in Fig. 4.36 [4.157]. This circuit includes polarization transformer, frequency shifter, and a coupler. A polarization transformer with feedback control insures that the received signal coming in one fiber is aligned in polarization with the local oscillator which is inserted in the other fiber. The frequency shifter is used to insure that the frequency of the local oscillator matches that of the received signal. Signal and local oscillator are then combined using a directional coupler. For ultimate sensitivity it is important that the coupling efficiency be exactly 50 %, which implies that electrically adjusted couplers will be important.

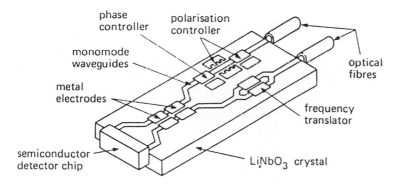

Fig. 4.36. Integrated coherent receiver circuit. (After [4.157])

After mixing, the signal is detected with two detectors to provide for the balanced receiver. A circuit of this type is currently under investigation [4.157].

4.8.2 Optical Switch Arrays

To date, integrated optic devices with the largest degree of integration have been optical switch arrays. These devices, which have potential application as space division switches for video switching and long-haul facility switching, have been fabricated using cascades of both directional coupler and intersecting waveguides 2 by 2 switches. The simplest and most general switch array is the crossbar shown schematically for the case of 4 by 4 in Fig. 4.37. Simple crossbar requires the maximum number of switches N^2

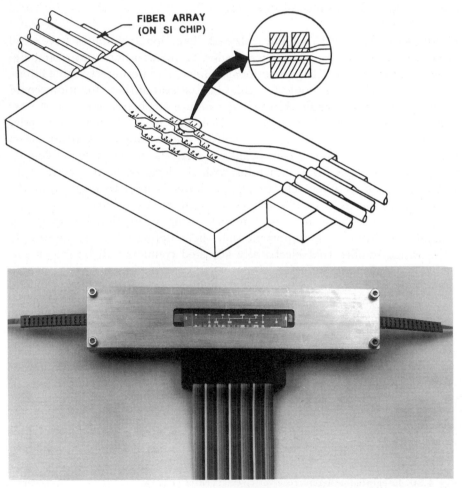

Fig. 4.37. (a) 4 × 4 crossbar switch. (b) Photograph of pigtailed switch. (After [4.164])

202

to achieve fully nonblocking switching in which each of N possible inputs can be connected to any one of N possible outputs. The number of stages required in the switch is $2N - 1$. Parameters of importance for the switch array are the total optical insertion loss, the fiber-to-fiber crosstalk and switch voltage. The device is an excellent test vehicle for both technology and circuit design in that uniformity of both insertion loss and switch voltage over 6 cm long crystals is important as is the need to design relatively low-loss bends using as little distance on the chip as possible [4.158–160a]. Because of the large number of switches, low switch voltage is essential, which makes tapered waveguide techniques important.

To date, several demonstrations of 4 by 4 crossbar switches have been reported [4.161–164]. Total crosstalk less than 35 dB has been achieved with average total fiber-coupled insertion losses, which are somewhat path dependent, of around 5 dB and required voltage less than 15 V to switch [4.164]. In addition, an 8 by 8 switch consisting of 64 directional couplers has been reported [4.165]. To achieve good crosstalk throughout the switch it is essential to use the reversed-delta beta electrodes when using directional couplers. These switch arrays have been used for wide-band video switching and time-division switching [4.166, 167].

4.9 Applications

Ti : LiNbO$_3$ waveguide devices appear attractive for application in communications, signal processing and sensing. Potential applications in second-generation optical communication systems like coherent, WDM, switching and local distribution seem especially attractive because the needs of these systems are well matched to the functionality possible with guided-wave devices. The coherent receiver chip previously described is a good example. Below we describe several applications where Ti : LiNbO$_3$ devices have already demonstrated strong potential.

4.9.1 External Modulators

Optical modulators employed for signal encoding have long been a principal objective of electro-optic device research. However, unlike early gas lasers, the semiconductor lasers employed in modern lightwave sysems can be modulated at multi-gigahertz rates. Nevertheless, external waveguide modulators are attractive for several application areas including very-high bit rate, long-haul systems, phase modulation in coherent systems and for inexpensive modulator arrays or remote modulators in local systems.

The lowest-loss wavelength of silica fiber is $\lambda = 1.55\,\mu\text{m}$. However, for this wavelength, silica fiber exhibits significant chromatic dispersion. Directly modulated lasers, even single-frequency distributed feedback lasers, exhibit a change in the output wavelength as the gain is turned on and off. This chirp together with fiber dispersion causes pulse spreading which can limit either the length or bandwidth of lightwave transmission systems. This chirp can be overcome by using external modulators. Indeed, even with its additional loss Ti:LiNbO$_3$ waveguide modulators have demonstrated superior performance to directly modulated lasers for data rates above 4 Gb/s [4.168, 169]. Indeed, at this writing the best performance, as measured by the bit rate distance product, for a unrepeatered single channel lightwave system is one in which 8 Gb/s of information was sent over 69 km of single-mode fiber with acceptably low error rate (10^{-9} errors/s). The system employed a Ti:LiNbO$_3$ traveling wave directional coupler switch [4.170]. Intensity modulators have also been used as remote modulators in local-distribution-system experiments [4.171–173]. High-speed switches and modulators have been used in a four-channel time division multiplexing system experiment to transmit an aggregate data rate of 16 Gb/s over 8 km of fiber [4.174].

In coherent lightwave systems, phase modulation encoding offers receiver sensitivity advantage over other techniques such as frequency or intensity modulation. It is not convenient to directly phase-modulate semiconductor lasers which makes external phase modulators attractive. While coherent-lightwave-system research is still evolving, maximum data rates are lower (2 Gb/s) than for the direct detection system. However, Ti:LiNbO$_3$ phase modulators have found extensive use in coherent system experiments reported to date including those that achieved the best bit rate distance figure of merit [4.175–180].

For these applications fiber pigtailed Ti:LiNbO$_3$ devices that are broadband, low-loss and can be driven with low drive power are required. A photograph of a fiber pigtailed phase modulator used for coherent trans-

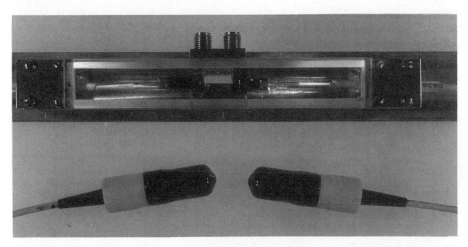

Fig. 4.38. Photograph of pigtailed traveling-wave modulator

mission experiments is shown in Fig. 4.38. This X-cut device operating at $\lambda = 1.55\,\mu$m, exhibits a total insertion loss of less than 2 dB, requires 8 V for a π phase shift and has a 3 dB optical modulation bandwidth of ~ 5 GHz [4.181].

4.9.2 High-Speed Analog to Digital Conversion

The periodic response of interferometric modulators has been employed to demonstrate megasample/second optical analog-to-digital conversion. The device, shown schematically in Fig. 4.39, consists of an array of Y-branch interferometers for which the drive electrode lengths differ by power of 2 [4.182]. Each interferometer also has a fixed-length electrode for biasing. The interferometer electrodes are driven in parallel by the input analog electrical signal. The optical input is a train of short pulses at a repetition rate corresponding to the sampling rate. One requires a number of interferometers equal to the number of digital bits of accuracy required. The interferometer with the longest electrode has a periodic output intensity with voltage with the shortest period while the short-electrode device takes the largest voltage change to control the output optical intensity. Accordingly, the bits for these interferometers correspond to the least significant and most significant bits, respectively. Comparators after detection from each interferometer determine whether the output is above or below a threshold value and set the bit to "one" or "zero" accordingly. Devices with 4 bit accuracy and 828 megasamples/s [4.183] as well as 2 bit, 2 gigasamples/s have been demonstrated [4.184].

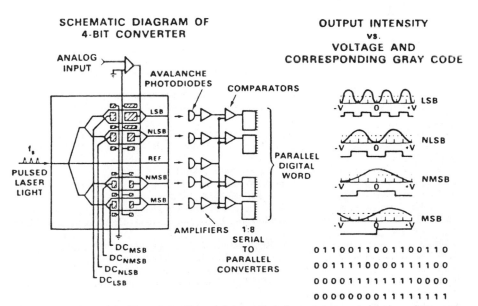

Fig. 4.39. Electro-optic waveguide analog-to-digital converter employing Y-branch interferometric modulators. (After [4.183])

4.9.3 Fiber Gyroscope Chip

Because of its low loss, optical fiber offers an attractive medium for rotational sensing via the Sagnac effect. To utilize and process this nonreciprocal phase shift in fiber, devices like splitters, polarizers, phase modulators and frequency shifters are required. While the first two are passive functions which can be performed with in-line fiber devices, $Ti:LiNbO_3$ chips that provide the active as well as passive functions are attractive to build compact fiber gyroscopes. Several such chips have been reported [4.185, 186].

References

4.1   T. Suhara, H. Nishihara: IEEE J. **QE-22**, 845–868 (1986)
4.2   I.P. Kaminow, J.R. Carruthers: Appl. Phys. Lett. **22**, 326–328 (1973)
4.3   R.V. Schmidt, I.P.Kaminow: Appl. Phys. Lett. **25**, 458–460 (1974)
4.4   M.L. Shah: Appl. Phys. Lett. **26**, 652–653 (1975)
4.5   J.L. Jackel, C.E. Rice, J.J. Veselka: Appl. Phys. Lett. **47**, 607–608 (1982)
4.6   K. Yamamoto, T. Taniuchi: Optoelectronics Conference, Tokyo, (1986) Paper B11-4
4.7   M.E. Lines, A.M. Glass: *Principles and Applications of Ferroelectrics and Related Materials* (Oxford Univ. Press, Oxford 1977)
4.8   R.V. Schmidt, L.L. Buhl: Private Communications
4.9   A. Askin, G.D. Boyd, J. Dziedzic, R.G. Smith, A.A. Ballman, J.J. Levinstein, K. Nassau: Appl. Phys. Lett. **9**, 72–74 (1966)
4.10 R.C. Alferness, L.L. Buhl: Opt. Lett. **5**, 473–475 (1980)
4.11 J.L. Jackel, V. Ramaswamy, S. Lyman: Appl. Phys. Lett. **38**, 509–511 (1981)
4.12 J.L. Jackel: Opt. Commun. **3**, 82–85 (1982)
4.13 S. Miyazawa, R. Guglielmi, A. Carenco: Appl. Phys. Lett. **31**, 842–744 (1977)
4.14 B. Chen, A.C. Pastor: Appl. Phys. Lett. **30**, 570–572 (1977)
4.15 R.J. Esdaile: Appl. Phys. Lett. **33**, 733–735 (1978)
4.16 R.L. Holman, P.J. Cressman, J.F. Revelli: Appl. Phys. Lett. **22**, 280–282 (1978)
4.17 J. Noda, M. Fukuma, A. Saito: Appl. Phys. Lett. **27**, 19–21 (1975)
4.18 M. Minakata, S. Saito, M. Shibata, S. Miyazawa: J. Appl. Phys. **49**, 4677–4682 (1978)
4.19 R.J. Holmes, D.M. Smyth: J. Appl. Phys. **55**, 3531–3535 (1984)
4.20 M. Minakata, S. Saito, M. Shibata, S. Miyazawa: J. Appl. Phys. **49**, 4677–4680 (1978)
4.21 K. Sugii, M. Kukuma, H. Iwasaki: J. Mater. Sci. **13**, 523–527 (1978)
4.22 M. De Michelli, J. Botineau, S. Neveu,P. Silbillot, D.B. Ostrowsky, M. Papupchon: Opt. Lett. **8**, 114–115 (1983)
4.23 A. Yi-Yan: Appl. Phys. Lett. **42**, 633–635 (1983)
4.24 K.K. Wong, R.M. De La Rue, S. Wright: Opt. Lett. **23**, 265–266 (1987)
4.25 R.A. Becker: Appl. Phys. Lett. **43**, 131–133 (1983)
4.26 J.J. Veselka, G.A. Bogert: Elect. Lett. **23**, 265–266 (1987)
4.27 E.Y.B. Pun, K.K. Wong, I. Andonovic, P. Laybourn, R.M. De La Rue: Elect. Lett. **18**, 740–742 (1982)
4.28 E.A.J. Marcatili, S.E. Miller: Bell System Tech. J. **48**, 2161–2188 (1969)
4.29 A. Mahapatra, W.C. Robinson: Appl. Opt. **24**, 2285–2286 (1985)
4.30 D.Y. Zang, C.S. Tsai: Appl. Phys. Lett. **46**, 703–705 (1985)
4.31 M. De Michelli, J. Botineau, P. Sibillot, D.B. Ostrowsky, M. Papuchon: Opt. Commun. **42**, 101–103 (1982)
4.32 I.P. Kaminow, W.L. Mammel, H.P. Weber: Appl. Opt. **13**, 396–405 (1974)

4.33   G.L. Tangonan, D.L. Persechini, J.F. Lotspeich, M.K. Barnowski: Appl. Opt. **17**, 3259–3263 (1978)

4.34   C.M. Gee, G.D. Thurmond, H. Blauvelt, H.W. Yen: Appl. Phys. Lett. **47**, 211–213 (1985)

4.35   L.L. Buhl: Electron. Lett. **19**, 659–660 (1983)

4.36   W.K. Burns, T.G. Giallorenzi, R.P. Moeller, E.J. West: Appl. Phys. Lett. **33**, 944–947 (1978)

4.37   S. Yamada, M. Minakata: Jpn. J. Appl. Phys. **20**, 733–737 (1981)

4.38   G. Eisenstein, S.K. Korotky, L.W. Stulz, J.J. Veselka, R. Jopson, K.L. Hall: Elect. Lett. **21**, 363–364 (1985)

4.39   M. Izutsu, Y. Yamane, T. Sueta: IEEE J. **QE-13**, 287–290 (1977)

4.40   R.C. Alferness, C.H. Joyner, L.L. Buhl, S.K. Korotky: IEEE J. **QE-19**, 1339–1340 (1983)

4.40a R.A. Becker: Opt. Lett. **10**, 417–419 (1985)

4.40b C.R. Giles, S.K. Korotky: Topical Meeting on Integrated and Guided-Wave Optics, Santa Fe (1988)

4.40c C.H. Bulmer, W.K. Burns, S.C. Hiser: Appl. Phys. Lett. **48**, 1036–1038 (1986)

4.40d A.R. Beaumont, B.E. Daymond-John, R.C. Booth: Elect. Lett. **22**, 262–263 (1986)

4.40e R.V. Schmidt, P.S. Cross, A.M. Glass: J. Appl. Phys. **51**, 90–93 (1980)

4.40f G.T. Harvey, G. Astfalk, A.Y. Feldblum, B. Kassahun: IEEE J. **QE-22**, 939–946 (1986)

4.40g A.R. Beaumont, C.G. Atkins, R.C. Booth: Elect. Lett. **22**, 1260–1261 (1986)

4.41   See, for example, I.P. Kaminow: *An Introduction to Electro-Optic Devices* (Academic, New York 1974)

4.42   D. Marcuse: IEEE J. **QE-18**, 393–398 (1982)

4.43   D.G. Ramer: IEEE J. **QE-18**, 386–392 (1982)

4.44   L. Thylen, P. Granestrand: J. Opt. Commun. **7**, 11–15 (1986)

4.45   R.E. Tench, J.-M.P. Delavaux, L.D. Tzeng, R.W. Smith, L.L. Buhl, R.C. Alferness: IEEE J. **LT-5**, 493–501 (1986)

4.46   F. Auracher, D. Imhof: Siemens Forsch.- u. Entwick. **15**, 19–22 (1986)

4.47   K. Kubota, J. Noda, O. Mikami: IEEE J. **QE-16**, 754–760 (1982)

4.48   K.C. Gupta, R. Garg, L.J. Bahl: *Microstrip Lines and Slotliners* (Artech Dedham, Mass. 1979)

4.49   There is a typographical error in [4.47] which has propagated to other papers. In [Ref. 4.47, Eqs. 27 and 30] G should be replaced by $2G$. The results of [4.47] then agree with (4.2.10)

4.50   R.V. Schmidt, P.S. Cross: Opt. Lett. **2**, 45–57 (1978)

4.51   R.A. Becker: IEEE J. **QE-20**, 723–727 (1984)

4.52   W.W. Rigrod, I.P. Kaminow: Proc. IEEE **51**, 137–140 (1963)

4.53   S.K. Korotky, G. Eisenstein, R.C. Alferness, J.J. Veselka, L.L. Buhl, G.T. Harvey, P.H. Read: IEEE J. **LT-3**, 1–5 (1985)

4.54   H. Kogelnik, V. Ramaswamy: Appl. Opt. **13**, 1857–1862 (1974)

4.55   H. Haga, M. Izutsu, T. Sueta: IEEE J. **QE-22**, 902–906 (1986)

4.56   M. White, C.E. Enderby: Proc. IEEE **51**, 214–220 (1963)

4.57   R.C. Alferness, S.K. Korotky, E.A.J. Marcatili: IEEE J. **QE-20**, 301–309 (1984)

4.58   R.S. Tucker, G. Eisenstein, S.K. Korotky, U. Koren, G. Raybon, J.J. Veselka, L.L. Buhl, B.L. Kasper, R.C. Alferness: Elect. Lett. **23**, 209–210 (1987)

4.59   H. Haga, M. Izutsu, T. Sueta: IEEE J. **LT-3**, 116–120 (1985)

4.60   H. Haus, S. Kirsch, K. Mathyssek, F.J. Leonberger: IEEE J. **QE-16**, 870–880 (1980)

4.61   E.A.J. Marcatili: Appl. Opt. **19**, 1468–1476 (1980)

4.62   A. Djupsjobacka: Elect. Lett. **21**, 908–909 (1985)

4.63   D. Erasme, M.G.F. Wilson: Elect. Lett. **22**, 1025–1026 (1986)

4.64   M. Nazarathy, D.Dolfi: IOOC '87, Reno, Nevada (1987) Paper. TUQ37

4.65   M. Fukuma, J. Noda: Appl. Opt. **19**, 591–597 (1980)

4.66   R.Keil, F. Auracher: Opt. Commun. **30**, 23–28 (1979)

4.67   C.H. Bulmer, S.K.Sheem, R.P. Moeller, W.K. Burns: Appl. Phys. Lett. **37**, 351–353 (1981)

4.68 O.G. Ramer, C. Nelson, C. Mohr: IEEE J. **QE-17**, 970–974 (1981)
4.69 V.R. Ramaswamy, R.C. Alferness, M.D. Divino: Elect. Lett. **10**, 30–31 (1982)
4.70 R.C. Alferness, V.R. Ramaswamy, S.K. Korotky, M.D. Divino, L.L. Buhl: IEEE J. **QE-18**, 1807–1813 (1982)
4.71 L. McCaughan, E.J. Murphy: IEEE J. **QE-19**, 131 (1983)
4.72 R.C. Alferness, L.L. Buhl, M.D. Divino: Elect. Lett. **18**, 490–491 (1982)
4.73 W.K. Burns, G.B. Hocker: Appl. Opt. **16**, 2048–2049 (1977)
4.74 H. Suche, B. Hampel, H. Seibert, W. Sohler: Proc. Conf. on Integrated Optical Circuit Engineering, Boston, SPIE **578**, 156–158 (1985)
4.75 R.C. Alferness, S.K. Korotky, L.L. Buhl, M.D. Divino: Electron. Lett. **20**, 354–355 (1984)
4.76 E.J. Murphy, T.C. Rice, L. McCaughan, G.T. Harvey, P.H. Read: IEEE J. **LT-51**, 795–799 (1985)
4.77 J.J. Veselka, S.K. Korotky: IEEE J. **QE-22**, 933–938 (1986)
4.78 J. Komatsu, M. Kondo, Y. Ohta: Elect. Lett. **22**, 881–882 (1986)
4.79 K.Komatsu, S. Yamazaki, M. Kondo, Y. Ohta: IOOC '87, Reno, (1987) Paper WK-5
4.80 E.A.J. Marcatili: Bell Syst. Tech. J. **48**, 2071–2080 (1969)
4.81 R.C. Alferness, R.V. Schmidt, E.H. Turner: Appl. Opt. **18**, 4012–4016 (1979)
4.82 J. Noda, M. Fukuma, O. Mikami: Appl. Opt. **20**, 2284–2290 (1981)
4.83 H.F. Taylor: J. Appl. Phys. **13**, 327–333 (1974)
4.84 H.F. Schlaak: J. Opt. Commun. **5**, 122–125 (1984)
4.85 W.H. Louisell: *Coupled-Mode and Parametric Electronics* (Wiley, New York 1960)
4.86 S.E. Miller: Bell System. Tech. J. **33**, 661 and 719 (1954)
4.87 S.E. Miller: Bell Syst. Tech. J. **48**, 2189–2219 (1969)
4.87a H. Kogelnik: Bell Syst. Tech. J. **55**, 109–126 (1976)
4.88 R.C. Alferness, P.S. Cross: IEEE J. **QE-14**, 843–847 (1978)
4.89 T. Findakly, Chin-Lin Chen: Appl.Opt. **17**, 769–773 (1978)
4.90 A.F. Milton, W.K. Burns: Appl. Opt. **14**, 1207–1212 (1975)
4.91 R.C. Alferness: Appl. Phys. Lett. **35**, 260–262 (1979)
4.92 H. Kogelnik, R.V. Schmidt: IEEE J. **QE-12**, 396–401 (1976)
4.93 R.V. Schmidt, H. Kogelnik: Appl. Phys. Lett. **26**, 503–505 (1976)
4.94 G.A. Bogert, E.J. Murphy, R.T. Ku: IEEE J. **LT-4**, 1542–1545 (1986)
4.95 F. Auracher, D. Schiketanz, K.-H. Zeitler: J. Opt. Commun. **5**, 7–9 (1984)
4.96 R.S. Tucker, S.K. Korotky, G. Eisenstein, U. Koren, G. Raybon, J.J. Veselka, L.L. Buhl, B.L. Kasper, R.C. Alferness: In *Photonic Switching*, ed. by T.K. Gustafson, P.W. Smith, Springer Ser. Electron. Photon. Vol. 25 (Springer, Berlin, Heidelberg 1988) p. 208
4.97 W.E. Martin: Appl. Phys. Lett. **26**, 562–563 (1975)
4.98 V. Ramaswamy, M. Divino, R.D. Standley: Appl. Phys. Lett. **32**, 644–646 (1978)
4.99 M. Minakata: Appl. Phys. Lett. **35**, 145–147 (1978)
4.100 0. Mikami, S. Zembutsu: Appl. Phys. Lett. **35**, 145–147 (1978)
4.101 M. Papuchon, A.M. Roy, B. Ostrowsky: Appl. Phys. Lett. **31**, 266–268 (1977)
4.102 A. Neyer: Elect. Lett. **19**, 553–554 (1983)
4.103 H. Nakajima, I. Sawaki, M. Seino, K. Asama: IOOC '83, Tokyo (June 1983) Paper 29C4-5
4.104 C.S. Tsai, B.Kim, F.R. El-Arkani: IEEE J. **QE-14**, 513–517 (1978)
4.105 R.A. Forber, E. Marom: IEEE J. **QE-22**, 911–919 (1986)
4.106 J. Ctyroky: J.Opt. Commun. **1**, 139–143 (1986)
4.107 R.C. Alferness, M.D. Divino: Elect. Lett. **20**, 760–761 (1984)
4.108 F.J. Leonberger: Opt. Lett. **5**, 312–314 (1980)
4.109 F. Auracher, R. Keil: Appl. Phys. Lett. **36**, 626–629 (1980)
4.110 R.A. Becker: IEEE J. **QE-20**, 723–727 (1984)
4.111 W.K. Burns, A.B. Lee, A.F. Milton: Appl. Phys. Lett. **29**, 790–792 (1976)
4.112 A. Neyer, W. Sohler: Appl. Phys. Lett.**35**, 256–258 (1979)
4.113 R.C. Alferness: Appl. Phys. Lett. **36**, 513–515 (1980)
4.114 R.C. Alferness, L.L. Buhl: Opt. Lett. **7**, 500–502 (1982)
4.115 C.H. Bulmer, W.K. Burns: IEEE J. **LT-2**, 512–515 (1984)

4.115a M. Izutsu, A. Enokihara, T. Sueta: Opt. Lett. **7**, 549–551 (1982)
4.116 S.K. Korotky, R.C. Alferness: In *Integrated Optical Circuits and Components*, ed. by L.D. Hutcheson (Dekker, New York 1987)
4.117 S.K. Korotky, R.C. Alferness: IEEE J. **LT-1**, 244–251 (1983)
4.118 S.K. Korotky: IEEE J. **QE-22**, 952–958 (1986)
4.119 C.M. Gee, G.D. Thurmond, H.W. Yen: Appl. Phys. Lett. **43**, 998–1000 (1983)
4.120 K.Kubota, J. Noda, O. Mikami: IEEE J. **QE-16**, 754–758 (1980)
4.121 R.A. Becker: Appl. Phys. Lett. **45**, 1168–1170 (1984)
4.122 S.K. Korotky, G. Eisenstein, R.S. Tucker, J.J. Veselka, G. Raybon: Appl. Phys. Lett. **50**, 1631–1633 (1987)
4.123 R.C. Alferness, L.L. Buhl: Elect. Lett. **19**, 40–41 (1983)
4.124 R.C. Alferness: IEEE J. **QE-17**, 946–959 (1981)
4.125 R.C. Alferness, L.L. Buhl: Appl. Phys. Lett. **38**, 655–657 (1981)
4.126 C. Mariller, M. Papuchon: In *Integrated Optics*, ed. by H.P. Nolting, R. Ulrich. Springer Ser. Opt. Sci., Vol. 48 (Springer, Berlin, Heidelberg 1985) pp. 174–176
4.127 S. Thaniyavarn: Appl. Phys. Lett. **47**, 674–677 (1985)
4.128 M. Haruna, J. Shimada, H. Nishihara:Trans. IECE Jpn. **69**, 418–419 (1986)
4.129 R.C. Alferness, L.L. Buhl: Appl. Phys. Lett. **47**, 1137–1139 (1985)
4.130 L.L. Buhl: Electron. Lett. **19**, 659–660 (1983)
4.131 J. Ctyroky, H.J. Henning: Elect. Lett. **22**, 756–757 (1986)
4.132 T. Findakly, B.U. Chen: Electron. Lett. **20**, 128–129 (1984)
4.133 J.J. Veselka, G.A. Bogert: Elect. Lett. **23**, 265–266 (1987)
4.134 M. Papuchon, S. Vatoux: Elect. Lett. **19**, 612–613 (1983)
4.135 H. Yajima: IEEE J. **LT-1**, 273–279 (1983)
4.136 H. Nakajima, T. Horimatsu, M. Seino, I. Sawaki: IEEE Trans. **MTT-30**, 617–621 (1982)
4.137 M. Masuda, G.L. Yip: Appl. Phys. Lett. **37**, 20–22 (1980)
4.138 O. Mikami: Appl. Phys. Lett. **36**, 491–493 (1980)
4.139 R.C. Alferness, L.L. Buhl: Opt. Lett. **10**, 140–142 (1984)
4.140 A. Neyer: Elect. Lett. **20**, 744–746 (1984)
4.141 R.C. Alferness, L.L. Buhl: Appl. Phys. Lett. **30**, 131–133 (1981)
4.142 R.C. Alferness, L.L. Buhl: Appl. Phys. Lett. **40**, 861–862 (1982)
4.143 F. Heisman, R.C. Alferness: IEEE J. **QE-24**, 83–93 (1988)
4.144 F. Heisman, L.L. Buhl, R.C. Alferness: Elect. Lett. **23**, 572–574 (1987)
4.145 H.F. Taylor:Opt. Commun. **8**, 421–425 (1973)
4.146 R.C. Alferness, R.V. Schmidt: Appl. Phys. Lett. **33**, 161–163 (1978)
4.147 R.C. Alferness, J.J. Veselka: Elect. Lett. **21**, 466–467 (1985)
4.148 R.A. Steinberg, T.G. Giallorenzi: Appl. Opt. **15**, 2440–2453 (1976)
4.149 W.K. Burns, T.G. Giallorenzi, R.P. Moeller, E.J. West: Appl. Phys. Lett. **33**, 944–947 (1978)
4.150 Y. Bourbin, M. Papuchon, S. Vatoux, J.M. Arnoux, M. Werner: Elect. Lett. **20**, 496–497 (1984)
4.151 R.C. Alferness: Appl. Phys. Lett. **35**, 748–750 (1979)
4.152 J.E. Watson: IEEE J. **LT-4**, 1717–1721 (1986)
4.153 Y. Silberberg, P. Perlmutter, J. Baran: Appl. Phys. Lett. **51**, 1230–1232 (1987)
4.154 W. Warzanskyi, F. Heisman, R.C. Alferness: Optical Fiber Commun. Conf., New Orleans (1988)
4.155 F. Heisman, R. Ulrich: Appl. Phys. Lett. **45**, 490–492 (1984)
4.156 M. Izutsu, S. Shikoma, T. Sueta: IEEE J. **QE-17**, 2225–2227 (1981)
4.157 W.A. Stallard, T.G. Hodgkinson, K.R. Preston, R.C. Booth: Elect. Lett. **21**, 1077–1079 (1985)
4.158 E.A.J. Marcatili, S.E. Miller: Bell Syst. Tech. J. **48**, 2161–2188 (1969)
4.159 L.D. Hutcheson, I.A. White, J.J. Burke: Opt. Lett. **5**, 276–278 (1980)
4.160 W.J. Minford, S.K. Korotky, R.C. Alferness: IEEE J. **QE-18**, 1802–1810 (1982)
4.160a S.K. Korotky, E.A.J. Marcatil, J.J. Veselka, R.H. Bosworth: Appl. Phys. Letts. **48**, 92–94 (1985)
4.161 M. Kondo, Y.Ohta, M. Fujiwara, M. Sakaguchi: IEEE Trans. **MTT-30**, 1747–1752 (1982)

4.162 L. McCaughan, G.A. Bogert: Appl. Phys. Lett. **47**, 348–350 (1985)
4.163 A. Neyer, W. Mevenkamp, B. Kretzschmann: Topical Meeting on Integrated and Guided-Wave Optics, Atlanta, GA (1986) Paper WAA2
4.164 G.A. Bogert, E.J. Murphy, R.T. Ku: IEEE J. **LT-4**, 1542–1545 (1986)
4.165 P. Granestrand, B. Stoltz, L. Thylen, K. Bergvall, W. Doldissen, H. Heinrich, D. Hoffmann: Elect. Lett. **22**, 816–818 (1986)
4.166 J.R. Erickson, et al.: Topical Meeting on Photonic Switching, Paper ThA5, Incline Village (March 18–20, 1987)
4.167 S. Suzuki et al.: IEEE J. **LT-4**, 894–899 (1986)
4.168 S.K. Korotky et al.: IEEE J. **LT-3**, 1027–1031 (1985)
4.169 T. Okyiyama et al.: European Conference on Optical Communication, (1987) Post-Deadline Paper
4.170 A.H. Gnauck, S.K. Korotky, B.L. Kasper, J.C. Campbell, J.R. Talman, J.J. Veselka, A.R. McCormick: Optical Fiber Conference, Atlanta, GA (1986)
4.171 P.J. Duthie, M.J. Wale, I. Bennion, J. Hankey: Elect. Lett. **22**, 517–518 (1986)
4.172 H. Kobrinski, S.S. Cheng: Elect. Lett. **23**, 943–944 (1987)
4.173 E.J. Murphy, J. Ocenasek, C.R. Sandahl, R.J. Lisco: IOOC '87, Reno (1986), Paper TuQ16
4.174 R.S. Tucker et al.: Elect. Lett. **23**, 1115–1116 (1987)
4.175 R. Wyatt et al.: Elect. Lett. **19**, 550–552 (1983)
4.176 Hooper, Midwinter, D.W. Smith: IEEE J. **LT-1**, 596–611 (1983)
4.177 R. Linke, B.L. Kasper, N.A. Olsson, R.C. Alferness: Elect. Lett. **22** , 30–31 (1986)
4.178 A.H. Gnauck et al.: Elect. Lett. **23**, 286–287 (1987)
4.179 S. Yamazaki et al.: IOOC '87, Reno (1987), Paper PDP-12
4.180 Y.K.Park et al.: Elect. Lett. **22**, 283–284 (1986)
4.181 R.C. Alferness, L.L. Buhl, M.D. Divino, S.K. Korotky, L.W. Stulz: Elect. Lett. **22**, 309–310 (1986)
4.182 H.F. Taylor: IEEE Proc. **63**, 1524–1525 (1975)
4.183 F.J. Leonberger, C.E. Woodward, R.A. Becker: Appl. Phys. Lett. **40**, 565–568 (1982)
4.184 R.A. Becker, C.E. Woodward, F.J. Leonberger, R.C. Williams: IEEE Proc. **72**, 802–819 (1984)
4.185 H.C. LeFevre, S. Vatoux, M. Papuchon, C. Puech: SPIE **25**, 717–719 (1986)
4.186 C.H. Bulmer, R.P. Moeller: Opt. Lett. **6**, 572–574 (1981)

5. Mode-Controlled Semiconductor Lasers

I.P. Kaminow and R.S. Tucker

With 55 Figures

Working with the bewildering variety of high-performance semiconductor lasers available today, it is startling to think back 20 years to the hopelessly crude devices being studied then. Occasionally one of these lasers would emit low-duty-cycle light pulses at very low temperatures and then only for seconds. Today we are still plagued by problems of yield and reliability but the performance criteria are orders of magnitude more demanding than simply convincing oneself that laser oscillation has indeed taken place. The early history and progress in lasers can be found in many excellent books and reviews [5.1–6, 14]. Our purpose here is to present, in a tutorial fashion, some special topics relevant to semiconductor lasers, amplifiers and superluminescent diodes, principally as they relate to high performance fiber optical communication systems. A review of optical telecommunications is given in [5.6].

5.1 Organization of the Chapter

Transverse-mode control is essential for realizing efficient coupling of the laser output into a single-mode fiber. The principles of optical waveguiding [5.6], as developed in Chap. 2 are fundamental to mode control. After reviewing some laser basics to set the stage, we discuss various laser structures that afford mode control by buried waveguides or by ridge-loaded guides. Since our main focus is fiber communications we will be concerned chiefly with InGaAsP devices that operate at wavelengths in the region of 1.3 and 1.5 μm, which are best suited to such systems. However, where appropriate, we will touch on AlGaAs lasers operating at 0.8 μm. We will also be concerned with the spectral purity and longitudinal mode control of these lasers. High-speed modulation is another important topic to be treated. Finally, we will discuss amplifiers and superluminescent diodes, which can be fabricated from these same laser structures by antireflection (AR) coating the output facets to eliminate the feedback.

It is not our intention in a brief chapter like this to be complete in the range of topics nor in the coverage of a specific topic. Rather, we will present a personal view.

5.1.1 Notation

Regretably, there are only 26 letters in the Roman alphabet. Even with the help of capitals, hats, bars and Greek letters, we have found it necessary, in the interest of conforming with conventions established in the literature, to use the same notation for more than one (unrelated) quantity. A glossary is provided in Appendix 5.A to help reduce confusion.

5.2 Laser Basics

5.2.1 Expitaxial Materials and Heterostructure

The basic geometry of the heterostructure laser is illustrated schematically in Fig. 5.1. It consists of an *active layer* with thickness h surrounded by *cladding layers.* The active layer medium has refractive index n_f and bandgap energy E_{gf}; its carrier type may be either n- or p-type and, frequently, it is not intentionally doped leaving a residual n-type "undoped" layer. The cladding layers have band-gap energy $E_{gs} > E_{gf}$ which ensures confinement to the active layer of the excited carriers. The refractive index of the cladding layers n_s is less than n_f in order to provide low-loss optical waveguiding in the region of the active layer, thereby ensuring effective interaction of the optical field with the gain produced by the excited carriers. In Fig. 5.1, the upper layer is p-type and the lower one n-type providing a p-n junction at the upper interface of the active layer. (We have used the planar waveguide notation of Chap. 2 for the subscripts, f: film, s: substrate, for consistency.) The length in the z direction is L and the width in the y direction is w giving a cross-sectional area $A = wL$ normal to the *injected current* into the active layer I_a and the associated *current density* $J = I_a/A$. The output xy facets are cleaved (110) crystallographic planes and the junction is in a (001) plane.

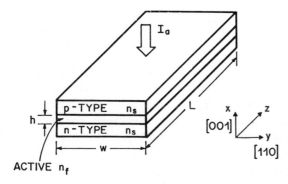

Fig. 5.1. Planar heterostructure laser. Refractive indices n are indicated for the active layer, f (film), and cladding layers, s (substrate)

The heterostructure laser has been the key to realizing efficient sources that operate at low current densities. In the earlier homojunction lasers, consisting of a doped pn junction in GaAs, dual confinement from two cladding layers was not present and these devices could only operate pulsed at low temperature [5.3]. Although we have indicated only confinement normal to the junction in Fig. 5.1, we will also be interested in confinement parallel to the junction for mode control and added efficiency.

Several alloy systems can provide the required conditions

$$n_{\mathrm{f}} > n_{\mathrm{s}} \quad , \quad E_{\mathrm{gf}} < E_{\mathrm{gs}} \quad , \tag{5.2.1}$$

for the epitaxially grown layers in the heterostructure laser. The two most common systems are $Al_y Ga_{1-y}As$, in which $y_{\mathrm{f}} < y_{\mathrm{s}}$ for the active and cladding layers, respectively; and $In_{1-x}Ga_x As_y P_{1-y}$, in which [5.7]

$$y = 2.2x \tag{5.2.2}$$

to assure a match of the lattice constant of the quaternary alloy to the usual binary substrate InP, and $x_{\mathrm{f}} > x_{\mathrm{s}}$. Figure 5.2 maps the lattice parameter versus band-gap energy in electron volts E_{g} and equivalent wavelength λ_{g}. The equivalent wavelength is given by

$$V_{\mathrm{g}}\lambda_{\mathrm{g}} = 1.24 \tag{5.2.3}$$

where $V_{\mathrm{g}} = E_{\mathrm{g}}/e$ is the energy gap in volts and λ_{g} is in micrometers. The boundary defines the lowest energy band gap with the solid line a direct gap and the dashed line an indirect gap. The lattice parameters of

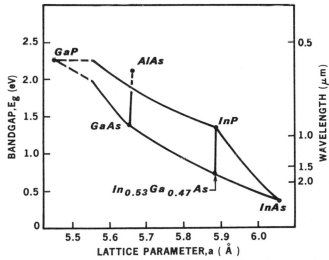

Fig. 5.2. Map of band-gap energy and wavelength versus lattice constant for AlGaAs and InGaAsP alloys. The dashed regions indicate indirect bandgaps

213

GaAs (5.654 Å) and AlAs (5.660 Å) differ by 0.1 % but the $Al_yGa_{1-y}As$ system can be reasonably well lattice matched to a GaAs substrate for $y < 0.3$ corresponding to a range of λ_g from 870 nm for GaAs to 760 nm. The $In_{1-x}Ga_xAs_yP_{1-y}$ system can be perfectly matched to InP, because of the extra degree of freedom, over $920 < \lambda_g < 1650$ nm, as y is varied with the constraint (5.2.2). The ranges of efficient laser action are somewhat less than this to ensure sufficiently large inequalities in (5.2.1).

The material properties of $Al_yGa_{1-y}As$ are thoroughly reviewed in [5.8]. The band-gap voltage varies as

$$V_g = 1.424 + 1.247y \quad (0 \leq y \leq 0.45) \quad . \tag{5.2.4}$$

The refractive index as a function of composition and photon energy below the band edge is given in Fig. 5.3. The curves are calculated from a semi-empirical model and give a good fit to experimental measurements [5.8]. These curves take on singular values near the band edges, where laser action takes place. Nevertheless, measurements of n through the band edge [Ref. 5.3 A, Fig. 2.5–4] show that n saturates at about the end points indicated in Fig. 5.3. In a typical case, the active layer would be GaAs and the cladding layers $Al_{0.3}Ga_{0.7}As$, corresponding to an index step of $\Delta n \approx 0.3$ at $\lambda \sim 870$ nm. To operate at shorter wavelengths, it is necessary to add Al to the active layer to increase the band-gap energy.

For the $In_{1-x}Ga_xAs_yP_{1-y}$ system, the band-gap voltage is given by [5.7]

$$V_g = 1.3 - 0.72y + 0.12y^2 \quad . \tag{5.2.5}$$

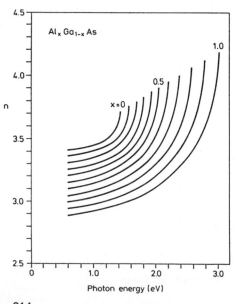

Fig. 5.3. Calculated refractive indices of $Al_xGa_{1-x}As$ [5.8]

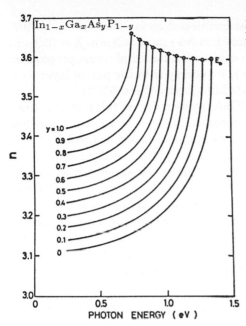

Fig. 5.4. Calculated refractive indices of $In_{1-x}Ga_xAs_yP_{1-y}$ [5.9]

Calculated refractive indices are given in Fig. 5.4 [5.9]. Other material parameters have been given in [5.10]. Refractive indices measured over a limited range of wavelengths can be found in [5.11], where an empirical relationship between the emission peak, λ_p, of the alloy and the laser wavelength, λ_L, was established; i.e., $\lambda_L - \lambda_p = 0.022\,\mu$m. This information is espcially important in determining the exact index for distributed-feedback lasers. Furthermore, the refractive index was also found to be reduced by -0.023 as the injected current increases from zero to threshold I_{th} due to the injected current density.

The carrier-induced reduction of refractive index is considerably larger than would be calculated by simple free-carrier effects. The major effect is due to a shift in the band edge to higher energy as the lower levels of the conduction and valence bands become full (Sect. 5.2.2). The Kramers-Kronig relation for the complex refractive index demands a corresponding reduction in index n due to an increase in carrier density N, or threshold injection current density J_{th}. Measured values of dn/dN are -1.2×10^{-20} cm^3 for GaAlAs lasers and -2.8×10^{-20} cm^3 for 1.3 μm InGaAsP lasers; the index change at threshold is (-0.03 to -0.06) for GaAlAs and (-0.04 to -0.10) for InGaAsP lasers, depending on layer thickness and doping [5.12]. Another measurement [5.13] gave an index change of -0.06 at threshold for 1.3 μm lasers. These induced index changes give an anti-guiding effect in the plane of the active layer that counteracts the positive waveguiding effects to be described in Sect. 5.3.2.

215

Typical values for the quantities mentioned so far are: $n_f - n_s \approx 0.2$, $E_{gs} - E_{gf} \cong 0.1\,\text{eV}$, $\lambda_{gf} - \lambda_{gs} \cong 200\,\text{nm}$, $h = 0.1\,\mu\text{m}$, $w = 5\,\mu\text{m}$, $L = 250\,\mu\text{m}$. With these values of $n_f - n_s$ and h, only the fundamental mode propagates with low loss, and approximately 30 % its energy is in the active layer with evanescent tails extending into the cladding layers (Sect. 5.2.2).

The short wavelengths of the AlGaAs system are useful in applications that require high-resolution focused spots and inexpensive, high numerical aperture (NA) optics to produce them; examples are laser printers and optical disc players. These systems generally do not operate at high modulation speeds. However, they do require a fundamental transverse mode output. Optical-fiber-communication applications require the longer wavelengths available with the InGaAsP system. The reason is the low-loss silica-based fibers. The lowest loss occurs at 1550 nm; and the region around 1300 nm is also interesting because the fiber dispersion vanishes there. Communication applications can be very demanding on laser performance in terms of mode control, linewidth and spectral purity, and high-speed operation.

5.2.2 Waveguide Propagation, Amplification and Oscillation

An amplifier element of length L containing an active layer of thickness h is illustrated in Fig. 5.5. For the present we consider a planar structure and ignore field confinement in the y direction. In Sect. 5.3 we will consider y-direction transverse-mode control. A fundamental transverse-mode propagates in the x direction with electric field

$$E(x,z,t) = E(x)e^{-j\overline{\beta}z} \tag{5.2.6}$$

where the time dependence $\exp(j\omega t)$ is understood and the modal propagation constant $\overline{\beta}$ may be complex. Note that the field $E(x)$ extends into the cladding layers on each side of the active layer, while the optical gain $g(x)$, which is due to carriers, is tightly confined to the active layer.

When the optical gain g is zero, the propagation constant is

$$\overline{\beta} = \beta - j\frac{\alpha}{2} \quad \text{where} \tag{5.2.7a}$$

$$\beta = k_0 \hat{n}_p \tag{5.2.7b}$$

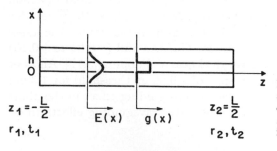

Fig. 5.5. Schematic drawing of the active layer waveguide illustrating confinement of the wavefunction $E(x)$ and the gain $g(x)$ due to injected carriers $N(x)$

is purely real, $\hat{n}_p$ (called N in Chap. 2), is the modal or waveguide refractive index for the pth mode for $g = 0$, $k_0 = 2\pi/\lambda$, where λ is the free-space wavelength, and α is the modal power attenuation coefficient. This attenuation includes effects such as scattering from imperfections in the bulk media and in the interface between the active and cladding layers, free carrier absorption in the active and cladding layers, and band-tail absorption in the cladding. It does not include band-edge absorption in the active layer. The factor 2 in the denominator of the α term arises because α is a *power* attenuation constant.

Band-edge absorption and amplification in the active layer are both represented by the gain g. Thus, the gain represents both stimulated emission and stimulated absorption in the active layer, and g can take positive or negative values. Figure 5.6 [5.14] shows curves of calculated gain g against photon energy for InGaAsP with a band gap corresponding to an emission wavelength of 1.3 μm. The curves are given for various electron densities N injected into the active layer. Further details of the gain process are given in Sect. 5.2.3.

The effect of electrons injected into the active layer can be represented by defining an increment in complex refractive index due to an increment in the carrier density δN. Separating the incremental index δn into real and imaginary parts as follows,

$$\delta n = \delta n' - j\delta n'' \quad , \tag{5.2.8}$$

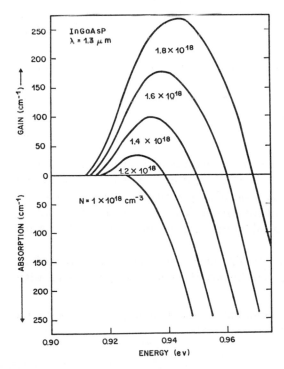

Fig. 5.6. Gain g (and absorption $-g$) near the band edge at 1.3 μm for InGaAsP [5.14]

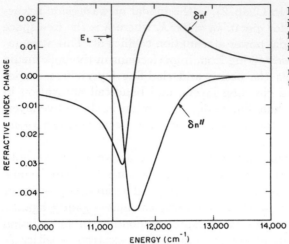

Fig. 5.7. Incremental refractive index change $\delta n = \delta n' - j\delta n''$ for a GaAs active layer for an increase of N from 0 to N_{th} versus photon energy. E_L designates the energy corresponding to the laser emission [5.15a,b]

the corresponding incremental power gain becomes

$$\delta g = -2k_0\delta n'' \quad . \tag{5.2.9}$$

According to the Kramers-Kronig relations, $\delta n''$ must be accompanied by a corresponding $\delta n'$. The calculated [5.15] increments for a GaAs active layer are plotted in Fig. 5.7 for an increment in N from $N = 0$ to approximately $N \sim 10^{18}\ \mathrm{cm}^{-3}$ (which is the electron density at laser threshold). The energy E_L corresponding to the laser emission wavelength occurs well below the peak in δg, or minimum in $\delta n''$, as indicated in the figure, since the net gain, g, which includes band-edge absorption as in Fig. 5.6, has its peak close to E_L. Note also that the injected carriers reduce the real part of the refractive index, chiefly due to the band-edge shift, which is apparent in Fig. 5.6, and to a small extent, due to the added free carriers. As we will see later, an important parameter is the ratio of any increment $\delta n'$ to the corresponding increment $\delta n''$ due to an arbitrary δN,

$$a \equiv \delta n'/\delta n'' \quad . \tag{5.2.10}$$

This "linewidth parameter" determines spectral linewidth and frequency chirp. It is often denoted by "α" in the literature.[1] It is evaluated at the laser energy and is defined here as a positive quantity. It is usually assumed that the ratio a is independent of the carrier density increment for small to moderate increments in N. Typical values of a are in the range 2 to 5 for InGaAsP lasers.

The quantities $\delta g/\delta N$ and $\delta n'/\delta N$ have been carefully measured as functions of the wavelength in an InGaAsP ridge waveguide laser with

---

[1] The sign convention used for a varies among authors. In the present chapter a is a positive number.

$\lambda = 1.53\,\mu m$ [5.16] and the behavior is qualitatively similar to that illustrated in Fig. 5.7. The value of $\delta g / \delta N$ at the lasing wavelength (1530 nm) is $2.7 \times 10^{-16}\,cm^2$ and at the peak (1500 nm) is $4.1 \times 10^{-16}\,cm^2$. The differential index $\delta n'/\delta N = -1.8 \times 10^{-20}\,cm^{-3}$ at 1530 nm and has a broad minimum of $-2 \times 10^{-20}\,cm^{-3}$ near 1520 nm. Thus, an increase of N by 1×10^{18} (below threshold) produces an index decrease of 2×10^{-2}. The linewidth parameter a varies from 3.8 at 1490 nm to 5.2 at 1530 nm to 11.2 at 1570 nm. Since, as shown in Sect. 5.5.2, line broadening is proportional to $(1 + a^2)$, the linewidth can be reduced substantially by "detuning" the laser wavelength from the gain peak toward shorter wavelengths to reduce a. This effect can be seen clearly in Fig. 5.7 by moving E_L to higher energy. This detuning can be achieved, at the expense of higher threshold current, by adjusting the grating period in a distributed feedback laser (Sect. 5.4.4).

Since the optical field is partly outside the active layer, the incremental *modal* refractive index is less than δn. This difference can be accounted for in terms of the mode confinement factor Γ which is defined below. Its value may lie between 0 and 1, and is typically in the range 0.3 to 0.5. Using this mode confinement factor, we are now in a position to generalize (5.2.7a) for arbitrary (non-zero) values of the gain g. To do this, we define at each wavelength a specific increment $\delta n_g = \delta n'_g - j\delta n''_g$ in the refractive index resulting from an increment in the carrier density from its zero-gain value to a value corresponding to g. For this increment, (5.2.9) becomes

$$g = 2k_0 \delta n''_g \quad . \tag{5.2.11a}$$

This gain g is the difference between any curve for given N in Fig. 5.6 and the zero baseline. Absorption corresponds to negative values of g. Note that Figs. 5.6, 7 are curves for different materials (InGaAsP and AlGaAs, respectively). However, they differ in a more fundamental way. Figure 5.6 gives the absolute gain g whereas Fig. 5.7 gives incremental quantities corresponding to an incremental change in carrier density. When band-edge effects are included (5.2.7a) becomes

$$\bar{\beta} = \beta + k_0 \Gamma \delta n'_g - j\left(\frac{\alpha}{2} + k_0 \Gamma \delta n''_g\right) \quad . \tag{5.2.11b}$$

From (5.2.10 and 11a), this can be written as

$$\bar{\beta} = \beta - \frac{a\Gamma g}{2} + j\left(\frac{\Gamma g - \alpha}{2}\right) \quad . \tag{5.2.11c}$$

The passive waveguide propagation constant β can be obtained from curves in Chap. 2, Fig. 2.8 using the refractive indices and thickness of the active and cladding layers. The incremental index δn is usually neglected when considering the waveguiding properties. If E is polarized along y (parallel to the plane of the guide), the wave is a transverse electric (TE) mode since there are no components of E along x or z. If E is polarized predominantly along x, a small z-component must also be present. However, the magnetic field of this mode is purely transverse (TM) along y.

Given k_0, $n_f - n_s$ and h, it is straightforward to calculate $E(x)$ for the generalized frequency or V-value as in Chap. 2,

$$V = k_0 h (n_f^2 - n_s^2)^{1/2}$$
$$\approx k_0 h (2 n_f \Delta n)^{1/2} \quad , \tag{5.2.12}$$

where the approximation holds for $\Delta n = n_f - n_s \ll n_f$. The number of modes that can propagate for the symmetric guide is the integer

$$M \leq 1 + V/\pi \quad . \tag{5.2.13}$$

In general, the fundamental mode has higher net modal gain ($\Gamma g - \alpha$) than the higher-order modes since it is concentrated in the high-gain active region and has less energy at the lossy interfaces. For the fundamental mode and n_f, n_s fixed, the effective width of the transverse wave function $E(x)$ decreases linearly with h for $V > 2$ and then increases rapidly as $V \to 0$ with a minimum near $V_m = 1.7$ (Chap. 2, Fig. 2.11). Thus, for $V < V_m$, a substantial fraction of the photon energy is outside the active region in which the stimulated emission gain takes place. The mode confinement factor, Γ, is defined as the ratio of optical power flow (or photon flux) in the active region to the total power flow in the mode. For TE modes

$$\Gamma = \frac{\displaystyle\int_0^h \eta_0 |E_y(x)|^2 dx}{\displaystyle\int_{-\infty}^{\infty} \eta_0 |E_y(x)|^2 dx} \quad , \tag{5.2.14}$$

with constant mode admittance,

$$\eta_0 = \beta \omega / \mu_0 \quad , \tag{5.2.15}$$

which cancels out in (5.2.14), where ω is the optical radian frequency and μ_0 is the magnetic vacuum permeability. The mode confinement factor is plotted in Fig. 5.8 [5.2] for asymmetric planar guides, where now the cladding media may be different on either side of the active layer with cladding indices n_s and n_c, and $n_c < n_s$. The asymmetry is characterized by the parameter

$$a' = \frac{n_s^2 - n_c^2}{n_f^2 - n_s^2} \quad . \tag{5.2.16}$$

Stronger confinement is obtained for lower asymmetry at a given V-value. For TM modes, E_y must be replaced by H_y in (5.2.14) and η_0 by mode impedance $\zeta_0 = \beta \omega / \varepsilon$, with ε the optical dielectric constant, where ζ_0 is now a funcion of x and does not drop out of (5.2.14).

220

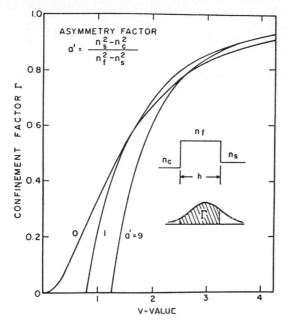

Fig. 5.8. Mode confinement factor Γ versus V-value for fundamental mode of slab guide with asymmetry factor a' [5.2]

Although $\Gamma \to 1$ gives efficient interaction between photons and excited carriers in the active region, a large mode width is sometimes more desirable, e.g. to reduce output beam divergence or power density at the mirror facet. This "large optical cavity" (LOC) behavior can be obtained by letting $V \to 0$ by means of thin active layers and/or small $n_f - n_s$.

We now consider a wave of amplitude E_i incident at $z = -L/2$ on the Fabry-Perot amplifier unit in Fig. 5.5 with amplitude transmission and reflection coefficients t_1, t_2, r_1, r_2 at the ends. The field E_o exiting at $z_2 = L/2$ is given by

$$\frac{E_o}{E_i} = \frac{t_1 t_2 e^{-j\overline{\beta}L}}{1 - r_1 r_2 e^{-j2\overline{\beta}L}} \quad . \tag{5.2.17}$$

This equation follows from $\sum_{n=0}^{\infty} x^n = (1-x)^{-1}$ for $x \le 1$. Transmission maxima occur at

$$\text{Re}\{\overline{\beta}_p\}L = p\pi \quad , \quad p \text{ integer} \tag{5.2.18a}$$

where, from (5.2.11)

$$\text{Re}\{\overline{\beta}_p\} = \beta - \frac{a\Gamma g}{2} \quad , \tag{5.2.18b}$$

and the solution $\overline{\beta}_p$ represents the pth *longitudinal mode*. The mode spacing ω_s between the p and $p+1$ longitudinal modes is

221

$$\omega_s = \frac{\pi}{L}\left(\frac{\partial \omega}{\partial \beta}\right) = \frac{\pi v_g}{L} \quad , \tag{5.2.18c}$$

where $v_g = \partial \omega / \partial \beta$ is the modal group velocity in the absence of gain. Equation (5.2.18c) follows from (5.2.18a) and $\mathrm{Re}\{\bar{\beta}_{p+1}\} = \mathrm{Re}\{\bar{\beta}_p\} + (\partial \beta / \partial \omega)\omega_s$, where it is assumed that g is a slowly-varying function of ω. Strictly, ω_s varies with ω due to dispersion in g. This effect can be neglected for present purposes.

It is clear from (5.2.18a and b) that the frequency of the longitudinal modes depends on gain. The change in mode frequency $\delta\omega$ for a change δg in gain is

$$\delta\omega = \frac{\partial \omega}{\partial \beta} \cdot \frac{\partial \beta}{\partial g} \delta g \quad . \tag{5.2.19a}$$

Using (5.2.18a), this becomes

$$\delta\omega = v_g a \Gamma \delta g / 2 \quad . \tag{5.2.19b}$$

When (5.2.18a) is satisfied, the peak power amplification (or gain) of the *Fabry-Perot amplifier* is

$$G_{\mathrm{FP}} \equiv \left(\frac{P_o}{P_i}\right) = \left(\frac{E_o}{E_i}\right)^2 = \frac{T_1 T_2 e^{(\Gamma g - \alpha)L}}{(1 - \sqrt{R_1 R_2} e^{(\Gamma g - \alpha)L})^2} \tag{5.2.20}$$

where T_1, T_2, R_1, R_2, are power transmission and reflection coefficients. In the limit of perfectly anti-reflection (AR) coated facets, the amplification of a *traveling-wave amplifier* is the single-pass gain[2]

$$G_0 = e^{(\Gamma g - \alpha)L} \quad . \tag{5.2.21}$$

With sufficient net exponential gain, the denominator in (5.2.20) vanishes, and output power is permitted with no input. Thus, the *idealized oscillation condition* is

$$1 - \sqrt{R_1 R_2} e^{(\Gamma g - \alpha)L} = 0 \quad , \quad \text{or} \tag{5.2.22a}$$

$$\tfrac{1}{2}\ln R_1 R_2 + (\Gamma g - \alpha)L = 0 \quad , \tag{5.2.22b}$$

together with the phase condition (5.2.18). This magnitude condition can be written in the following form: modal gain = total cavity losses, i.e.

$$\Gamma g = \alpha + \alpha_{\mathrm{m}} \quad , \tag{5.2.22c}$$

---

[2] Note that G_{FP} and G_0 are dimensionless amplifier power gains and g is an exponential gain coefficient with a dimension of nepers/length. It is common practice to use the term "gain" for all these parameters.

where α is the internal loss and $\alpha_m = L^{-1}\ln{(R_1 R_2)^{-1/2}}$ is the output coupling loss or mirror loss.

The loss in the "cold cavity", i.e. for $g \equiv 0$, can be expressed by a cavity or photon lifetime τ_p. In the special case of low-loss resonators

$$R_1 R_2 e^{-2\alpha L} \approx 1 \quad , \tag{5.2.23}$$

the fractional energy loss per round-trip in the resonator is

$$-\frac{\Delta W}{W} = 1 - R_1 R_2 e^{-2\alpha L} \tag{5.2.24a}$$

$$\approx -\ln R_1 R_2 + 2\alpha L \quad , \tag{5.2.24b}$$

using the approximation $-\ln x = 1 - x$, $x \approx 1$, for very low-loss resonators. The round trip time is

$$\Delta t = 2L/v_g \quad . \tag{5.2.25}$$

Assuming an exponential decay of energy,

$$W = W_0 e^{-t/\tau_p} \quad , \tag{5.2.26}$$

and small incremental loss, we find

$$\tau_p^{-1} = (v_g/2L)(1 - R_1 R_2 e^{-2\alpha L}) \tag{5.2.27a}$$

$$\approx v_g \left(\alpha - \frac{\ln R_1 R_2}{2L} \right) \quad . \tag{5.2.27b}$$

The photon lifetime τ_p can be viewed as the average time a photon spends in the cavity before it is lost by internal absorption or transmission through the facets. Although the exponential decay time is a convenient concept and (5.2.27) a concise means to represent cavity losses, the approximation (5.2.23) is not very accurate for semiconductor lasers, where $R_1 R_2 \approx (0.3)^2$ and $2\alpha L \approx 1$.

For an infinite plane wave, the reflectivity of a cleaved mirror in fact would be

$$R = \left(\frac{\hat{n} - 1}{\hat{n} + 1} \right)^2 \quad , \tag{5.2.28}$$

where $\hat{n}$ is the modal refractive index. This gives a value for R of 0.31 for a typical modal refractive index of 3.5. However, for a waveguide mode confined to a thickness on the order of a wavelength, (5.2.28) no longer holds. The modal reflectivity is a function of the polarization and mode width. One can think of the mode as composed of a group of rays, over a range of angles, bouncing between the active layer boundaries until they strike the cleaved facet. The range of angles, with respect to the facet normal, increases as the mode width decreases. For the TM polarization, R decreases with angle as the Brewster angle, at $16°$, is approached, but for the TE polarization, R increases monotonically with angle. Thus, for a small fundamental mode width, $R(\text{TE})$ is considerably greater than $R(\text{TM})$. A numerical analysis indicates typical values: $R(\text{TM}) \approx 0.24$ and $R(\text{TE}) \approx 0.40$ [5.17]. Since the semiconductors of interest have cubic symmetry (except when they are subjected to external stress), g should be polarization independent. Also α and Γ do not depend strongly on polarization. Thus, the higher $R(\text{TE})$ causes the TE mode to have the lower threshold in strain-free lasers. It has also been shown [5.18] that confinement parallel to the junction plane, for widths normally encountered in lateral confinement lasers to be discussed later, does not vary the reflectivity significantly from that for broad area lasers.

5.2.3 Laser Gain

A concise and elementary description of gain and absorption in semiconductors has been given in [Ref. 5.1, Chap. 15], [Ref. 5.4, Chap. 2], and [Ref. 5.14, Chap. 3]. To summarize: we consider a medium with the band structure illustrated in Fig. 5.9 [5.1], which is a plot of electron energy E versus electron momentum κ. We assume that all transitions conserve momentum, i.e., are vertical, and that the transitions are purely radiative. The origin of the E axis need not be specified, as only energy differences are relevant. The conduction-band minimum and valence-band maximum occur at $\kappa = 0$, indicating a direct-band-gap medium. Electrons can be injected into the conduction band and holes into the valence band by means of optical pumping or by means of a forward-biased p-n junction, as will be described shortly. The electrons and holes recombine spontaneously between bands with a characteristic time τ_n on the order of 10^{-9} s, and emit photons of energy $\hbar\omega \geq E_g$, where E_g is the energy gap between the band extrema. Electrons within the bands re-establish equilibrium with an intraband relaxation time τ_i on the order of 10^{-12} s by means of electron-electron and electron-phonon collisions. For sufficiently large injection current I_a into the active region, the density of holes N_v and electrons N_c will be the same and the carrier density $N = N_c \approx N_v$ will be given by

$$\frac{N}{\tau_n} = \frac{I_a}{v} \ , \tag{5.2.29}$$

where $v = whL$ is the volume of the active medium and, for now, we assume no appreciable photon density present to produce stimulated emission.

The carrier densities can also be represented by the quasi-Fermi energies E_{Fc} and E_{Fv}, indicated in Fig. 5.9. Physically, E_{Fc} and E_{Fv} are chemical potentials representing the energy required to remove an electron from the

224

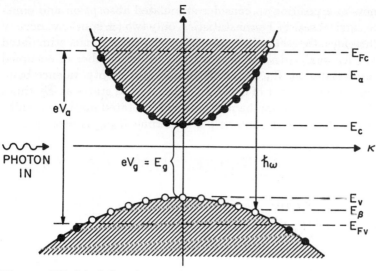

Fig. 5.9. Filled (*solid*) and vacant (*open*) valence and conduction band states at electron energies E and momenta κ. The quasi-Fermi levels are E_{Fc} and E_{Fv} [5.1]

conduction band or add a hole to the valence band. In equilibrium, (i.e., with no thermal or charge flux), $E_{Fc} = E_{Fv}$, so that an electron and hole at the Fermi level would recombine without loss of energy. The situation depicted in Fig. 5.9 is non-equilibrium, however, and

$$E_{Fc} - E_{Fv} = eV_a \qquad (5.2.30)$$

where e is the electron charge and V_a is the forward bias voltage applied across the junction through which the carriers are injected.

The relation between carrier density and Fermi level is

$$N_{c,v} = \int_0^\infty \varrho_{c,v}(E) f(E - E_{Fc,v}) dE \quad , \qquad (5.2.31)$$

where $\varrho_{c,v}(E)$ is the density of discrete states available at energy E and $f(E, E_{Fc,v})$ is the Fermi function, describing the probability of electron occupation of each state, for conduction and valence bands, respectively. Thus, $E_{Fc,v}$ are obtained from (5.2.31 and 29), with

$$\varrho_{c,v}(E) = \frac{1}{2\pi^2} \left(\frac{2m_{c,v}}{\hbar^2} \right)^{3/2} E^{1/2} \qquad (5.2.32)$$

$$f(E - E_{Fc,v}) = \{\exp[(E - E_{Fc,v})/kT] + 1\}^{-1} \quad , \qquad (5.2.33)$$

where $m_{c,v}$ are effective masses, $\hbar$ and k are Planck and Boltzmann constants, and T is the absolute temperature.

225

We are now in a position to consider stimulated absorption and emission. Since the carriers satisfy Fermi statistics, only two ($\pm$ spin) can occupy each state. Therefore, the net gain — stimulated emission minus stimulated absorption — at $\hbar\omega = E_\alpha - E_\beta$ is proportional to the number of occupied conduction band states at E_α times the number of empty valence-band states at E_β minus the number of occupied valence-band states at E_β times the number of empty conduction band states at E_α. Stated mathematically, the net number of stimulated photons per unit time R_{stim} is proportional to

$$
\begin{aligned}
R_{\text{stim}} &= v_g \Gamma g(\omega) I \sim v \int_E \varrho(E) dE \{ f(E_\alpha - E_{Fc})[1 - f(E_\beta - E_{Fv})] \\
&\quad - f(E_\beta - E_{Fv})[1 - f(E_\alpha - E_{Fc})] \} \\
&= v \int_E \varrho(E) dE [f(E_\alpha - E_{Fc}) - f(E_\beta - E_{Fv})] \quad .
\end{aligned}
\tag{5.2.34}
$$

The first equality in (5.2.34) defines the relationship between R_{stim} and $g(\omega)$, where v_g is the group velocity, I is the total number of photons in the mode in the modal volume v/Γ and $v = whL$ is the active volume. In the ideal case, $\varrho(E) \equiv 0$ for $E < E_g$, so that

$$
g(\omega) = 0 \quad \text{for} \quad \hbar\omega < E_g = eV_g \quad ,
\tag{5.2.35}
$$

although band-tail effects make it difficult to define E_g in practice. In addition, the factor in the bracket in the last equality in (5.2.34) taken with (5.2.30 and 33) shows that

$$
g(\omega) \leq 0 \quad \text{for} \quad \hbar\omega > eV_a \quad ,
\tag{5.2.36}
$$

i.e. the medium is lossy. Hence, the necessary condition for positive gain, first stated by Bernard and Durafforg, is

$$
E_{Fc} - E_{Fv} > E_g
\tag{5.2.37a}
$$

or

$$
V_a > V_g \quad ,
\tag{5.2.37b}
$$

i.e. the junction bias voltage must be greater than the energy-gap voltage. The condition (5.2.37) for positive gain in semiconductors is equivalent to the condition of population inversion of atomic levels for gain in gas lasers.

In the range

$$
eV_g < \hbar\omega < eV_a \quad ,
\tag{5.2.38}
$$

the gain is positive and has a peak that increases in magnitude and shifts toward higher frequency (energy), as N increases (Fig. 5.6). If ω_p is the frequency of the gain peak and the carrier density at which $g(\omega_p) = 0$,

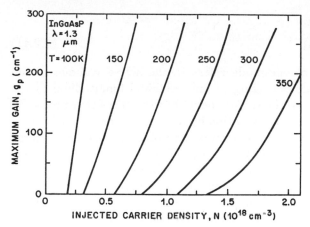

Fig. 5.10. Gain at the peak of the gain spectrum for InGaAsP at $1.3\,\mu$m versus current density illustrating their approximately linear relationship [5.14]

making the medium transparent at ω_p, is N_t, then

$$g(\omega_\mathrm{p}) \approx \overline{g}_0(N - N_t) \qquad (5.2.39)$$

where $\overline{g}_0$ is a *spatial gain constant* and the linear approximation is reasonably good for $N \approx N_t$ as illustrated in Fig. 5.10 [5.14]. Note that the transparency condition, $N = N_t$, corresponds to the difference in Fermi energies being equal to the band-gap energy. Typical values for $\overline{g}_0$ and N_t are 1.5×10^{-16} cm^2 and 1.5×10^{18} cm^{-3} for AlGaAs lasers, and 1.0×10^{-16} cm^2 and 1.0×10^{18} cm^{-3} for InGaAsP lasers. Since the oscillation condition (5.2.22) for lasers demands that Γg is just equal to the cavity losses, N is almost clamped at the laser threshold value determined by (5.2.39 and 22). In the usual case, N is near N_t and (5.2.39) holds.

5.2.4 Spontaneous Emission

As we noted in (5.2.29), electrons and holes will recombine spontaneously with N/τ_n transitions per unit volume per unit time, assuming purely radiative recombination. In terms of the Fermi statistics of Sect. 5.2.3, the spontaneous emission rate is proportional to the number of occupied conduction-band states times the number of empty valence-band states. The number of spontaneous photons per unit time R_spon is independent of the photon number and is proportional to (with the same proportionality factor as in (5.2.34)),

$$R_\mathrm{spon} \sim v \int_E \varrho(E)dE\{f(E_\alpha - E_{\mathrm{F}c})[1 - f(E_\beta - E_{\mathrm{F}v})]\} \quad . \qquad (5.2.40)$$

Einstein noted that in a closed system at equilibrium, the total number of downward transitions, due to stimulated and spontaneous emission, must equal the number of upward transitions, due to stimulated absorption; and furthermore that the average number of photons in *one mode* at equilibrium

227

be given by

$$I = [\exp(\hbar\omega/kT) - 1]^{-1} \quad . \tag{5.2.41}$$

Note that the condition of one mode refers to the active volume, i.e. an individual transverse mode belonging to an individual longitudinal mode. These two conditions lead to the requirement that the transition probabilities for stimulated emission and absorption be the same (Einstein coefficients $B_{21} = B_{12}$) and that the spontaneous emission probability is proportional to the stimulated emission probability (Einstein coefficients $A_{21} = B_{21}/v$) [Ref. 5.4, p. 49]. The latter constraint defines a relationship between spontaneous emission and gain that will become clear shortly.

The ratio of stimulated and spontaneous rates in (5.2.34 and 40) can be obtained by recognizing that the bracketed factors are independent of the integration over energy in the conduction and valence bands. Then

$$\frac{R_{\mathrm{stim}}}{R_{\mathrm{spon}}} = I\{1 - \exp[(\hbar\omega - eV_{\mathrm{a}})/kT]\} \quad , \tag{5.2.42}$$

where we have made use of (5.2.30) and (5.2.41). Since the spontaneous emission is always positive, we see that the net stimulated emission is positive when $eV_{\mathrm{a}} > \hbar\omega$, as specified by (5.2.38). The stimulated absorption coefficient $A(\omega)$ is equivalent to $-g(\omega)$ when the active region is unpumped, i.e., when $eV_{\mathrm{a}} < \hbar\omega$ and $(\hbar\omega - eV_{\mathrm{a}}) \gg kT$. Under these conditions (5.2.42 and 34) can be combined to give

$$\Gamma A(\omega) = -R_{\mathrm{stim}}/v_{\mathrm{g}}I = (R_{\mathrm{spon}}/v_{\mathrm{g}})\exp[(\hbar\omega - eV_{\mathrm{a}})/kT] \quad , \tag{5.2.43}$$

providing a relationship between the absorption spectrum $A(\omega)$ (see Fig. 5.6) and the spontaneous emission spectrum $R_{\mathrm{spon}}(\omega)$, which describes the emission from a light-emitting diode. Rewriting (5.2.43) as

$$R_{\mathrm{spon}}(\omega) = v_{\mathrm{g}}\Gamma A(\omega)\exp[(eV_{\mathrm{a}} - \hbar\omega)/kT] \tag{5.2.44}$$

and inserting (5.2.44 into 5.2.42) we obtain, using the definition of $R_{\mathrm{stim}} = v_{\mathrm{g}}\Gamma gI$,

$$\Gamma g(\omega) = R_{\mathrm{stim}}/v_{\mathrm{g}}I = \Gamma A(\omega)\{\exp[(eV_{\mathrm{a}} - \hbar\omega)/kT] - 1\} \quad . \tag{5.2.45}$$

Then, with (5.2.44 and 45), the total number of spontaneous photons *with a given polarization coupled into a single cavity mode per unit time* is given by [5.19, 5.15]

$$R_{\mathrm{spon}}(\omega)/v_{\mathrm{g}} = \Gamma\{g(\omega) + A(\omega)\} = \Gamma g(\omega)n_{\mathrm{sp}}(\omega) \quad , \tag{5.2.46}$$

with a spontaneous emission factor n_{sp} defined as

$$n_{\mathrm{sp}} \equiv \frac{R_{\mathrm{spon}}}{v_{\mathrm{g}}\Gamma g} = \left\{1 + \frac{A(\omega)}{g(\omega)}\right\} = \left\{1 - \exp\left[\frac{\hbar\omega - eV_{\mathrm{a}}}{kT}\right]\right\}^{-1} \quad , \tag{5.2.47a}$$

where $A(\omega)$ is the absorption associated with the band edge and should not be confused with α representing extraneous losses. For $eV_a \gg \hbar\omega$, $n_{sp} \to 1$ corresponding to complete inversion in a two-level laser; for $eV_a \gtrsim \hbar\omega$, $n_{sp} \gtrsim 1$ corresponding to incomplete inversion and excess spontaneous emission for a given gain; and, for $eV_a \lesssim \hbar\omega$, $n_{sp} < 0$ corresponding to no inversion. It is worth re-emphasizing the single-mode, single-polarization definitions of the parameters in (5.2.46 and 47). The spontaneous emission power generated in a laser amplifier is calculated in Sect. 5.7.3.

In the unpumped case, $V_a = 0$, n_{sp} is negative and $[\exp(\hbar\omega/kT) - 1]^{-1} = -n_{sp}$, which is the (positive) number of photons in an individual mode of the active volume at equilibrium as specified in (5.2.41). Also, $-v_g \Gamma g = -R_{stim}/I$ in (5.2.46 and 47a) represents the rate of absorption by the medium of photons in the mode. Then, for $V_a = 0$, (5.2.46 and 47a) merely state that the rate of spontaneous photon generation is just balanced by the rate of stimulated photon absorption, which assures that the system is in thermal equilibrium with its surroundings, i.e.

$$R_{spon} = v_g \Gamma g [\exp(\hbar\omega/kT) - 1]^{-1} \ , \quad V_a = 0 \ . \tag{5.2.47b}$$

Put another way, (5.2.47b) is a statement of the *fluctuation-dissipation* theorem of statistical mechanics or the Nyquist theorem of electrical engineering, namely that a dissipative medium in thermal equilibrium will be subject to a fluctuating force that radiates the thermal energy absorbed by the medium [5.20,21].

Equation (5.2.46) summarizes important relationships among $A(\omega)$ and $R_{spon}(\omega)$, which are readily measured quantities, and $g(\omega)$, which is not. Figure 5.11 [5.15] illustrates these relationships for an experimental 1.3 μm InGaAsP laser at threshold ($I_a = 80$ mA, solid curves) and below threshold ($I_a = 2.6$ mA, dashed curve). To obtain these curves, the spontaneous emission $R_{spon}(\omega)$ was first measured for different I_a. Then (5.2.44) was used to compute $A(\omega)$. It was found that $A(\omega)$ is reduced slightly as I_a increases but, above threshold, $A(\omega)$ is effectively clamped. The gain curve $g(\omega)$ was calculated using (5.2.45). The voltage V_a in these expressions can be measured as the voltage across the pn-junction (taking care to eliminate any voltage drop across the series resistance), or V_a can be estimated by trial-and-error to satisfy the requirement that the gain be a maximum at the laser energy, $E_L = \hbar\omega_L$, as indicated in the figure. Note that $g(\omega)$ abruptly drops to zero (and $n_{sp} \to \infty$) for $\hbar\omega = eV_g$ as expected from (5.2.35,47). Note also that at low energies $g(\omega)$ and $R_{spon}(\omega)/v_g\Gamma$ coincide as expected from (5.2.46,47) as $A(\omega) \to 0$ and $n_{sp} \to 1$. Typical derived values of n_{sp} at E_L are 2.6 [5.19] for AlGaAs lasers and 1.7 [5.15] for 1.3 μm InGaAsP lasers, indicating stronger inversion for the latter. Since the inversion is nearly clamped above threshold, n_{sp} is approximately constant in that regime.

Near the oscillation condition, $A(\omega)$ does not vary appreciably with ω, or with V_a, due to clamping. If we put $A(\omega_L) = A$ (constant), then (5.2.45)

229

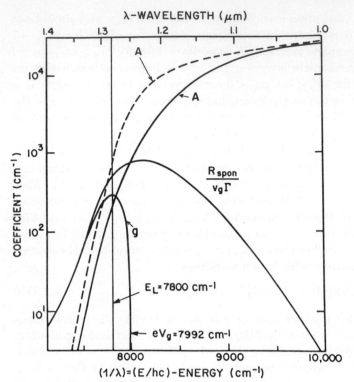

Fig. 5.11. Relationships for a 1.3 μm InGaAsP laser among experimental spontaneous emission spectra $R_{\text{spon}}(\omega)/v_g\Gamma$, and calculated absorption $A(\omega)$ and gain spectra $g(\omega)$. eV_a is adjusted to align the gain peak and laser energy $E_L = \hbar\omega_L$. Solid curves at threshold ($I_a = 80$ ma), dashed curve at $I_a = 2.6$ ma [5.15a,c]

becomes

$$g(\omega_L) = A\{\exp[(eV_a - \hbar\omega_L)/kT] - 1\} \qquad (5.2.48)$$

and (5.2.44) becomes

$$R_{\text{spon}}(\omega_L) = v_g\Gamma A\exp[(eV_a - \hbar\omega_L)/kT] \quad . \qquad (5.2.49)$$

Comparing (5.2.48, 49) with (5.2.39), we obtain

$$\Gamma\overline{g}_0 N_t = A \quad \text{and} \qquad (5.2.50)$$

$$R_{\text{spon}} = v_g\Gamma\overline{g}_0 N \quad . \qquad (5.2.51)$$

Thus the spontaneous emission coupled into the lasing mode is porportional to N. This spontaneous emission is usually written in the form

$$R_{\text{spon}} = \gamma\frac{Nv}{\tau_n} \qquad (5.2.52)$$

where τ_n is the spontaneous recombination lifetime, v is the volume of the

230

active region, and γ is the fraction of spontaneous photons that are coupled into the lasing mode.

If the laser oscillates close to the frequency of maximum gain, then $\omega \simeq \omega_p$ and the spontaneous emission factor can be obtained by inserting (5.2.39 and 51) into (5.2.47):

$$n_{sp} = \frac{N}{N - N_t} \quad . \tag{5.2.53}$$

This is consistent with the model of a two-level system such as in a gas laser, in which $n_{sp} \to 1$ for large N. In a semiconductor laser the inversion is not complete and $n_{sp} > 1$.

5.2.5 Photon Rate Equation

The total rate of change $\dot{I}$ of the photon number is obtained by subtracting the rate of cavity loss I/τ_p from the stimulated and spontaneous injection rates:

$$\dot{I} = R_{stim} + R_{spon} - \frac{I}{\tau_p} \quad . \tag{5.2.54}$$

This rate equation can be written as

$$\dot{I} = \Gamma g_0 (N - N_t)I + \gamma \frac{Nv}{\tau_n} - \frac{I}{\tau_p} \quad \text{where} \tag{5.2.55}$$

$$g_0 = v_g \bar{g}_0 \quad . \tag{5.2.56}$$

Equation (5.2.55) can be converted to a rate equation in the photon density S by dividing by the modal volume v/Γ :

$$\dot{S} = \Gamma g_0 (N - N_t)S + \Gamma \gamma \frac{N}{\tau_n} - \frac{S}{\tau_p} \quad . \tag{5.2.57}$$

A physical interpretation of this type of rate equation is given in Sect. 5.2.8.

5.2.6 Spectral Hole Burning

Spectral hole burning is a small localized reduction in gain at the energy corresponding to the wavelength of an oscillating mode. It is caused by a localized reduction in the number of occupied conduction-band states and empty valence band states. Each time an electron and hole recombine, the number of conduction-band states at this energy is temporarily reduced. The electron is replenished from the large pool of electrons in the conduction band by electron-electron and electron-phonon collisions. These scattering events do not occur instantaneously and so the net effect is a slight decrease in the average number of occupied states at the energy level in question. There are a number of different relaxation times that characterize the various scattering mechanisms, but the overall scattering rate can be approximated in terms of a single intra-band relaxation time τ_i on the order of 1.0 ps [5.22].

Considering just one oscillating mode, the reduction in gain due to spectral hole burning is roughly proportional to the net stimulated emission rate $\overline{g}_0(N - N_t)S$ and also to the intra-band relaxation time. Thus the gain reduction can be written as

$$\Delta g \approx \overline{g}_0(N - N_t)S\varepsilon \tag{5.2.58}$$

where ε is a parameter which is proportional to τ_i. With spectral hole burning included, the net gain becomes

$$\begin{aligned} g &= \overline{g}_0(N - N_t) - \Delta g \\ &= \overline{g}_0(N - N_t)(1 - \varepsilon S) \quad . \end{aligned} \tag{5.2.59}$$

This simple model of gain compression is reasonably accurate for small εS. For large εS the model can be generalized to include higher-order terms in S

$$g = \overline{g}_0(N - N_t)(1 - \varepsilon S - \varepsilon' S^2 - \varepsilon'' S^3 - \ldots) \quad . \tag{5.2.60}$$

If there are two or more modes oscillating simultaneously, then additional perturbations to the gain can arise due to beating between the modes. Small-signal variations in the electron density at the beat frequency give rise to frequency mixing terms due to the product of N and S in (5.2.57). These terms lead to a small component of gain at each mode, which is porportional to the photon density in another mode [5.23, 24]. The mixing of the beat-frequency component of carrier density with the central laser mode has a phase-modulation component which gives an in-phase boost to the upper-frequency longitudinal side-mode and a decrement to the lower side-mode. These asymmetrical gain perturbations can have a significant influence on the longitudinal-mode spectrum of multi-mode lasers [5.25].

5.2.7 Carrier Injection in a Heterojunction

The energy-band structure of a forward-biased $In_{1-x}Ga_xAs_yP_{1-y}$ heterojunction is illustrated in Fig. 5.12, where band-bending at the hetero-interfaces is ignored. The difference in band-gap energies between cladding and active layers, $E_{gs} - E_{gf}$, may be unequally divided between conduction and valence bands so that $\Delta E_c \neq \Delta E_v$. With the forward bias voltage V_a applied, electrons are injected into the active $In_{1-x}Ga_xAs_yP_{1-y}$ layer (with $x = x_f$), from the n-$In_{1-x}Ga_xAs_yP_{1-y}$ cladding (with $x = x_s$), and holes are injected from the p cladding. From (5.2.38), stimulated emission occurs if

$$eV_{gf} \leq \hbar\omega \leq eV_a \quad . \tag{5.2.61}$$

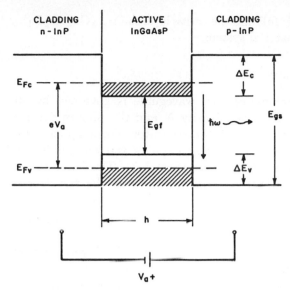

Fig. 5.12. Energy band structure at a forward biased heterojunction in InGaAsP [5.1]

The holes and electrons are well-confined to the active layer if both ΔE_v and ΔE_c are several times greater than kT. This condition is satisfied by making x_s sufficiently small compared with x_f.

For the material systems InGaAsP and AlGaAs, the refractive index n_f of the active layer will always be greater than that of the cladding when $E_{gf} < E_{gs}$. Thus the optical mode will also be confined to the neighborhood of the active region, although a fraction $(1 - \Gamma)$ of the mode will extend into the cladding regions. Nevertheless, only the fraction Γ of the mode within the active region will be subject to gain. Since, from (5.2.39) the gain g is proportional to the density $(N - N_t)$, the gain in the active layer is inversely proportional to its thickness h, for given injection current I_a or current density J above threshold. However, as h is reduced, the modal confinement Γ is also reduced so that at some value of h, the *modal* gain Γg reaches a maximum value.

A further limitation on gain versus h is incorporated by non-radiative recombination at the heterointerfaces. When the epitaxial layers are grown, the lattice constants may not match perfectly or impurities may be incorporated at the interface, leading to surface states that permit holes and electrons to recombine without producing stimulated radiation, thereby reducing the active layer gain. These interface effects become a more significant factor in reducing modal gain as the active layer thickness is decreased. Thus, the gain maximum, and consequently the laser threshold current density J_{th} minimum, will occur at larger h for poor epitaxial growth.

The preceding discussion for $In_{1-x}Ga_xAs_yP_{1-y}$ materials holds for $Al_yGa_{1-y}As$ materials provided $y_f < y_s$. In limiting cases, the active layer is binary GaAs and the cladding is $Al_yGa_{1-y}As$ with $y = y_s$ for the latter sys-

233

tem, and the cladding is binary InP and the active layer is $In_{1-x}Ga_xAs_yP_{1-y}$ with $y = y_f$, for the former material system.

5.2.8 Modal Rate Equations

The gain process in a semiconductor-laser waveguide is governed by an interaction between the excited carrier density N and the photon density S_m in each of the longitudinal modes of the resonator. For simplicity, we assume that only the fundamental transverse mode of the waveguide can propagate due to an index guided structure (Sect. 5.3). The longitudinal mode spacing is given by (5.2.18c) in terms of ω_s or as the wavelength spacing

$$\lambda_s = \lambda^2/2n_g L \quad , \tag{5.2.62}$$

where $n_g = c/v_g$ is the group refractive index in the active layer. Because L is the largest cavity dimension, λ_s is the smallest allowed modal wavelength increment. Below threshold, the number and relative strengths of these longitudinal modes is defined by a normalized distribution function $F(\lambda)$ proportional to the spontaneous emission spectrum. If $F(\lambda)$ has a Lorentzian shape with half-width at half-maximum $\Delta\Lambda$, then

$$F(\lambda_m) = \frac{1}{1 + [(\lambda_m - \lambda_0)/\Delta\Lambda]^2} \tag{5.2.63}$$

where the function peaks at λ_0 and λ_m is the wavelength of the mth mode. The normalization is such that

$$\int_{-\infty}^{\infty} F(\lambda)d\lambda = M\lambda_s \tag{5.2.64}$$

where M is the mean number of λ_s increments, or longitudinal modes, under $F(\lambda)$. If, for example, the distribution had been the rectangular function

$$\begin{aligned}
F'(\lambda) &= 1 \quad , \quad \lambda_0 - \Delta\Lambda' < \lambda < \lambda_0 + \Delta\Lambda' \\
&= 0 \quad \text{otherwise} \quad ,
\end{aligned} \tag{5.2.65}$$

then we would have had

$$M' = 2\Delta\Lambda'/\lambda_s \quad . \tag{5.2.66}$$

For the Lorentzian case, we have instead

$$M = \pi\Delta\Lambda/\lambda_s = 2\pi n_g L\Delta\Lambda/\lambda^2 \quad . \tag{5.2.67}$$

If the carrier density N, injected into the active layer, is independent of position within the layer thickness h (Fig. 5.1), and similarly the photon density S_m is constant within the active layer, then the time rates of change,

$\dot{N}$ and $\dot{S}_m$, are given by the *multimode rate equations* [5.26],

$$\dot{N} = \frac{J}{eh} - \frac{N}{\tau_n} - g_0(N - N_t)(1 - \varepsilon S_0)S_0$$
$$- \sum_{m \neq 0} g_m(N - N_t)(1 - \varepsilon S_m)S_m \qquad (5.2.68a)$$

$$\dot{S}_m = \Gamma g_m(N - N_t)(1 - \varepsilon S_m)S_m - \frac{S_m}{\tau_p} + \Gamma\gamma\frac{N}{\tau_n} \quad . \qquad (5.2.68b)$$

The first term in the carrier density rate equation (5.2.68a) gives the rate of carrier injection due to the current density J. The second term is the rate of loss of carriers due to spontaneous emission, which is proportional to N and the spontaneous emission rate τ_n^{-1}, which in general is a function of N. Since, as we note later, N is approximately clamped near its threshold value in laser operation, it is often a good approximation to take τ_n as a constant. The third term is the rate of reduction or increase in N due to stimulated emission or absorption, respectively, of the central mode. Gain compression is included in this term, but gain changes due to beating between modes are ignored. The summation term represents changes in N due to stimulated emission and absorption of the side modes. The coefficients g_m are proportional to the gain at λ_m.

The first term in the photon-density rate equation (5.2.68b) is the rate of increase or decrease of photon density in the active layer due to stimulated gain or absorption, respectively. Each carrier lost due to stimulated electron-hole recombination in (5.2.68a) appears as a photon in 85.2.68b), but, unlike the carriers, which are confined to the active layer, the photon is distributed according to the modal wavefunction such that only the fraction Γ appears in the active layer. Since we assume only the fundamental transverse mode, Γ is approximately independent of m. The second term in (5.2.68b) gives the rate of loss of photons, which is proportional to S_m and the inverse of the photon lifetime τ_p in the cavity. The last term gives the rate of increase of S_m due to spontaneous emission in the active layer. Here, γ is the fraction of the isotropically radiated spontaneous emission N/τ_s coupled into the mth longitudinal mode (Sects. 5.2.4, 5). Only the fraction Γ of the mode appears in the active region.

The factor γ is usually estimated [5.26] by counting the number of discrete electromagnetic normal modes in an ideal box with conducting walls and dimensions approximating the modal volume, whL/Γ, and within the spectral width of the emission. Then it is assumed that the spontaneous emission is distributed uniformly over all modes of the box, one of which is the mode m, to obtain the fraction coupled into that mode. The number of modes in the range λ to $\lambda + \lambda_s$ is calculated as

$$4\pi n^2 wh/\Gamma\lambda^2 \quad , \qquad (5.2.69)$$

where n is the phase refractive index, by counting the number of discrete standing waves in the box and allowing for two orthogonal polarizations. The number of modes in the wavelength range of the spectral distributions $F(\lambda)$ is found by multiplying the number of standing waves in λ_s in (5.2.69) by M, the mean number of λ_s increments, in (5.2.67). Then γ is the reciprocal of this total number of resonator modes [5.26]

$$\gamma = \frac{\Gamma \lambda^4}{8\pi^2 n^2 n_g whL\Delta\Lambda} \quad , \tag{5.2.70a}$$

where γ is the same for each mode λ_m near λ_0. For modes far removed from the peak, γ will decrease proportionally with $F(\lambda)$.

In the case of an actual single-transverse-mode laser, the count of available modes in volume whL is not very precise for the following reasons. (a) The cavity boundary conditions are determined by refractive index steps rather than conducting walls, resulting in a poorly defined resonator volume, especially in the transverse dimensions which are comparable in size with λ/n_f. (b) Some of the spontaneous emission goes into radiation modes which are not counted in (5.2.65). (c) In gain-guided or partially gain-guided lasers, the modal wavefront may no longer be planar. If this is the case, then a given laser longitudinal mode into which the spontaneous emission couples is actually a linear combination of K ideal cavity modes so that the effective γ in (5.2.65) must be multiplied by K, where $K = 1$ for an ideal index-guided laser and $K > 1$ for gain-guided lasers [5.27]. The factors (a) and (b) above can be represented by an increased cavity volume and, therefore, have the effect of reducing γ. If we take $\Gamma = 0.3$, $K = 1$, $\lambda = 1.5\,\mu m$, $n_g = 4$, $n = 3.5$, $\Delta\Lambda = 0.05\,\mu m$, $w = 2\,\mu m$, $h = 0.1\,\mu m$, $L = 250\,\mu m$, then (5.2.70a) gives $\gamma = 1.5 \times 10^{-4}$.

An alternative method for estimating γ comes directly from (5.2.51 and 52). Combining these with (5.2.56) yields

$$\gamma = \frac{\Gamma g_0 \tau_n}{whL} \quad . \tag{5.2.70b}$$

If we take $g_0 = 4.5 \times 10^{-12}\,s^{-1}\,m^3$, $\tau_n = 3\,ns$, then for the same dimensions as used above, (5.2.70b) gives $\gamma = 0.81 \times 10^{-4}$, which is in reasonable agreement with the value obtained from (5.2.70a). The ambiguity associated with γ can be avoided entirely by employing the quantities R_{stim}, R_{spon} and I, as in Sects. 5.2.4, 5, that refer to the total number of photons in a cavity mode.

In cases where all but the $m = 0$ mode are suppressed by one of the methods to be discussed in Sect. 5.4, the multimode equations (5.2.68) are reduced to *single-mode rate equations* simply by setting $S_m = 0$, $m \neq 0$:

$$\dot{N} = \frac{J}{eh} - \frac{N}{\tau_n} - g_0(N - N_t)(1 - \varepsilon S)S \quad , \tag{5.2.71a}$$

$$\dot{S} = \Gamma g_0(N - N_t)(1 - \varepsilon S)S - \frac{S}{\tau_p} + \Gamma\gamma\frac{N}{\tau_n} \quad . \tag{5.2.71b}$$

In these equations, S can also be taken as an approximation to the total photon density in a multimode laser if suitable effective values are used for γ, ε and g_0. The rate equations (5.2.71) will be used in Sect. 5.6 to explore the high-speed modulation behavior. The photon number rate equation (5.2.54) has the advantage over (5.2.71b) that it sidesteps the need to define the spontaneous coupling factor γ. If the spontaneous emission factor n_{sp} is known, R_{spon} in (5.2.54) can be obtained directly from (5.2.46). For this and other reasons, (5.2.54) will be used as a basis for the analysis of spectral linewidth in Sect. 5.5.

5.2.9 Longitudinal Variation of Photon Density

The distribution of photon density along the z axis of a laser cavity with end mirrors at $z = \pm L/2$ having power reflectivities R_2 and R_1, respectively, is given by the sum of two counter-propagating fluxes

$$S(z) = S_+(0)e^{(\Gamma g - \alpha)z} + S_-(0)e^{-(\Gamma g - \alpha)z} \quad , \tag{5.2.72}$$

where $(\Gamma g - \alpha)L \approx -\frac{1}{2}\ln R_1 R_2$ is determined by the oscillation condition (5.2.22). At the mirrors

$$S_-(L/2) = R_2 S_+(L/2) \quad , \tag{5.2.73a}$$

$$S_+(-L/2) = R_1 S_-(-L/2) \tag{5.2.73b}$$

and at the center

$$S_+(0) = (R_1/R_2)^{1/2} S_-(0) \quad . \tag{5.2.73c}$$

For the symmetrical case, $R_1 = R_2 = R$, the photon density is minimum at $z = 0$ and maximum at $z = +L/2$, with [5.26]

$$\frac{S(L/2)}{S(0)} = \frac{1}{2}R^{-1/2}(1 + R) \quad . \tag{5.2.74}$$

For $R = 0.3$, the above ratio is 1.19 so that there is significant variation in $S(z)$ in a typical laser. For a traveling wave amplifier in which $R_1 = R_2 \approx 10^{-3}$, the ratio is approximately 500 (Sect. 5.7.3). Furthermore, $S(z)$ represents only the envelope of the optical intensity. In general, there will be a standing wave with periodicty $\lambda/2n_g$.

The photon flux passing through each mirror in the general case is

$$F_1 = v_g(1 - R_2)S_+(L/2) \tag{5.2.75a}$$

$$\begin{aligned} F_1 &= v_g(1 - R_1)S_-(-L/2) \\ &= v_g(1 - R_1)(R_2/R_1)^{1/2}S_+(L/2) \quad . \end{aligned} \tag{5.2.75b}$$

237

The power out of each end is related to $F_{1,2}$ by [5.26]

$$P_{1,2} = (\hbar\omega)\left(\frac{wh}{\Gamma}\right)F_{1,2} \quad . \tag{5.2.76}$$

In the symmetrical case, (5.2.76) becomes

$$P_1 = P_2 = \frac{v_g}{2}(\hbar\omega)\left(\frac{wh}{\Gamma}\right)\frac{(1-R)}{R^{1/2}}S(0) \quad . \tag{5.2.77a}$$

An alternative way to determine P_1 and P_2 is to count the number of photons emitted from the mirror. In a period equal to the photon lifetime τ_p, each one of the I photons in the modal volume is lost by internal absorption or mirror losses. The number of photons lost through one mirror in this time is

$$I\frac{\alpha_m}{2(\alpha_m + \alpha)}$$

where $\alpha_m = L^{-1}\ln(1/R)$ is the mirror loss and $\alpha_m + \alpha$ is the total loss. Using (5.2.27b) for τ_p, the output power per facet (for $R_1 = R_2 = R$) becomes

$$P_1 = P_2 = \frac{v_g}{2}(\hbar\omega)\alpha_m I \quad \text{or} \tag{5.2.77b}$$

$$P_1 = P_2 = \frac{v_g}{2}(\hbar\omega)\left(\frac{wh}{\Gamma}\right)\ln(1/R)S \tag{5.2.77c}$$

where S is the average photon density in the modal volume. Equation (5.2.77c) is approximately the same as (5.2.77a) since

$$\ln(1/R) \approx \frac{(1-R)}{R^{1/2}}$$

for $R \approx 1$.

The ratio of powers from each mirror in the asymmetrical case is, from (5.2.76a)

$$\frac{P_1}{P_2} = \left(\frac{R_2}{R_1}\right)^{1/2}\left(\frac{1-R_1}{1-R_2}\right) = \frac{\eta_1}{\eta_2} \quad , \tag{5.2.78}$$

where η_1 and η_2 are the *external quantum efficiencies* for mirrors 1 and 2 (defined below). As indicated in (5.2.72), Eq. (5.2.78) holds only above threshold. The second equality holds since $P_{1,2}$ increases linearly with J above threshold and the external quantum efficiency $\eta_{1,2}$ is proportional to $P_{1,2}/(J - J_{th})$. A more rigorous derivation is given below. Equation (5.2.78) is sometimes a convenient expression for measuring an unknown reflectivity R_1 when the other reflectivity R_2 is known, as in the case of an anti-reflection (AR) coating applied to a cleaved cavity laser, and when the P versus J curve is near ideal (i.e., just above threshold). Another method for finding R_1 given R_2, that is better suited for $R_1 \to 0$, is given in Sect. 5.2.11.

The total number of stimulated photons generated per unit time in the resonator can be obtained by integrating $v_g g S(z)$ over the modal volume whL/Γ, with $S(z)$ given by (5.2.72). Above threshold, each additional electron injected into the active region produces a stimulated photon. If we allow for a fraction $(1 - \eta_\mathrm{I})$ of the external drive current I_d at the device terminals to leak around the active region or to be lost by nonradiative recombination and the remaining fraction η_I to pass through the active region, then the current I_a injected into the active region is $I_\mathrm{a} = \eta_\mathrm{I} I_\mathrm{d}$. The number of electrons lost or stimulated photons generated per unit time in the modal volume is

$$\frac{\eta_\mathrm{I}}{e}(I_\mathrm{d} - I_\mathrm{th}) = 2\Gamma g v_g \left(\frac{whL}{\Gamma}\right)\left(\ln\frac{1}{R_1 R_2}\right)^{-1}$$

$$\times (1 - \sqrt{R_1 R_2})\left(1 + \sqrt{\frac{R_1}{R_2}}\right) S_+(L/2) \tag{5.2.79}$$

where e is the electronic charge. The *external quantum efficiency* is defined as the ratio of the number of emitted photons *per facet*, as obtained by multiplying (5.2.75) by (wh/Γ), to the number of injected electrons per unit time:

$$\eta_1 = \eta_\mathrm{I}\frac{(1 - R_1)(R_2/R_1)^{1/2}\ln(1/R_1 R_2)}{2\Gamma g L[1 - (R_1 R_2)^{1/2}][1 + (R_2/R_1)^{1/2}]}, \tag{5.2.80a}$$

$$\eta_2 = \eta_\mathrm{I}\frac{(1 - R_2)\ln(1/R_1 R_2)}{2\Gamma g L[1 - (R_1 R_2)^{1/2}][1 + (R_2/R_1)^{1/2}]}. \tag{5.2.80b}$$

The ratio η_1/η_2 is as given in (5.2.78). For $R_1 = R_2 = R$, the total quantum efficiency for both facets is [5.15]

$$\eta_\mathrm{T} \equiv \eta_1 + \eta_2 = \eta_\mathrm{I}\frac{\ln 1/R}{\Gamma g L}. \tag{5.2.81a}$$

From (5.2.22c)

$$\eta_\mathrm{T} = \eta_\mathrm{I}\frac{\alpha_\mathrm{m}}{\alpha_\mathrm{m} + \alpha} \tag{5.2.81b}$$

$$= \eta_\mathrm{I} v_g \alpha_\mathrm{m} \tau_\mathrm{p} \tag{5.2.81c}$$

and, with $\Gamma g = \alpha_\mathrm{m} + \alpha$,

$$\eta_\mathrm{T} = \eta_\mathrm{I}\frac{\Gamma g - \alpha}{\Gamma g}. \tag{5.2.81d}$$

Equation (5.2.81a) shows that Γg can be calculated from a measurement of η_T. Equation (5.2.81b) illustrates that η_T is proportional to the ratio of the output coupling loss to the total loss, and that η_T may be improved by reducing both current leakage and the ratio of total losses to the coupling loss. From (5.2.77b) and (5.2.81c), the output power per facet can be written as

$$P = \frac{\eta_T I \hbar \omega}{2\eta_I \tau_p} = \frac{\eta_T S \hbar \omega w h L}{2\eta_I \tau_p \Gamma} . \tag{5.2.82}$$

When only one output is required, η_1 can be increased (and η_T reduced) by placing a highly-reflective coating on facet 2 such that $R_2 \approx 1$, $R_1 = R$. Then

$$\eta_T = \eta_1 = \eta_I \frac{\ln 1/R}{2\Gamma g L} = \eta_I \frac{\alpha_m}{2(\alpha_m + \alpha)} \quad \text{and} \tag{5.2.83a}$$

$$\eta_2 = 0 . \tag{5.2.83b}$$

If $\eta_1(RR)$ is the quantum efficiency per facet for $R_1 = R_2 = R$ and $\eta_1(R)$ is the efficiency for the output facet for $R_1 = R$, $R_2 = 1$, then

$$\eta_1(R) = \frac{\eta_1(RR)}{1 - \eta_1(RR)} . \tag{5.2.84}$$

5.2.10 Steady-State Solution of Rate Equations

The rate equations (5.2.68, 71) can be employed to analyze the steady-state behavior of a laser near threshold [5.28, 29]. In the steady state, $\dot{N} = \dot{S} = 0$. We consider the single-mode case (5.2.71), for simplicity, and obtain

$$N = \frac{S/\Gamma g_0 \tau_p + N_t(1 - \varepsilon S)S}{(1 - \varepsilon S)S + \gamma/g_0 \tau_n} \tag{5.2.85}$$

$$S = \Gamma \tau_p \left[\frac{J}{eh} - \frac{N}{\tau_n}(1 - \gamma) \right] . \tag{5.2.86}$$

Combining (5.2.85 and 86) yields

$$S = \frac{\Gamma \tau_p J}{eh} - \frac{1}{g_0 \tau_n} \left[\frac{1 + \Gamma g_0 N_t \tau_p(1 - \varepsilon S)}{(1 - \varepsilon S) + \gamma/S g_0 \tau_n} \right] (1 - \gamma) . \tag{5.2.87}$$

Above threshold, the $\gamma/S g_0 \tau_n$ term is small. If gain compression is also small ($\varepsilon S \ll 1$) then (5.2.87) becomes for $J > J_{th}$

$$S \cong \frac{\Gamma \tau_p}{eh}(J - J_{th}) \quad \text{where} \tag{5.2.88}$$

$$J_{th} = \frac{N_{th} eh}{\tau_n} \tag{5.2.89a}$$

240

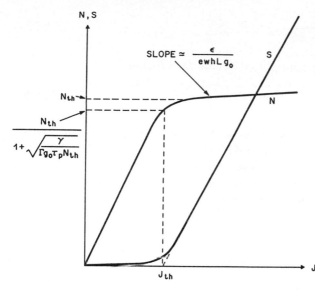

is the threshold current density and

$$N_{\text{th}} = (N_t + 1/\Gamma g_0 \tau_{\text{p}})(1 - \gamma) \tag{5.2.89b}$$

is the nominal carrier density at threshold. The photon density versus drive current density characteristic is shown in Fig. 5.13. The approximate characteristic (5.2.88) is a straight line which intercepts the horizontal axis at $J - J_{\text{th}}$, while the full characteristic (5.2.87) has a knee in this region. The two curves are indistinguishable for J more than about 10 % above threshold. If the expression for S is converted to output power by (5.2.77) and J to drive current I_{d}, (5.2.88) gives an idealized power versus current curve

$$P = 0 \quad , \quad I_{\text{d}} < I_{\text{th}} \quad , \tag{5.2.90a}$$

$$P = \zeta(I_{\text{d}} - I_{\text{th}}) \quad , \quad I_{\text{d}} > I_{\text{th}} \tag{5.2.90b}$$

with *slope efficiency*

$$\zeta = \left. \frac{dP}{dI_{\text{d}}} \right|_{I_{\text{d}} > I_{\text{th}}} = \frac{\eta_{\text{I}} v_{\text{g}}(1 - R)\hbar\omega}{2R^{1/2} e \Gamma g_0 (N_{\text{th}} - N_t)L} \tag{5.2.91a}$$

$$= \frac{\eta_{\text{T}} \hbar\omega}{2e} \quad . \tag{5.2.91b}$$

Equation (5.2.85) can be used to demonstrate that the electron density is not completely clamped above threshold. Putting $S \to \infty$ in (5.2.85) yields, for γ small,

$$N = N_t + \frac{1}{\Gamma g_0 \tau_p (1 - \varepsilon S)} \quad . \tag{5.2.92a}$$

With $\varepsilon S \ll 1$, this can be approximated (for $I_d > I_{th}$) as

$$N \cong N_{th} + \frac{\varepsilon S}{\Gamma g_0 \tau_p} \quad . \tag{5.2.92b}$$

The first term in (5.2.92b) is constant, but the second term increases with S. Complete clamping of N occurs only if $\varepsilon = 0$. Figure 5.13 also shows N plotted against the current density J. Note that the carrier density at threshold ($J = J_{th}$) is slightly less than the nominal value N_{th}. From (5.2.89a) in (5.2.87 and 85) it can be shown that the actual carrier density at threshold is

$$N|_{J=J_{th}} = \frac{N_{th}}{1 + \sqrt{\gamma / \Gamma g_0 \tau_p N_{th}}} \quad . \tag{5.2.93}$$

For small S (below threshold), $S \ll \gamma / \tau_n g_0$, and (5.2.85 and 86) reduce to

$$N = \frac{g_0 N_{th} \tau_n}{(1 - \gamma)\gamma} S \quad , \tag{5.2.94a}$$

$$\frac{J}{eh} = \left(\frac{N_{th} g_0}{\gamma} + \frac{1}{\Gamma \tau_p} \right) S \quad , \tag{5.2.94b}$$

indicating that the density of spontaneous photons coupled into the laser mode increases directly with J and the device acts like a light emitting diode (LED). Although we will see later that g_0 can be eliminated from these expressions, stimulated emission does contribute to output in the LED regime, i.e. below threshold.

With (5.2.89b and 93) substituted into (5.2.53) and for small γ and ε, the spontaneous emission factor n_{sp} becomes

$$n_{sp} = 1 + \Gamma g_0 N_t \tau_p \quad . \tag{5.2.95}$$

The spectral behavior of a *multimode* laser can be explained approximately in terms of the steady-state photon rate equation. With $\dot{S}_m = 0$ in (5.2.68b) we obtain

$$S_m = \frac{\gamma \Gamma N / \tau_n}{1/\tau_p - \Gamma g_m (N - N_t)(1 - \varepsilon S_m)} \quad , \tag{5.2.96}$$

where, as explained earlier, the gain factor g_m is maximum for the central mode and decreases for $|m| > 0$, following a nearly parabolic gain curve; but the loss term $1/\tau_p$ in the denominator is constant. For the central mode, the modal gain term $\Gamma g_m (N - N_t)(1 - \varepsilon S_0)$ is almost equal to the loss term $1/\tau_p$, and S_0 is large. Since g_m is smaller for the side modes, the amplitudes of the side modes are less than the central mode. For a given

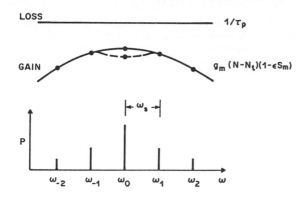

Fig. 5.14. Loss and gain spectra, including spectral hole burning. Laser longitudinal-mode spectrum is also shown

value of S_0, this discrimination against the side modes is greater when the spontaneous emission factor in the numerator is less and also when the mode separation is larger, as for L small [5.26]. The situation is illustrated in Fig. 5.14 where the loss and gain curves are plotted against frequency. For the case $S_0 = 1$, the spacing between the curves at ω_0 is just the spontaneous emission factor in the numerator. When S_0 is large, the gain compression term $(1 - \varepsilon S_0)$ due to spectral hole burning reduces the gain for S_0, as indicated by the dashed curve, and allows other modes to achieve comparable photon densities. Thus, efforts to reduce γ and L are not always successful in achieving single longitudinal-mode operation.

5.2.11 Measurement of Modal Reflectivity and Laser Gain

When a laser is operated below threshold, the output spectrum consists of the spontaneous emission modulated by the Fabry-Perot resonances due to the end mirrors. The measured modulation depth can be employed to calculate the below-threshold laser gain [5.30] or the modal reflectivity of an end mirror [5.31].

Consider the laser structure shown in Fig. 5.5. Assume that a spontaneous emission event at $z = L/2$ produces an electric field $E(0)$ that is reflected at $L/2$, travels to $-L/2$, is reflected, and returns to $L/2$. The power transmitted through $R_2 \ (= r_2^2)$ after an infinite number of bounces is $|E(0)|^2 (1 - R_2)|1 - \hat{a}|^{-2}$, where

$$\hat{a} = \sqrt{R_1 R_2} \exp[(\Gamma g - \alpha)L + \mathrm{j}2\beta L] \quad . \tag{5.2.97}$$

If we integrate the effect of spontaneous emission events over $-L/2 \le z \le L/2$, then the output power is

$$P = \frac{P'(1 - R_2)}{|1 - \hat{a}|^2} \quad , \tag{5.2.98}$$

where P' is the spontaneous emission integrated over the active volume. The spectrum $P(\lambda)$ has maxima and minima due to the phase factor in (5.2.97).

243

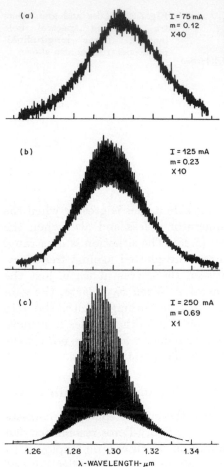

(a) I = 75 mA
m = 0.12
X40

(b) I = 125 mA
m = 0.23
X10

(c) I = 250 mA
m = 0.69
X1

1.26   1.28   1.30   1.32   1.34
λ-WAVELENGTH-μm

Fig. 5.15. Modulated spontaneous emission spectra from an antireflection-coated laser operating below threshold as an edge emitting LED [5.31]

The modulation index m of the output spectrum, which is independent of P', is

$$m = \frac{P_{\max} - P_{\min}}{P_{\max} + P_{\min}} \quad . \tag{5.2.99}$$

A modulated spectrum is shown in Fig. 5.15. $P_{\max}$ and $P_{\min}$ are most easily measured near the peak of the spectral envelope. Using (5.2.98), we find

$$m = \frac{2|\hat{a}|}{1 + |\hat{a}|^2} \tag{5.2.100a}$$

$$\approx 2|\hat{a}| \quad , \quad |\hat{a}| \ll 1 \quad . \tag{5.2.100b}$$

Thus, measurement of m also yields a measurement of $|\hat{a}|$.

If we have a symmetric laser with cleaved mirror reflectivities $R_1 = R_2 = R_l$, then just below threshold, $I_d \lesssim I_{th}$, $|\hat{a}| \approx 1$, and, from (5.2.97),

$$R_l^2 = \exp[-2(\Gamma g - \alpha)L] \quad , \quad I_d \lesssim I_{th} \quad . \tag{5.2.101}$$

If, now, the mirrors are coated so that the reflectivities are R_1 and R_2, then the new $|\hat{a}|$ is given by (5.2.97). With a drive current I_d equal to I_{th} for the uncoated laser, (5.2.101) holds and

$$R_1 R_2 = (|\hat{a}| R_l)^2 \quad , \quad I_d \lesssim I_{th} \quad . \tag{5.2.102}$$

Thus, a measurement of $|\hat{a}|$ for the coated laser using the modulated spontaneous spectrum at $I_d \lesssim I_{th}$, yields $R_1 R_2$ when R_l is known for the cleaved laser. The reflectivity R_2 can be determined if (a) $R_1 = R_l$ (uncoated), (b) $R_1 = R_2$ (both coated) or (c) $R_1 \approx 1$ (high reflection coating). If $R_2 \ll 1$, then (c) will give the most accurate result and (b) the least since it is difficult to obtain a reliable measurement for $m < 0.1$.

Gain clamping near threshold has been neglected in the combination of (5.2.101 and 102). For the usual case in which $R_1 R_2 \ll R_l^2$, the threshold of the coated laser is much larger than I_{th} for the original laser, so that the gain factor at I_{th} is unclamped for the measured $|\hat{a}|$ in (5.2.102). On the other hand, the gain factor in (5.2.101) occurs near threshold where the photon density may be substantial so that g at I_{th} may be smaller in (5.2.101) than in (5.2.102), and

$$R_1 R_2 = (f|\hat{a}| R_l)^2 \tag{5.2.103}$$

where f is a factor larger than unity. Thus, neglecting clamping effects gives a value $R_1 R_2$ smaller than the true value. Nevertheless, we expect that, at least for single-mode lasers in which γ is small, the photon density below threshold as it appears in (5.2.94) will be very small (see also [5.28]). Then clamping can be ignored to the precision realizable in the measurement of m.

The above method is especially suitable for low $R_1 R_2$ because the coated device has a very high threshold current. Heating effects may make it impossible to use the method discussed in connection with (5.2.78), which requires operation above threshold of the coated device. A third method for measuring facet reflectivity by measuring the finesse of an external cavity consisting of the laser facet and an optical fiber facet is described in [5.32].

5.3 Structures for Transverse-Mode Control

The lasers we have considered so far are two-dimensional (planar) structures of the type shown in Figs. 5.1, 3, where the field is assumed to be uniform in the y direction. In this section we describe lasers with mode control in the y transverse direction. A large number of laser structures providing transverse-mode control have been proposed for both the GaAlAs and In-GaAsP material systems. After a brief discussion of gain guiding, we will be concerned with a few representative structures that afford fundamental transverse-mode operation by virtue of a built-in optical waveguide or index-guiding. The various epitaxial growth methods available for preparing the wafer and overgrowing (where necessary) have been reviewed in [Ref. 5.5., Pt. A]. Some additional structures have been discussed in [5.33, 35].

5.3.1 Stripe Geometry Laser, Blocking Layer

The broad-area-contact laser shown in Fig. 5.1 is the simplest laser structure. The simplest structure that provides optical field confinement in the transverse direction is the stripe geometry laser, shown in Fig. 5.16. Confinement occurs because gain is provided only under the narrow stripe contact. Typical dimensions are length: 200–400 μm, width: 300–500 μm, stripe width W : 5–10 μm, active layer thickness h : 0.1–0.2 μm. The mirror facets are usually cleaved and the width of the chip defined by sawing. The insulating layer of Si_3N_4 or SiO_2 that defines the stripe is typically 0.1 to 0.2 μm thick may cover most of the area of the chip. The contact above this dielectric layer gives rise to a large shunt capacitance which, together with the contact resistance and bulk resistance from stripe to active layer, may limit the modulation response (Sect. 5.6). A mesa may be etched near the contact stripe to reduce the capacitance.

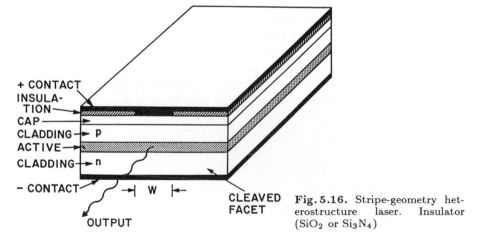

Fig. 5.16. Stripe-geometry heterostructure laser. Insulator (SiO_2 or Si_3N_4)

The current from the contact stripe spreads to a width greater than W due to the thickness of the cladding layer ($\sim 2\,\mu$m) and the high conductivity of the cap layer. (The heavily-doped cap layer is used to facilitate a low resistance ohmic contact.) Therefore, the gain may be positive but poorly defined over the excited width of the active layer. The modal properties in the junction plane are also poorly defined and the output is generally multimode in the longitudinal and transverse directions with a curved wavefront at the mirror facet. Such a laser is said to be *gain-guided* in that the transverse extent of the beam is defined solely by the injected current. Gain-guided lasers are less efficient than devices in which the injected current, the optical mode and the excited carriers are all confined to the same narrow, pencil-like region. Furthermore, transverse confinement of the optical beam can assure a fundamental transverse mode, in the plane of the junction and normal to it, for efficient coupling of the laser into a single-mode fiber. We will discuss laser structures for realizing this three-way confinement in the next subsection. First, we mention a method for achieving better current confinement than that from a narrow contact stripe.

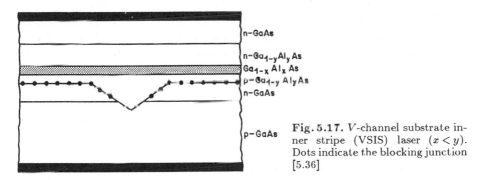

n–GaAs

n–Ga$_{1-y}$Al$_y$As
Ga$_{1-x}$Al$_x$As
p–Ga$_{1-y}$Al$_y$As
n–GaAs

p–GaAs

Fig. 5.17. V-channel substrate inner stripe (VSIS) laser ($x < y$). Dots indicate the blocking junction [5.36]

The cross-section of a VSIS (*V*-channel substrate inner stripe) laser [5.36] is shown in Fig. 5.17. Unlike most other lasers described here, this laser is made on a *p*-GaAs substrate, despite the fact that *n*-type substrates are more readily available, because it is thought that the epitaxial growth is of better quality and the stripe *n*-type ohmic contact has lower resistance. The active Ga$_{1-x}$Al$_x$As layer has a higher refractive index and lower bandgap energy than the Ga$_{1-y}$Al$_y$As cladding layers ($x < y$). The *n*-GaAs *blocking layer* serves to confine the current to the $\sim 2\,\mu$m-wide V-channel aperture by introducing a reverse-biased *pn*-junction (indicated by dots in Fig. 5.17 and subsequent structure diagrams) that is within $\sim 0.6\,\mu$m of the active layer. Thus, only a narrow current stripe is injected into the active layer, giving low threshold current and tending to limit the lateral extent of the transverse mode. Further mode confinement is provided indirectly because the evanescent field is only weakly guided by the 0.08 μm-thin active layer and, therefore, penetrates into the GaAs blocking layer. Since GaAs is lossy at the laser wavelength ($\sim 720\,\mathrm{nm}$) fields outside the V-channel are absorbed, and, since GaAs has a higher refractive index than the cladding $\mathrm{Ga}_{1-y}\mathrm{Al}_y\mathrm{As}$, there is an (anti-guiding) index discontinuity at the channel boundary that partially reflects light back into the channel region. These lasers generally operate with a near-fundamental transverse mode and are widely used in optical compact disc players.

5.3.2 Buried Heterostructure Lasers

The three-way confinement of current, photons and carriers in both transverse dimensions mentioned above can be realized in several buried heterostructure (BH) designs. Lasers that incorporate a waveguide to confine photons to a narrow strip within the junction plane are said to be *index-guided*, as opposed to gain-guided lasers. The techniques involve finding processing steps compatible with the epitaxial growth method to achieve an active core region of higher index and smaller bandgap energy surrounded on all sides by a cladding region. The conduction and valence band steps ΔE_c and ΔE_v should each be several times kT to confine the carriers.

If the effective index step between core and cladding regions in the plane of the active layer is Δn_eff, then the active region width for fundamental mode operation is limited by [see (5.2.12) and Chap. 2]

$$W \le \frac{\lambda}{2(2n\Delta n_\mathrm{eff})^{1/2}} \;. \tag{5.3.1}$$

In practice, a fundamental mode laser will result even with an actual width several times larger than the value given by (5.3.1), since ($\Gamma g - \alpha$) is lower for the higher-order transverse modes.

The InGaAsP etched-mesa buried heterostructure (EMBH) laser [5.33, 37], shown in Fig. 5.18, can be fabricated using liquid phase epitaxy (LPE) [Ref. 5.5, Pt. A]. A planar wafer containing the active layer surrounded by p- and n-InP cladding layers is etched to form a mesa having width W at

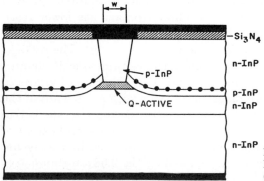

Fig. 5.18.
Etched-mesa buried heterostructure (EMBH) laser. Q =InGaAsP

the active layer. Additional p- and n-InP layers are regrown over the mesa to form the lateral cladding and a current-blocking layer.

A typical mesa width for a fundamental-transverse-mode laser is 1.5 μm. If the active region is 1.3 μm-bandgap InGaAsP material and the cladding is InP, then the core-cladding index difference is 0.4, from Fig. 5.4. If the active layer thickness is 0.15 μm, then $V \approx 1$ and $\Gamma \approx 0.3$ for the planar waveguide, using Fig. 5.8. A rough estimate of Δn_{eff} for the effective lateral index step is $\Delta n_{\text{eff}} \approx 0.3 \times 0.4 = 0.12$, using Γ and the planar index step, and (5.3.1) gives an ideal $W \leq 0.75$ μm. On the other hand, the effective V-value in the lateral dimension for the actual $W = 1.5$ μm is, see (5.2.12),

$$V_{\text{eff}} = \frac{2\pi}{\lambda} W (2n \Delta n_{\text{eff}})^{1/2} \approx 6.4 \quad , \tag{5.3.2}$$

indicating, see (5.2.13), that 3 transverse modes could be supported by such a guide, although in practice only the fundamental mode has sufficient gain to oscillate.

Several critical processing steps make it difficult to produce the EMBH laser with good yields: (a) The etched width of the mesa must be carefully controlled to assure fundamental mode operation, (b) the regrowth over the mesa must not introduce substantial non-radiative recombination centers at the edge of the mesa, (c) the position of the blocking junction (again indicated by dots in Fig. 5.18) must be nearly coincident with the active layer to avoid current leakage [5.33] through forward-biased pn junctions near the active region, and (d) the regrown layers should be uniform in thickness in order to avoid shorts or leaks outside the active region and to provide a planar top surface for further processing.

The double-channel, planar buried heterostructure (DCPBH) laser shown in Fig. 5.19 relieves some of these fabrication difficulties. The structure provides InGaAsP lasers of superior performance in terms of high output power, high operating temperature and low threshold current at 1.3 μm [5.33, 38] and 1.55 μm [5.33, 39] using LPE growth and regrowth. Rather

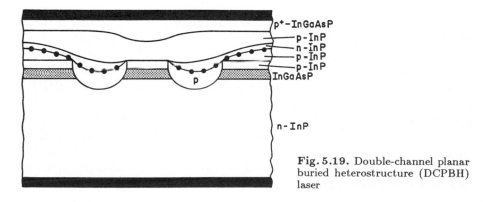

Fig. 5.19. Double-channel planar buried heterostructure (DCPBH) laser

249

than form a mesa by etching a wide-area plain as in the EMBH laser, two narrow ($\sim 10\,\mu$m-wide) channels define the active region in the planar epitaxial wafer. Then the regrowth needs to nucleate only in a narrow region adjacent to the active region, giving good reproducibility and quality. The outlying regions remain nearly planar, and the leakage current through the pnp transistor in this region is low, even at high drive currents.

The blocking junctions can be eliminated altogether in the EMBH and DCPBH structures if Fe-doped semi-insulating (SI) InP is deposited in the regrowth step, by VPE or MOVPE [5.40]. This can then be followed by a highly conducting p-InGaAsP cap layer. The reduced leakage and reduced capacitance, due to the thick SI layer in place of a thin Si_3N_4 layer, allow high-power and high-speed operation.

The channel-substrate buried heterostructure (CSBH) (or V-groove) laser [5.33, 34, 41] shown in Fig. 5.20 is less critical to process than the preceding devices but may have poorer leakage current characteristics. Selective chemical etching is used to define an accurate, self-limiting V-groove with (111A) crystal facets in the substrate containing a p-InP epitaxial layer on an n-InP substrate. Successive regrowth steps by LPE fill the V groove and form a lens-like, buried active region with surrounding blocking junction. The top surface is reasonably planar. However, a rather large forward-biased leakage path remains at the mouth of the V groove.

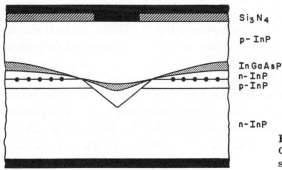

Si$_3$N$_4$

p- InP

InGaAsP
n- InP
p- InP

n- InP

Fig. 5.20.
Channel-substrate buried heterostructure (CSBH) laser

The three preceding BH structures were originally based on LPE growth. However, the planar epitaxial material can be made by any method, such as molecular beam epitaxy (MBE), vapor phase epitaxy (VPE) or organometallic vapor phase epitaxy (OMVPE). But the regrowth steps have usually been restricted to LPE. The constricted-mesa buried heterostructure (CMBH) laser shown in Fig. 5.21 lends itself to vapor phase regrowth and can also provide reduced shunt capacitance. A mesa, about $10\,\mu$m wide, is etched into a planar wafer by means of side channels. A selective chemical etch is then used to undercut the active layer to a $1.5\,\mu$m width. A variety of schemes have been employed to fill in the open regions, including mass transport of InP from the surroundings [5.42], vapor phase regrowth of InP [5.43], or

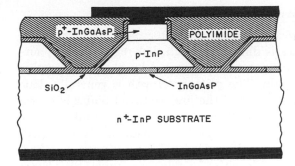

Fig. 5.21.
Constricted-mesa buried heterostructure (CMBH) laser [5.44]

vapor deposition of SiO₂ followed by polyimide [5.44]. The advantages of the latter structure are the much reduced shunt capacitance due to the thick polyimide dielectric, the current confinement by SiO₂ without capacitative reverse-bias junctions, and the possibility of reduced width of the active region due to large lateral index step. These devices have yielded exceptionally high-speed modulation characteristics (Sect. 5.5). In those devices in which the regions laterally adjacent to the active stripe are InP, the current confinement is by virtue of the higher resistance of the path through the larger bandgap-energy pn-junction.

Some of these BH structures have been grown on SI *substrates* in order to reduce the shunt capacitance and also allow the possibility of an *integrated optoelectronic circuit* containing a field effect transistor (FET) on the same chip [5.28]. The three-channel buried crescent (TCBC) laser shown in Fig. 5.22 is an example of such a device [5.45]. The buried crescent region is formed in the three-channel substrate in a manner similar to the CSBH laser. The injection current flows from the upper n contact, through the n-InP, InGaAsP crescent, and p-InP to the Zn-diffused area that reaches the lower corner of the p InP. This constricted current path limits the effective width of the Si₃N₄ capacitor, thereby reducing the parasitic capacitance. Current leakage is limited by the SI channels. An FET circuit can be added to the SI substrate.

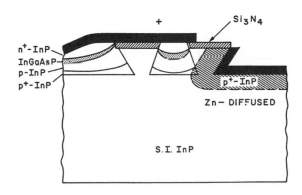

Fig. 5.22. Three-channel buried crescent (TCBC) laser [5.45]

5.3.3 Ridge Waveguide Lasers

BH lasers are difficult to fabricate because of the regrowth step. Epitaxial growth on a non-planar surface is difficult enough, but regrowth on a surface that has been exposed to air and contaminated by photolithographic processing is even more uncertain. The primary growth is generally done on an etched and cleaned substrate, and the first epitaxial buffer is grown thick enough to reach equilibrium; these precautions cannot be taken in the regrowth.Ridge-loading techniques can be used to provide lateral waveguiding without regrowth. However, lateral carrier confinement is also sacrified. Nevertheless, the threshold current densities for ridge waveguide (RW) lasers are similar to the current densities for BH lasers, presumably because non-radiative recombination at the regrown interfaces and current leakage in the BH devices compensate for the lack of good carrier confinement.

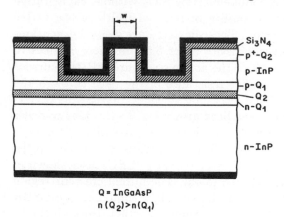

Fig. 5.23. Ridge waveguide (RW) laser [5.47]

Ridge-waveguide lasers were first described in GaAlAs [5.46]. The In-GaAsP system allows a variation in the RW structure that is especially simple to fabricate employing a stop-etch technique. This structure, which is shown in Fig. 5.23 [5.47–49], was used to demonstrate the first cw, fundamental mode laser at 1.5 μm [5.50]. The starting wafer contains an InGaAsP active layer about 0.15 μm thick surrounded by p- and n-cladding layers. In the symmetrical case, these are both of lower refractive index InGaAsP; and, in the asymmetrical case, the p-layer is lower-index InGaAsP and the n-layer is PInGaAsP. The p-cladding is about 0.15 μm-thick InGaAsP. The next p-layer is 2 μm-thick InP, which is followed by a heavily-doped p^+-InGaAsP cap layer.A ridge, 3 to 5 μm-wide, is generated by etching two 10 μm-wide channels. A selective chemical etch for InP that automatically stops at the top of the p-InGaAsP cladding layer and simultaneously gives a vertical $(1\bar{1}0)$ ridge wall is used. Contacts are applied on the top of the ridge through a window in the Si_3N_4 insulating layer. Since the ridge confines the current to within about 0.15 μm of the active layer, the current spreading is negligible and no blocking junction is needed.

The lateral waveguiding occurs because the planar optical wavefunction under the ridge has a larger guide index than the wavefunction in the channel region, the reason being that the evanescent tail of the wavefunction in the former region extends into the p-InP and, in the latter, it extends into the lower index Si_3N_4. The effective lateral index step can be estimated from the curves in Chap. 2 for the two types of planar guide by the equivalent index method [Chap. 2, 5.47]. Typically, the lateral index step is on the order of ~ 0.02, which gives a V value of 8 at $1.5\,\mu m$ for a $5\,\mu m$-wide ridge. The nominal number of allowed transverse modes is then 3, although only the fundamental mode has sufficient gain to oscillate even at twice the threshold current. Since the effective lateral guiding step is of the same order of magnitude as the reduction in index due to current injection, one might expect even less lateral guiding and lower V value at threshold. Indeed, one may wonder if the index-guiding disappears above threshold giving rise to a gain-guided laser. Nevertheless, measurements of the output beam profile show that the longitudinal beam waist is at the mirror, indicating index-guiding, rather than inside the laser, which would indicate gain-guiding [5.47]. A factor that may help maintain the index step is some spreading of current through the cladding layer just outside the ridge width.

The weak lateral guiding allows an active width about three times that of the BH devices, which means that tolerances are reduced and also the contact and bulk series resistances are smaller compared to BH lasers. The smaller resultant RC time constant reduces parasitic rolloff at higher modulation frequencies but the increased active width lowers the maximum achievable modulation bandwidth for a given output power (Sect. 5.6).

Recent reports on RW lasers with narrow ridges, between 1.5 and $3.5\,\mu m$ wide, indicate thresholds comparable to BH structures at $1.3\,\mu m$ [5.51] and at $1.5\,\mu m$ [5.52]. At $\lambda = 1.5\,\mu m$, a cw threshold of $18\,mA$ at $20°$ C was reported for a ridge width of $3.5\,\mu m$. These lasers operated for over 10,000 h at $50°$ C with an output of $5\,mW$. The absence of lateral carrier confinement in the RW laser is apparently balanced by non-radiative recombination at the edges of the BH active layer and current leakage around the blocking layer.

The hetero-epitaxial ridge-overgrown (HRO) laser [5.53, 54] shown in Fig. 5.24 is a ridge-waveguide laser in which the ridge is grown during a second LPE growth, rather than being etched on the final epitaxial wafer. To fabricate the HRO laser, a SiO_2 layer is deposited on a wafer containing the InGaAsP active layer and an upper InGaAsP cladding layer. Window stripes, 3 to $5\,\mu m$ wide, are opened in the SiO_2 and p-InP is grown over the wafer by LPE. Nucleation takes place only in the window, and the InP grows rapidly in thickness and mushrooms out over the SiO_2, forming a ridge several μm thick. An InGaAsP contact layer may be added and the surface covered with metallization. The SiO_2 window provides current confinement self-aligned with the ridge.

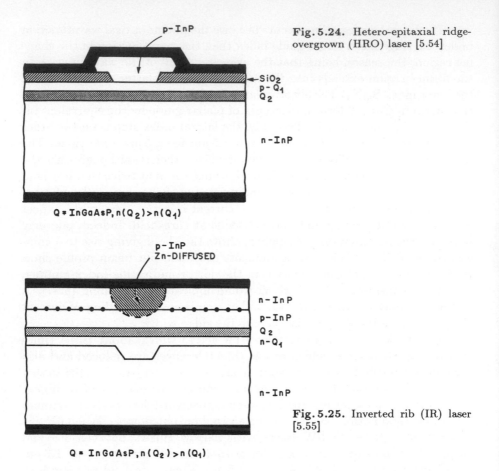

p-InP

Fig. 5.24. Hetero-epitaxial ridge-overgrown (HRO) laser [5.54]

←SiO₂
p- Q₁
Q₂

n-InP

Q = InGaAsP, n(Q₂)>n(Q₁)

p-InP
Zn-DIFFUSED

n-InP
p-InP
Q₂
n-Q₁

n-InP

Fig. 5.25. Inverted rib (IR) laser [5.55]

Q = InGaAsP, n(Q₂)>n(Q₁)

Another variation of the ridge-loaded laser is the inverted rib (IR) laser [5.55, 56] shown in Fig. 5.25. In this device, an InGaAsP layer is grown over a channeled InP substrated to form the inverted rib. Then the active and cladding layers are grown. The current confinement can be provided by a stripe contact or, as in Fig. 5.25, by a Zn-diffused path through a blocking junction. This structure gives somewhat weaker guiding, due to the smaller index step between InGaAsP and InP versus Si₃N₄ and InP, and poorer current confinement than the RW laser since the stripe contact or blocking junction is several μm from the active layer.

5.4 Longitudinal Mode Control

As indicated in (5.2.18c), a laser with Fabry-Perot (FP) mirrors can oscillate on several longitudinal modes spaced apart by angular frequency $\omega_{\mathrm{s}} = \pi v_{\mathrm{g}}/L$ and covering a spectral width of ~ 20 nm. Although a cw spectrum might show a distribution of mode strengths that is roughly defined by the gain profile as discussed in Sect. 5.2.10, a time resolved spectrum might show a random distribution of mode strengths with time. And if the laser is driven by a pulsed current, the number and power distribution of modes will vary randomly from pulse to pulse. Since the transit time of pulses through a dispersive fiber is a function of the mean oscillating wavelength, these fluctuations in modal distribution give rise to a *modal partition noise* in the receiver due to deviations in arrival times of the pulse sequence in a pulse-code modulated (PCM) system. Furthermore, if each pulse contains several longitudinal modes, the fiber dispersion will spread the pulses and limit the resolvable bit-rate. For these reasons, a laser that is stabilized to operate on only one longitudinal mode is desirable.

In coherent systems, single-longitudinal-mode behavior is essential to the heterodyne or homodyne detection process. In addition, the intrinsic linewidth of the mode limits the system performance. In this section, we will consider structures to ensure single-longitudinal-mode behavior and wavelength tunability. In Sect. 5.5 we will consider linewidth control.

As noted in Sect. 5.2.10, the discrimination against side modes can be enhanced by using a short FP resonator with effective length adjusted, by temperature or drive current, see (5.2.18), to put the central mode at the gain peak. The larger mode spacing enhances the gain difference between central- and side-modes. However, the resultant side-mode suppression is usually not adequate for many applications nor is the inherent temperature and current sensitivity satisfactory.

Strictly speaking, if the gain profile were *homogeneously broadened,* i.e., if the distribution of excited states were tightly coupled with intraband time constant $\tau_i \rightarrow 0$, then there would be no spectral hole burning and the longitudinal mode closest to the gain peak would uniformly deplete the gain profile and no other mode would ever achieve enough gain to oscillate, despite increasing drive current. Thus, spectral hole burning or, equivalently, gain compression is responsible for multiple longitudinal mode behavior. Single-longitudinal-mode behavior is more often observed in AlGaAs lasers than in InGaAsP lasers of the same length. Also, single-longitudinal-mode behavior is sometimes observed when a laser chip with its facets AR coated is inserted into a much longer air-filled resonator (*extended-cavity* laser). These observations are not well-understood, although they seem to indicate reduced spectral and spatial hole-burning. Nevertheless, reliable and effective longitudinal mode control requires the introduction of a frequency sensitive element in the laser resonator as discussed below.

5.4.1 Three- and Four-Mirror Resonators

If a third mirror is incorporated in the conventional FP resonator structure, as illustrated in Fig. 5.26a [5.57], the added constraint on the boundary

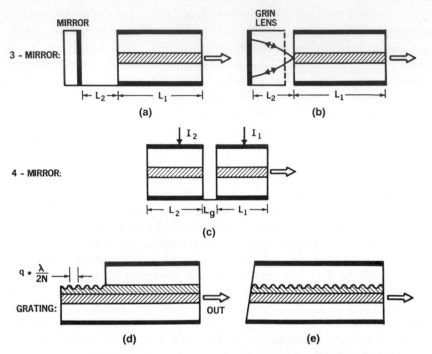

Fig. 5.26a–e. Longitudinal mode control with **(a,b)** three- and **(c)** four-mirror (C^3) resonators, and with **(d)** distributed Bragg reflector (DBR) and **(e)** distributed feedback (DFB) gratings. For first and second-order gratings, $q = 1$ and 2, respectively

conditions severely limits the number of wavelengths with constructive feedback. The internal mirror is assumed to be semi-transparent. One can then think of two independent FP resonators, one of length L_1 containing the gain element and the other of length L_2 filled with air. Only modes common to both will oscillate, and if L_1 and L_2 are incommensurate, only one mode will satisfy the oscillation condition. Typically, $L_2 > L_1 \approx 250\,\mu$m. It may be necessary to tune the L_1 resonator by changing the temperature or drive current to achieve an overlap. However, the cavities are not totally independent and an injection-locking effect will allow some deviation from overlap and even allow single-mode pulses to be realized. The external resonator L_2 will have a greater locking range if the coupling to the gain element is stronger, as provided by the graded-index lens in Fig. 5.26b [5.58]. In this case, a high-reflection mirror is coated on one side and an antireflection (AR) coating on the other side of the lens, which has focal length $L_2/2$.

The four-mirror resonator in Fig. 5.26c is more complicated to understand because we now have three coupled resonators, including the gap L_g which is usually only a few wavelengths long. The currents I_1 and I_2 can be adjusted independently, along with temperature, to achieve a single-mode regime over a reasonable drive current range that permits single-mode pulsed operation when L_g is properly adjusted. The coupled cavities may be formed by chemical etching or a FP laser chip may be recleaved, and held in alignment by its thick metal contact, to form a cleaved-coupled-cavity (C^3) laser. The theory and operation of coupled-cavity lasers is reviewed in detail in [5.59]. These devices are critical to fabricate and to control at present.

5.4.2 Distributed Bragg Gratings

Single-longitudinal-mode operation can be achieved by incorporating a frequency-selective element in the laser resonator. A Bragg grating provides frequency-selective feedback in the distributed Bragg reflector (DBR) laser and the distributed feedback (DFB) laser, which are illustrated schematically in Figs. 5.26d and e. A DBR laser was first demonstrated in poly(methyl methacrylate) (PMMA) doped with Rhodomine 6G dye as the active medium [5.60]. Holographic phase gratings were introduced in the PMMA block at either end to provide feedback as in a laser with external, dielectric-coated mirrors. The DFB laser was first demonstrated [5.61] in dichromated gelatin, doped with Rhodomine 6G in which a continuous holographically-induced grating provided feedback throughout the length of the laser. In this subsection, we will outline the performance of Bragg gratings and, in Sects. 5.4.3, 4 discuss their application in semiconductor DFB and DBR lasers.

Consider a medium of refractive index n containing a region of length L in which the perturbed refractive index varies periodically with spatial period Λ and indices in the range $n \pm n_1$, for $n_1/n \ll 1$. This region comprises a *distributed Bragg phase grating* that couples oppositely traveling waves with propagation constants $\pm \beta$ at angular frequency ω. The *Bragg condition* is satisfied when $\beta = \beta_B$:

$$2\beta_B = K \equiv 2\pi/\Lambda \quad . \tag{5.4.1}$$

with K the grating wavenumber. The Bragg wavelength is

$$\lambda_B = 2\hat{n}\Lambda \quad , \tag{5.4.2a}$$

and the Bragg frequency is

$$\omega_B = cK/2\hat{n} \quad , \tag{5.4.2b}$$

where $\hat{n}$ is the modal index for the planar guide. At this frequency the coupling between oppositely traveling waves is maximized, and the transmission is minimized as the many small reflections from each phase perturbation add constructively in the backward direction.

The ω-β diagram for an infinitely long grating is shown in Fig. 5.27. The slopes are given by the group velocity $v_g(\omega)$, and, in the nearly linear regions are given approximately by $\pm n = \pm \omega/\beta$ for forward and backward

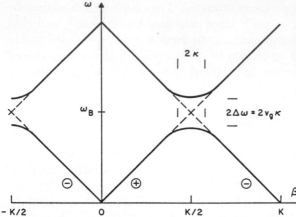

Fig. 5.27. ω-β diagram for an infinitely long grating, showing Bragg coupling between forward (+) and backward (−) waves. The stop-band width, $2\Delta\omega$, at the Bragg frequency, ω_B, is proportional to the coupling coefficient κ [5.63, p. 238]. Adapted by permission of Prentice Hall Inc., Englewood Cliffs, New Jersey

waves, respectively. Since the medium is translationally invariant for displacements of integer multiples of Λ, the ω-β curves are invariant when β is increased or decreased by increments of K, as illustrated in Fig. 5.27. Physically, this means that, for a driving frequency ω, the fields in the grating region are represented by a Fourier series of traveling waves with spatial harmonics $\beta \pm pK$ (with p an integer) in which the strength of the pth harmonic depends upon the profile of the periodic index perturbation. The Bragg coupling between forward and backward waves at the cross-overs has a strength determined by the coupling coefficient κ, which is a function of n_1/n. For a grating with sinusoidal index, $n(z) = n + n_1 \cos Kz$, and gain, $g(z) = g + g_1 \cos Kz$, the coupling coefficient is [5.62]

$$\kappa = (\pi n_1/\lambda_B) + jg_1/2 \quad . \tag{5.4.3}$$

The coupling produces a transmission stop-band, centered at ω_B, in which transmission is forbidden (Fig. 5.27). The stop-band has a width [5.63]

$$2\Delta\omega = 2v_g|\kappa| \quad , \tag{5.4.4}$$

where v_g is the group velocity of the unperturbed medium at ω_B. The actual group velocities in the periodic medium at the stop-band edges ($K/2$, $\omega_B \pm \Delta\omega$) approach zero, as indicated in Fig. 5.27. The deviation of β from β_B is defined by the detuning parameter

$$\delta = \beta - K/2 \approx \frac{\omega - \omega_B}{v_g} \quad , \tag{5.4.5}$$

since $\partial\beta/\partial\omega = v_g^{-1}$ at ω_B in the absence of coupling.

The more practical case of a Bragg grating of finite length is analyzed in [5.63] where it is shown that inside the stop-band the wave decays exponentially according to $\exp(\pm\gamma z)$, with

$$\gamma = \sqrt{|\kappa|^2 - \delta^2} \quad . \tag{5.4.6}$$

It is also shown that the amplitude reflection coefficient $r(-L)$, defined as the ratio of amplitudes of the negative-going to positive-going waves, for a wave incident at $z = -L$ on a grating of length L and terminated at $z = 0$ by a medium with reflection coefficient $r(0)$ is

$$r(-L) = -\frac{(\kappa^*/\kappa) + jr(0)[(\gamma/\kappa)\coth\gamma L - j(\delta/\kappa)]}{r(0) - j[(\gamma/\kappa)\coth\gamma L + j(\delta/\kappa)]} \; , \tag{5.4.7}$$

where κ is allowed to be complex corresponding to a periodically varying gain or loss along with the periodic index variation as in (5.4.3). Note that the definition of κ used here [5.62] differs from that of [5.63] by a factor j.

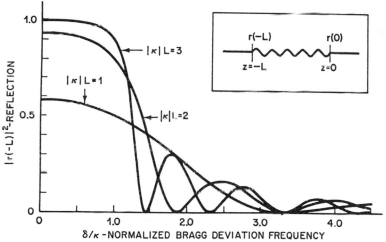

Fig. 5.28. Power reflectivity $R = |r(-L)|^2$ from Bragg grating shown in inset with matched output, $r(0) = 0$, versus normalized deviation frequency [5.63, p. 240]. Adapted by permission of Prentice Hall Inc., Englewood Cliffs, New Jersey

The power reflection coefficient $R = |r(-L)|^2$ is plotted in Fig. 5.28 for a matched or AR-coated output, $r(0) = 0$, and different values of $|\kappa|L$. The *Bragg length* is defined as $|\kappa|^{-1}$; nearly 100 % reflection is realized for a grating that is three Bragg lengths long, $|\kappa|L \geq 3$. Note that there is a stop-band for $|\delta/\kappa| \lesssim 1.5$. Unlike the infinite grating, the transmission through the finite grating

$$T = 1 - |r|^2 \tag{5.4.8}$$

is not forbidden in the stop-band but is decreased in value relative to the transmission for larger $|\delta/\kappa|$. The bandwidth of the stop-band is defined by the normalized frequencies $\delta_{1/2}$ at which T goes to twice its value at $\delta = 0$.

In the limit of large $|\kappa|L$ and small $\delta/|\kappa|$ we have [5.63]

$$\left|\frac{\delta_{1/2}}{\kappa}\right| = \sqrt{\frac{\ln 2}{|\kappa|L}} \qquad (5.4.9a)$$

and, with (5.4.5), we define a stop-band width,

$$2\Delta\omega_{1/2} = 2v_g\delta_{1/2} = 2v_g\sqrt{\frac{|\kappa|\ln 2}{L}} \quad . \qquad (5.4.9b)$$

In the limit $\coth \gamma L = 1$ with κ real, (5.4.7) reduces to

$$r(-L) \approx \frac{-\mathrm{j}}{1+\mathrm{j}(\delta/\kappa)} \approx \mathrm{e}^{-\mathrm{j}(\pi/2+\delta/\kappa)} \qquad (5.4.9c)$$

for $\delta/\kappa \ll 1$. Thus, the phase shift on reflection decreases linearly with δ from its Bragg value of $-\frac{\pi}{2}$. The effective mirror position may be taken as approximately one half a Bragg length, $\sim 1/2\kappa$, inside the grating.

A set of reflection zeros (transmission maxima) occur outside the stop-band, where γ is purely imaginary, for

$$\mathrm{j}\gamma L = \pm p\pi \quad , \quad p = 1, 2, 3, \ldots \quad . \qquad (5.4.10)$$

The first zeros occur at

$$(\beta_{\pm 1} - K/2) = \pm(\pi^2 + \kappa^2)^{1/2} \quad . \qquad (5.4.11)$$

Far from the stop-band edges, the frequencies of the reflection zeros approach those of a Fabry-Perot resonator of length L.

5.4.3 Semiconductor DFB Lasers

A Bragg grating can be incorporated in most of the laser structures described in Sect. 5.3 by interrupting the wafer growth to allow a grating to be etched into the surface of an appropriate epilayer and then continuing the growth of epilayers. The wafer processing can then proceed as with normal multilayer lasers.

In the case of DFB lasers, the grating can be etched into a layer on either side of the active layer. Figure 5.29a illustrates the case of a sinusoidal grating etched into the substrate with index n_s, which might be n-InP. An n-InGaAsP cladding layer (n_c) is grown over the grating, followed by the active (n_f), p-cladding (n_c) and p-InP layers (n_s), with $n_f > n_c > n_s$. The strength of the coupling coefficient κ is determined by the evanescent field that interacts with the grating, by the index step $n_c - n_s$, and by the grating amplitude h_g [5.64]. The grating can be formed by holographically exposing photoresist with two UV laser beams separated by an angle θ, and chemical etching. Precise control of θ can give a grating period Λ with

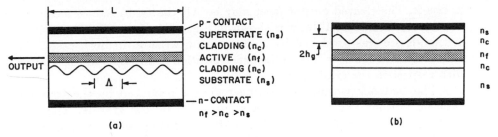

Fig. 5.29a,b. Bragg gratings (a) etched into the substrate cladding and (b) etched into the superstrate cladding

a precision better than $\pm 10\,\text{Å}$. Both first-order $(\Lambda = \lambda_B/2\hat{n})$ and second-order $(\Lambda = \lambda_B/\hat{n})$ gratings are used. An electron beam machine can also be employed to write gratings in the photoresist. The chemical etching process and melt-back during overgrowth limit the grating amplitude to $h_g \lesssim \Lambda/2$ with a corresponding coupling $\kappa \lesssim 5\,\text{mm}^{-1}$.

If the grating period Λ is specified in the substrate, then the refractive indices, layer thicknesses and bandgap of the active layer must all be carefully controlled in the subsequent growth steps in order to assure that the modal index $\hat{n}$ will produce a Bragg wavelength λ_B in the center of the gain profile.

Alternatively, the epilayers may be grown up to the top cladding layer. Then the band gap of the active layer can be determined by optical pumping and the layer thicknesses can be measured with an electron microscope. Then the required Λ can be computed, taking into consideration the reduction in refractive index due to the injection current at threshold. The grating-layer structure for this type of device is shown in Fig. 5.29b.

A DCPBH laser with DFB grating etched in the p-cladding layer is illustrated in Fig. 5.30a [5.65]. The back face is etched with a slope to avoid

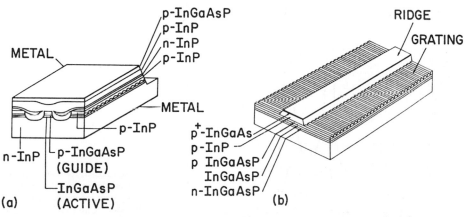

Fig. 5.30. (a) DCPBH-DFB laser [5.65]. (b) Ridge waveguide DFB laser [5.66]

feedback. The cleaved output facet may be AR coated to reduce feedback. A ridge-waveguide DFB laser is shown in Fig. 5.30b [5.66, 67]. The grating in this device was produced by electron-beam writing. In both cases, the output spectra are sensitive to the exact position of the ends of the grating with respect to the grating phases [5.68].

The feedback within the grating region is produced by the coupling between forward and backward waves provided by the grating. The gain in the active region in combination with the grating allows a number of longitudinal modes to oscillate [5.62]. The two modes with highest gain are just outside the stop-band, near the reflection minima in Fig. 5.28, where low-transmission-loss traveling waves can propagate and where the small group velocity allows a long interaction time, or equivalently strong coupling between forward and backward waves.

Gain can be incorporated in the reflection coefficient expression (5.4.7) by means of a periodic gain (gain grating) leading to a complex κ, or by means of uniform gain represented by a complex $\overline{\beta} = \beta - j\Gamma g/2$ and substituting $\overline{\delta} = \delta - j\Gamma g/2$ and a corresponding $\overline{\gamma}$ in (5.4.5–7). In the latter case, which is the usual one, the reflection coefficient for waves incident from the left at $z = -L$ for a grating with matched termination, $r(0) = 0$, becomes [5.69], for κ real,

$$r(-L) = -\frac{j(\kappa/\overline{\gamma})\sinh \overline{\gamma}L}{\cosh \overline{\gamma}L + j(\overline{\delta}/\overline{\gamma})\sinh \overline{\gamma}L} \ . \tag{5.4.12}$$

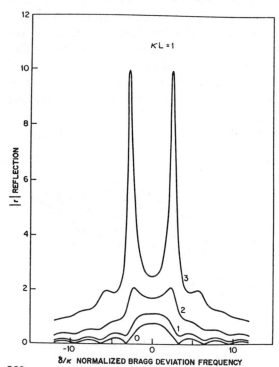

Fig. 5.31. Absolute reflectivity $|r|$ versus deviation frequency for Bragg grating with gain, i.e. a DFB laser. The gain $\Gamma g L$ is a parameter [5.69]

The absolute value of $|r|$ for $\kappa L = 1$ is plotted versus δ/κ in Fig. 5.31 [5.69]. As the gain $\Gamma g L$ increases through values 0, 1, 2, 3, the oscillating modes take shape and threshold occurs when $|r| \to \infty$, so that an infinitesimal input wave gives a finite reflected wave. For values of $\Gamma g L$ that give $|r|$ peaks greater than unity, the transmission function $|t|$ will exhibit similar peaks, i.e. similar laser output appears at 0 and $-L$. These oscillating modes correspond to zeros of the denominator of (5.4.12). The threshold gain for the two fundamental modes is [5.69], $\Gamma g_{th} L = 3.5$ for $\kappa L = 1$ as in Fig. 5.31 and,

$$\Gamma g_{th} = 2\pi^2/\kappa^2 L^3 \tag{5.4.13}$$

for large κL. Note that the two oscillation modes lie outside the stop-band for $g = 0$, as given by (5.4.96). This double-mode behavior, which ideally gives two identical modes separated by about $10\,\text{Å}$, is not satisfactory for system applications that require single frequency output. In practice, asymmetries in mirrors and grating phases usually favor one mode or the other in actual devices and a random selection of DFB lasers will usually contain a substantial fraction of devices with one dominant mode. Note that a gain grating, as opposed to a phase grating, intrinsically produces a dominant mode at the center of the stop-band [5.62].

5.4.4 DBR and Phase-Slip DFB Lasers

A DBR laser is illustrated schematically in Figs. 5.32a and b. Two passive Bragg gratings of lengths L_1 and L_3 with periodic index $\Delta n(z)$ are separated by a gain region of length L_2. The gratings serve as frequency-

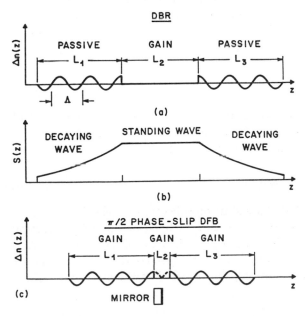

(a)

(b)

(c)

Fig. 5.32. (a) Schematic of a DBR laser showing grating profile $\Delta n(z)$. (b) A Fabry-Perot standing wave is in the central section and decaying waves in the grating sections. (c) A $\frac{\pi}{2}$ phase-slip DFB with $L_2 = \Lambda/2$. A symmetrically placed mirror can achieve the same result

263

selective mirrors with reflectivity bands given by (5.4.9) or Fig. 5.28. As with any Fabry-Perot, the transmission linewidth of the resonator can be much narrower than that of the reflectors [5.63]. A number of longitudinal modes with spacing ω_s (5.2.18c) can be supported within the grating band. The standing wave for each mode has uniform intensity $S(z)$ in L_2, but the intensity decays exponentially in the passive gratings. As L_2 is reduced, ω_s increases until one and only one longitudinal mode is allowed at ω_B. If gain is added to the grating regions L_1 and L_3, the device becomes a phase-slip DFB as illustrated in Fig. 5.32c.

The chief advantage of the DBR laser over the DFB is that the grating region can be separated from the gain region, which might be sensitive to defects introduced while forming the grating or growing the epilayers over it. This anticipated difficulty has proved not to be a real limitation for semiconductor DFB lasers. The chief difficulty for the DBR laser has been the coupling of the waveguide regions containing the passive gratings with the gain region. Since the grating regions are not usually excited, they would be strongly absorbing if they contained the same waveguide layer as the gain region. Therefore, a layer of wider bandgap material must be provided for the grating regions and it must be coupled to the active layer [5.70].

Another important advantage of DBR lasers that has only recently been utilized is the wavelength tunability it offers. If the grating regions are excited by separate electrodes, the Bragg frequency ω_B can be tuned by the injected carriers which alter $\hat{n}$ in (5.4.2) [5.71]. The variation of ω_B with grating current introduces a corresponding deviation frequency δ, according to (5.4.5), that causes the Fabry-Perot modes defined by L_2 to scan with the grating phase shift (5.4.9c).

Unlike the uniform DFB laser, the DBR oscillates at frequencies within the stop-band of the gratings. The exact resonance condition must include the phase change on reflection from the gratings. Examination of (5.4.12) in the absence of gain and for κ real shows that the reflection coefficient at the Bragg frequency is

$$r(\omega_B) = -\mathrm{j}\tanh\kappa L \quad , \tag{5.4.14}$$

independent of position z. Hence, a phase shift of $-\frac{\pi}{2}$ between incident and reflected waves occurs for each Bragg mirror at $\omega = \omega_B$. Therefore, for Fabry-Perot resonance the round-trip phase shift over $2L_2$ must be given by

$$2\beta_B L_2 = \pi + p2\pi \quad , \quad p \text{ integer} \tag{5.4.15}$$

for constructive feedback. The phase shift on reflection for the L_1 grating is the same when observed from the boundary with L_2 as for the L_3 grating when observed from its boundary with L_2. In the case of a continuous DFB grating, which has no phase slip ($L_2 = 0$), the phase condition for resonance cannot be satisfied at ω_B but can only be satisfied at the stop-band edges.

The shortest L_2 that permits DBR oscillation at the center of the stop-band is given by ($p = 0$),

$$\beta_B L_2 = \tfrac{\pi}{2} \quad , \tag{5.4.16}$$

corresponding to a *phase slip* of $\tfrac{\pi}{2}$ or a quarter of a Bragg wavelength in the waveguide L_2. Mathematically, this means that if the L_1 grating is defined by $n(z) = n + n_1 \cos(2\beta_B z)$, then the L_3 grating is defined by $n(z) = n + n_1 \cos[2(\beta_B z - \tfrac{\pi}{2})] = n - n_1 \cos(2\beta_B z)$, for the $\tfrac{\pi}{2}$ phase slip, as illustrated in Fig. 5.32c. Physically, one can imagine cutting a continuous grating at its even symmetry point and separating the two segments by

$$L_2 = \lambda_B/4\hat{n}_2 = \Lambda/2 \quad , \tag{5.4.17}$$

or a half-wave of the grating period. Since the region L_2 is generally much smaller than L_1 and L_3 for the phase-slip DFB laser, the details of the grating profile in L_2 cannot have a significant effect and the two grating segments L_1 and L_3 could as well continue into the L_2 region as indicated by the dashed lines. In fact, the $\tfrac{\pi}{2}$ phase slip could take place continuously over the total grating length. This effect may well occur inadvertently in practical lasers as a result of small variations in layer thickness or composition. The phase slip could also be realized by cleaving a continuous grating at the center of L_2 and applying a high-reflector mirror, as indicated in Fig. 5.32c. Random cleaving would achieve the approximate mirror position almost 50 % of the time.

If gain is provided in the grating regions as well as the gap region for a $\tfrac{\pi}{2}$ phase slip, the reflection coefficient for a wave incident from the left of L_1 is [5.69], for $L_1 = L_3 = L/2$,

$$r(-L) = \frac{(2\kappa\bar{\delta}/\bar{\gamma})\sinh^2(\bar{\gamma}L/2)}{(\kappa^2/\bar{\gamma}) - (\bar{\delta}^2/\bar{\gamma})\cosh\bar{\gamma}L - j\bar{\delta}\sinh\bar{\gamma}L} \quad . \tag{5.4.18}$$

The absolute reflectivity is plotted in Fig. 5.33 versus δ/κ for $\kappa L = 1$ and increasing values of $\Gamma g L$. For $g = 0$, there are zeros equivalent to those in (5.4.10) for the continuous Bragg grating. However, an additional zero at the center of the stop-band is also present. As the gain increases, the Bragg mode has the greatest gain and reaches resonance first, with $\Gamma g_{th} = 3.1$ for the conditions of Fig. 5.33, and

$$\Gamma g_{th} \approx 4\kappa e^{-\kappa L} \tag{5.4.19}$$

for large κL. Thus, the $\tfrac{\pi}{2}$ phase-slip DFB has two advantages over the continuous DFB: firstly, it produces only one longitudinal mode in the stop-band rather than a pair of modes outside the stop-band; and, secondly, this mode has a lower threshold gain at large κL, with $\Gamma g_{th} L$ reduced by half at $\kappa L = 3$ [5.69].

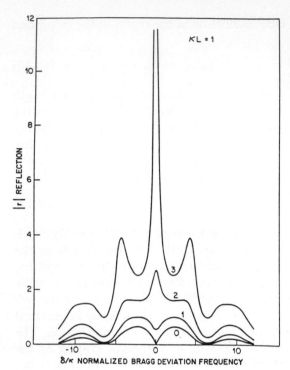

Fig. 5.33. Absolute reflectivity $|r|$ versus deviation frequency for two Bragg gratings separated by the phase slip $\pi/2$ with gain, i.e. a phase slip DFB laser. The gain ΓgL is a parameter [5.69]

As the phase slip varies between 0 and π for L_2 between 0 and Λ, the central mode moves from right to left in Fig. 5.33 and the outer modes increase in size at the boundaries of the region. Furthermore, residual end-mirror reflectivities would affect the symmetry of the three modes [5.72].

5.5 Linewidth

In a true single-frequency laser the optical power is confined to one longi-tudinal mode. The spectral linewidth of this mode is determined by phase noise in the optical field. In some applications, such as intensity modulated, direct detection communications systems, the spectral linewidth is of little practical concern. For coherent systems, however, the linewidth is a critical parameter affecting system performance. We present here a simple theory of linewidth in semiconductor lasers and identify the main factors that con-trol the linewidth. Section 5.5.1 is concerned with the linewidth of a simple Fabry-Perot laser, while Sect. 5.5.2 briefly covers linewidth reduction using external resonators and extended cavities.

5.5.1 Linewidth of Fabry-Perot Laser

Phase noise in the optical field of a laser is due to spontaneous photons coupled into the lasing mode. There are two main mechanisms by which these spontaneous photons give rise to fluctuations in the phase of the optical field:

i)   When a spontaneous photon is injected into the lasing mode, both the amplitude and phase of the optical field in the mode undergo changes. These changes can be described in terms of the field associated with the injected photon combined by phasor addition with the field of the lasing mode. The fluctuations in phase of the total field due to a large number of random spontaneous emission events give rise directly to phase noise.

ii) The amplitude fluctuations in (i) above couple into fluctuations in the carrier density via the rate equations. These fluctuations in carrier density N (and gain g) in turn cause fluctuations in the refractive index of the active region and the frequency of the Fabry-Perot mode. This second component of phase noise is affected by the dynamics of the rate equations and, in general, is enhanced at frequencies at or near the relaxation oscillation resonance frequency [5.73]. In the simple theory presented here [5.19], this enhancement is neglected. A more complete analysis has been presented in [5.73].

The optical field E' in the laser can be written, in phasor form, as

$$E' = I^{1/2}e^{j\phi}$$

where I is the average photon number, and the units of the field E' are scaled so that the magnitude of E' is simply the square root of I. The phase of the field is ϕ.

The phasor diagram in Fig. 5.34 shows how the ith spontaneous photon produces a change ΔI_i in the photon number and a change $\Delta\phi'_i$ in the phase, by the mechanism described in (i) above. Using simple geometrical

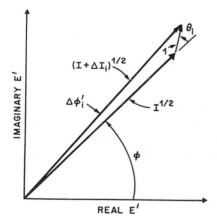

Fig. 5.34. Phasor diagram showing change in magnitude and phase of optical field due to a spontaneous photon [5.19]

267

considerations, it can be shown from Fig. 5.34 that these changes are

$$\Delta\phi_i' = I^{-1/2}\sin\theta_i \quad \text{and} \tag{5.5.1a}$$

$$\Delta I_i = 1 + 2I^{1/2}\cos\theta_i \tag{5.5.1b}$$

where θ_i is the (random) phase of the ith spontaneous photon. Note that the average phase change over a large number of spontaneous photons is zero. In this simple analysis, one photon is added to the lasing mode per spontaneous emission event.

In order to calculate the component of phase change described in (ii) above, it is necessary to determine the coupling between intensity and phase of the optical mode. This is achieved using the photon rate equation (5.2.54) and the expression (5.2.19b) for the deviation $\delta\omega$ in oscillation frequency from its steady-state value. Assuming that ΔI_i is small compared with the steady-state photon number I and that the resulting change ΔN in carrier density is small compared with N, it follows from (5.2.19) and (5.2.54) that

$$\dot\phi = \frac{a\dot I}{2I} \tag{5.5.2}$$

where $\dot\phi = \delta\omega$. The phase change $\Delta\phi_i''$ resulting from the change ΔI_i in photon number is obtained by integrating (5.5.2) with I assumed constant and equal to the steady-state value. The initial condition is $I(0) = I + \Delta I_i$ and the final condition (after the relaxation oscillations have died out) is $I(\infty) = I$. The integration results in

$$\Delta\phi_i'' = -\frac{a\Delta I_i}{2I} = -\frac{a}{2I}(1 + 2I^{1/2}\cos\theta_i) \quad . \tag{5.5.3}$$

The total phase change $\Delta\phi_i$ for one spontaneous event is the sum of $\Delta\phi_i'$ and $\Delta\phi_i''$:

$$\Delta\phi_i = \Delta\phi_i' + \Delta\phi_i'' = -\frac{a}{2I} + I^{-1/2}(\sin\theta_i - a\cos\theta_i) \quad . \tag{5.5.4}$$

The total phase fluctuation $\Delta\phi$ in time t for $M = R_{\mathrm{spon}}t$ spontaneous emission events is obtained by summing (5.5.4) over the M events:

$$\Delta\phi = -\frac{aR_{\mathrm{spon}}t}{2I} + \sum_{i=1}^{M} I^{-1/2}(\sin\theta_i - a\cos\theta_i) \quad . \tag{5.5.5}$$

The first term gives a linear increase in phase with time and therefore gives rise to a frequency offset. It can be ignored without loss of generality. The mean-squared phase deviation $\langle\Delta\phi^2\rangle$ resulting from the second term in (5.5.5) is easily computed because, for random θ_i, the average of all cross terms vanishes. Thus

$$\langle\Delta\phi^2\rangle = \frac{R_{\mathrm{spon}}(1 + a^2)t}{2I} \quad . \tag{5.5.6}$$

The phase ϕ has a Gaussian probability distribution and the power spectrum of the laser is Lorentzian with a full width at half maximum of [5.19]

$$\Delta\omega = \frac{\langle\Delta\phi^2\rangle}{t} \quad . \tag{5.5.7}$$

From (5.5.6), the linewidth becomes

$$\Delta\omega = \frac{R_{\text{spon}}(1 + a^2)}{2I} \tag{5.5.8}$$

Equation (5.5.8) shows that the linewidth is proportional to $1 + a^2$ and inversely proportional to the photon number I. Typically $a \sim 5$, and the a^2 term dominates, indicating that the main mechanism causing the linewidth is carrier density fluctuations ΔN, due to coupling from ΔI_i. For this reason a is often called the "linewidth enhancement factor". A more useful expression for the linewidth can be obtained by expressing I in terms of the output power per facet P, using (5.2.77b) and by expressing R_{spon} in terms of the gain g, using (5.2.46). For $R_1 = R_2 = R$ the linewidth becomes

$$\Delta\omega = \frac{v_g^2 \hbar\omega \Gamma g n_{\text{sp}} \ln R^{-1}}{4PL}(1 + a^2) \quad . \tag{5.5.9}$$

Thus the linewidth is inversely proportional to the output power. Typical values for $\Delta\omega/2\pi$ are on the order of tens to hundreds of MHz.

The gain term can be eliminated from (5.5.9) using (5.2.22c). The linewidth then becomes

$$\Delta\omega = \frac{v_g^2 \hbar\omega n_{\text{sp}} \ln R^{-1}(\alpha + L^{-1}\ln R^{-1})}{4PL}(1 + a^2) \quad . \tag{5.5.10}$$

This shows that for long lasers $(\alpha \gg L^{-1}\ln R^{-1})$ the linewidth is proportional to L^{-1} and for short lasers $(\alpha \ll L^{-1}\ln R^{-1})$ the linewidth is proportional to L^{-2}. In conventional devices, where $L \sim 250\,\mu$m the α and $L^{-1}\ln R^{-1}$ terms in the numerator are roughly equal in magnitude.

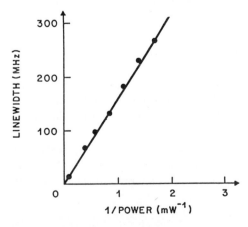

Fig. 5.35. Measured linewidth of In-GaAsP laser [5.74]

The linewidths given by (5.5.9 and 10) are strictly valid only for Fabry-Perot lasers, but can also be used to estimate the linewidths in DFB lasers if an effective value for R is selected. Figure 5.35 shows the measured linewidth of an InGaAsP DFB laser [5.74] as a function of P^{-1}. The linewidth closely follows the inverse power dependence predicted by the theory.

5.5.2 Linewidth Reduction Using Extended Cavities

Coherent optical communications demand stable lasers with linewidths which are lower than can be achieved with an isolated semiconductor laser chip. Injection locking [5.75] to an external narrow-linewidth source is one solution to this problem, but is cumbersome. A preferable solution is to modify the operating conditions of the laser itself in order to reduce the linewidth. This can be achieved by applying some form of feedback to the device. The feedback can take the form either of an electrical feedback signal that controls the linewidth by employing the direct frequency modulation properties of the laser [5.76, 77], or an optical signal reflected from a reflector external to the semiconductor chip [5.78–84]. We concentrate here on the optical-feedback method.

There are two different approaches to linewidth reduction by optical feedback. In the first approach, the strength of the feedback signal from the external cavity is low [5.76–80] (feedback ratio on the order of −40 to −50 dB). The laser becomes phase-locked to the reflected signal and resembles a conventional injection-locked oscillator. An important characteristic of this scheme is that the linewidth is a strong function of the phase of the reflected signal and can be either greater or less than the linewidth of the original laser, depending on the feedback path length. In addition, undesirable effects such as line splitting [5.81] or excessive broadening by "coherence collapse" [5.82] may occur if inappropriate values of feedback ratio and path length are used. When the feedback is adjusted correctly for minimum linewidth, the linewidth decreases with increasing length of the external cavity [5.80].

The second approach to linewidth reduction makes use of very strong feedback levels [5.83–85] (feedback ratio of −10 to −5 dB). To achieve this strong feedback it is usually necessary to anti-reflection coat one laser facet and to couple this facet very strongly to the external reflecting element. The reflector therefore becomes part of an *extended cavity* as opposed to an external cavity reflecting a relatively small amount of light as in the approach described above. Since the oscillation properties of the laser are determined by a single (extended) cavity, the exact feedback path length is not critical.

Figure 5.36 shows the basic structure of an extended-cavity laser. We restrict the present analysis to a cavity with a simple mirror reflector. The more general case of a cavity with a separate Bragg grating is treated in

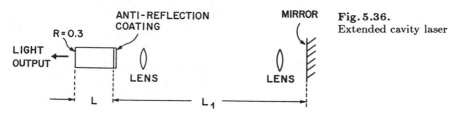

ANTI-REFLECTION COATING

R=0.3

MIRROR

Fig. 5.36.
Extended cavity laser

LIGHT OUTPUT

LENS

LENS

L

L_1

[5.84]. The extended cavity is shown as being air-filled, but other configurations such as fiber cavities [5.85] and integrated semiconductor waveguide cavities [5.86] are also possible. The extension length of the cavity is L_1 and the group velocity in the extension is v_{g1}. An expression for the linewidth of the extended cavity laser can be obtained by modifying the argument used to obtain (5.5.9 and 10). For simplicity, we assume that the effective reflectivity of the extended section of the cavity, as seen by the laser chip at the AR coated facet, is the same as the reflectivity R of the uncoated facet. The general conclusions are similar for other reflectivities, but the analytical results are more complicated.

For a given output power P, the photon number I and the group velocity v_g in the semiconductor portion of the cavity are the same as for the uncoated semiconductor laser without feedback. The photon number I_e in the extended (passive) section of the cavity is

$$I_e = I \cdot \frac{L_1 v_g}{L v_{g1}} \tag{5.5.11}$$

and the total photon number is

$$I_t = I + I_e = I\left(1 + \frac{L_1 v_g}{L v_{g1}}\right) \quad, \tag{5.5.12}$$

which is larger than I. On the other hand, the spontaneous emission rate R_{spon} into the lasing mode decreases because the number of cavity modes is proportional to the cavity length. Therefore, for the extended cavity laser, the spontaneous emission rate becomes

$$R_t = \frac{R_{spon}}{[1 + (L_1 v_g / L v_{g1})]} \quad, \tag{5.5.13}$$

and the linewidth $\Delta\omega_t$ of the extended-cavity laser is obtained from (5.5.8) by replacing R_{spon} with R_t and by replacing I with I_t, i.e.,

$$\Delta\omega_t = \frac{\Delta\omega}{[1 + (L_1 v_g / L v_{g1})]^2} \quad, \tag{5.5.14}$$

where $\Delta\omega$ is the linewidth of the laser without feedback as given in (5.5.8–10). Typically, $L_1 v_g \gg L v_{g1}$ and (5.5.14) becomes

271

$$\Delta\omega_t \approx \Delta\omega \left(\frac{Lv_{g1}}{L_1 v_g}\right)^2 \ , \qquad\qquad\qquad (5.5.15)$$

which shows that the linewidth is inversely proportional to the square of the length of the passive section of the cavity. For $\Delta\omega/2\pi = 100\,\text{MHz}$, $L_1 = 15\,\text{cm}$, $L = 300\,\mu\text{m}$, and $v_{g1}/v_g = 4$ the linewidth is reduced by a factor of 1.56×10^4 to $6.4\,\text{kHz}$.

In the above analysis it is implicitly assumed that the phase varies slowly during the external cavity round trip time [5.79]. If coherence time effects are taken into account, the linewidth may decrease less rapidly with L_1 than is given by (5.5.15) [5.87].

One problem with linewidth reduction using an extended cavity is that the longitudinal mode spacing $\omega_s \simeq \pi v_{g1}/L_1$ decreases with cavity length and single-mode operation is more difficult to realize. A practical solution is to insert frequency selective elements such as a diffraction grating into the cavity and to use these elements to select the desired mode. In air-filled extended cavities the diffraction grating can be used as the external reflecting element [5.83]. Alternatively, an AR-coated DFB laser can be used as the active medium, thereby using the DFB grating as the mode-selective element [5.85].

5.6 High-Speed Modulation

The simplest technique for impressing a signal on the output of a semiconductor laser is by current modulation. In this method, the laser is biased above threshold and a modulation signal is superimposed on the drive current. The time-varying component of the output power becomes an analog of the modulation waveform. Modulation bandwidths well in excess of $1\,\text{GHz}$ can be routinely achieved in semiconductor lasers, while bandwidths of $\sim 20\,\text{GHz}$ are possible in specially designed devices [5.88]. It is the purpose of this section to describe the general form of the direct modulation characteristics and to show how details of the laser structure affect the modulation bandwidth.

In most applications of directly modulated semiconductor lasers it is the modulated envelope of the output optical power or intensity which is of prime interest. This is the intensity modulation (IM) component of the output signal. It is important to recognize that the output optical frequency also varies in response to the modulation component of the drive current. This frequency chirp or frequency modulation (FM) is potentially useful as a modulation format in frequency-shift-keyed (FSK) communications systems [5.6, Chaps. 21, 26]. In high-bit-rate intensity-modulated systems, however, chirp combined with chromatic dispersion in the fiber can become a major obstacle to achieving long spans between repeaters.

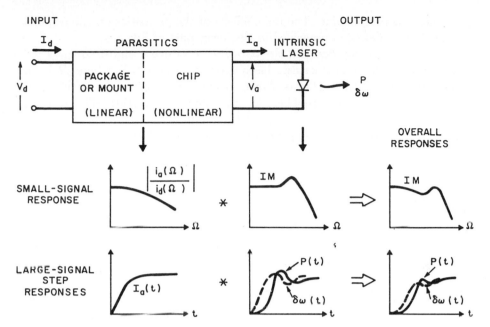

Fig. 5.37. Two-port model of semiconductor laser, showing influence of parasitics on modulation response

5.6.1 Modulation Response

In common with other types of semiconductor device, the response of the semiconductor laser may be affected by circuit parasitics at very high frequencies. The influence of circuit parasitics on direct modulation response is summarized in Fig. 5.37, which shows a packaged laser subdivided into a parasitic circuit and the intrinsic laser (active layer and cavity). The parasitics are further divided into package parasitics and chip parasitics. The package parasitics depend on construction details, but generally include a bond wire inductance and one or more stray capacitances to ground. These parasitics can be almost completely eliminated if the laser is monolithically integrated with the electrical drive circuit. The chip parasitics are stray capacitance and the resistance associated with the semiconductor material surrounding the active region. Chip parasitics are also strongly structure-dependent and in most lasers the chip parasitics are more troublesome than package parasitics. The most significant effect of chip parasitics is high-frequency shunting of the modulation current around the active region.

The drive current and voltage at the input terminals of the laser in Fig. 5.37 are I_d and V_d, respectively. These transform to I_a and V_a at the intrinsic laser due to the filtering effect of the parasitics, dc leakage around the active region, and nonradiative recombination. The output variables of interest are the optical output power P and the optical frequency shift $\delta\omega$

273

from some nominal value. The main effects of the parasitics on the response of these variables is summarized in the lower part of Fig. 5.37.

In the frequency domain, the parasitics produce a high-frequency roll-off in the small-signal intrinsic laser drive current $i_a(\Omega)$ where Ω is the modulation frequency. This is shown in the upper left curve in Fig. 5.37. (Lower-case symbols are used here for small-signal quantities.) This response is combined with the IM response of the intrinsic laser, which is shown in the center curve. The magnitude of the response of the intrinsic laser is independent of frequency at low frequencies but rises to a relaxation oscillation resonance peak [5.2] before rolling off at higher frequencies. (Details of the intrinsic laser response are given in Sect. 5.6.4.) The overall small-signal frequency response of the parasitics and intrinsic laser is shown in the upper right curve. The resonance peak remains in the response, but the high-frequency portion of the response is depressed due to the parasitic roll-off. This results in a dip in the IM response at frequencies below the resonance peak. If the dip is large it can drastically reduce the useful modulation bandwidth of the overall device.

In the time domain, the parasitics lead to a slowing-down of fast transitions of the drive current waveform $I_a(t)$ to the intrinsic laser. This waveform is illustrated in Fig. 5.37 for an $I_d(t)$ with a step waveform of zero rise-time. The time response of the intrinsic laser output power $P(t)$ shows a turn-on delay, followed by overshoot and ringing. Similar behavior is associated with the chirp $\delta\omega(t)$. As shown in the lower right curve of Fig. 5.37, the effect of the parasitics on the time domain response is to increase the overall turn-on time and to reduce the overshoot. Although not shown explicitly in Fig. 5.37, the turn-off time is also increased. Further discussion of these effects is given in Sect. 5.6.8.

5.6.2 Origin of Chip Parasitics

The details of chip parasitic resistance and capacitance depend strongly on device structure. To illustrate some common sources of chip parasitics, Fig. 5.38 shows the cross section of the etched mesa buried heterostructure (EMBH) laser, from Fig. 5.18, with the parasitics included. The resistance R_{sp} of the contact and the p region above the active region combined with the resistance R_{ss} of the substrate below the active region gives a total resistance of $R_s = R_{sp} + R_{ss}$ in series with the intrinsic laser. The small-signal input resistance of the intrinsic laser forward-biased junction is much less than $1\,\Omega$ at currents above threshold due to near clamping of the Fermi levels. On the other hand, R_s is usually of the order of $3\text{--}10\,\Omega$, and dominates the total series resistance. The main sources of shunt capacitance are as follows.

1) Reverse-Biased Blocking Junction. Under normal operating conditions, with the intrinsic device forward-biased, the blocking junction between the upper n layer and the p isolation layer is weakly reverse-biased. The space-charge capacitance C_L of this junction can be quite large, as it usually extends across the entire area of the chip. Typically, C_L is of the order of 100 pF. One parasitic high-frequency current path follows the upper n-layer from the top contact and passes through the blocking junction capacitance to ground. In Fig. 5.38 it is represented as a lumped approximation to a distributed network, with coupling resistances R_{Ni}.

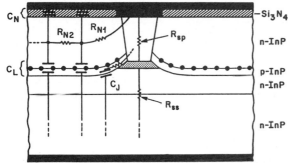

Fig. 5.38. EMBH laser of Fig. 5.18 with parasitics included

2) MIS Isolation Layer. The insulator layer under the top metal contact in Fig. 5.38 provides a metal-insulator-semiconductor (MIS) capacitance C_N distributed across the chip surface. A typical value for this capacitance is approximately 10 pF for a 500 μm $\times$ 250 μm chip area and a 0.2 μm-thick silicon nitride insulator layer. The MIS capacitance C_N appears in series with the blocking junction capacitance C_L, which is usually much larger than C_N. The total capacitance from the top contact through C_N to ground is therefore approximately equal to C_N. This is the dominant parasitic capacitance in many types of semiconductor laser.

3) Forward-Biased Junction Adjacent to Active Region. The third high-frequency parasitic current path follows the thin resistive p layer and the capacitance C_J associated with the forward-biased region of the p-n junction adjacent to the active layer. This forward-biased junction also provides a dc leakage path in shunt with C_J, but this is omitted in Fig. 5.38, for simplicity. Since C_J is the capacitance of a forward biased junction, it can take on large values (up to ~ 1000 pF). It is proportional to the magnitude of the dc leakage current. C_J may be important in devices such as channeled substrate planar lasers, which exhibit significant amounts of leakage around the active region.

5.6.3 Evaluation of Parasitics

The parasitic paths described in (1–3) above are reasonably complex and may be difficult to characterize precisely. A simplified circuit model of the

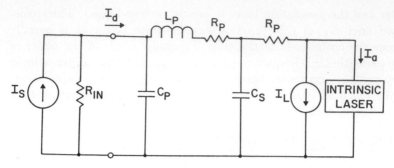

Fig. 5.39. Simplified circuit model of laser and its parasitics

chip parasitics is given in Fig. 5.39, combined with the package parasitics (inductance L_p and loss resistance R_p due to the bondwire, and standoff capacitance C_p associated with the ceramic contact pad) [5.89–91]. In a given laser, one of the parasitic paths (1–3) above will usually dominate. Therefore, Fig. 5.39 shows just one chip parasitic capacitance. Where necessary, the model can be generalized to include additional shunt paths. The capacitance C_s is the effective shunt parasitic capacitance and R_s is the resistance in series with the intrinsic laser. The current generator I_L models dc leakage around the active region.

The signal generator in Fig. 5.39 has a source resistance R_{IN} which is typically 50 or 100 Ω in systems using conventional microwave test instrumentation. If the laser is driven directly from an active device such as a transistor (as in the case of an integrated optoelectronic circuit), R_{IN} may be as high as 500 Ω or more. The parasitic circuit is terminated at its output by the intrinsic laser. Since the small-signal input resistance of the intrinsic laser is small it can be replaced by a short circuit with negligible loss of accuracy in calculations of the high-frequency parasitic roll-off.

A convenient method for determining values of the parasitics in Fig. 5.39 is to fit the model to measured electrical scattering parameters over a range of frequencies. The transfer function of the model in Fig. 5.39 can then be calculated and the influence of the parasitics thereby evaluated. Measurements of the parasitics by scattering parameter techniques have been described in [5.89–91]. Similarly, calculations of the parasitic roll-off have been described in [5.89–91]. The main findings of these studies are summarized below.

i)   The stand-off capacitance C_p has little effect on the response for low R_{IN} ($\leq 50\,\Omega$) but can resonate with L_p to give a weakly enhanced response for large R_{IN} ($> 100\,\Omega$).

ii)   If the bondwire inductance L_p is small ($\leq 0.2\,\text{nH}$), it has little effect on the response up to 20 GHz. Low inductances of this order can be achieved using short ($< 0.5\,\text{mm}$) wire and/or with tape or mesh replacing the usual thin wire. If $L_p \geq 1\,\mathrm{nH}$, then L_p can cause significant roll-off in the response above about $6\,\mathrm{GHz}$ with $R_{IN} = 50\,\Omega$. If $R_{IN} > 50\,\Omega$, this inductive roll-off occurs at lower frequencies.

iii)   For small L_p, the dominant elements affecting the high-frequency parasitic roll-off are the $R_s C_s$ parasitics in the chip.

5.6.4 Dependence of Parasitics on Device Structure

All laser structures which employ a reverse-biased blocking junction for current confinement suffer from the high-frequency shunting effects of the space-charge capacitance associated with this junction. Like the EMBH laser, the DCPBH laser of Fig. 5.19 also has a relatively large shunt capacitance due to the blocking junction. In the EMBH laser, the influence of the total shunt capacitance can be reduced to a degree by decreasing the series capacitance C_N of the insulator layer. This can be achieved by increasing the thickness of the insulator and reducing the area of the top contact. A similar reduction in total capacitance of the DCPBH laser can be obtained by adding an insulator layer (not shown in Fig. 5.19). The space-charge capacitance of the blocking junctions in both devices can be reduced by reducing the doping in the cladding layers.

The ridge laser (Fig. 5.23) does not have a blocking junction capacitance, but a similar capacitance is provided by the junction associated with the unpumped region of the active layer. However, most of this capacitance is electrically isolated from the active region by the thin ($\sim 0.2\,\mu\mathrm{m}$) p-type quaternary cladding layer in the channels on each side of the ridge and the ridge laser is therefore a low capacitance device, provided the top contact is small. The ridge width is generally at least a factor of two wider than the active region of an EMBH laser and the series parasitic resistance R_s is correspondingly smaller, further reducing $R_s C_s$.

Lasers fabricated on semi-insulating substrates, such as the TCBC laser shown in Fig. 5.22 also have low parasitic capacitance. In the TCBC laser the two contact pads are almost co-planar and are separated by the semi-insulating substrate. This leads to a large resistance in series with C_N, which greatly reduces the high-frequency shunting effect of this capacitance.

The constricted mesa laser (Fig. 5.21) and lasers with semi-insulating blocking layers [5.40] can be fabricated with exceptionally low shunt parasitic capacitance. There are a number of reasons why this is so. First, there is no reverse-biased blocking junction. Second, since there is very little quaternary material on each side of the active region, the capacitance C_J associated with any forward biased junction in this area is minimized. Finally, the top contact pad capacitance can be made small ($\sim 1\,\mathrm{pF}$) by minimizing the pad area and using a relatively thick ($\sim 1.5\,\mu\mathrm{m}$) polyimide layer under the contact. Parasitic roll-off frequencies as high as $24\,\mathrm{GHz}$ have been achieved with this structure [5.88].

5.6.5 The Intrinsic Laser –
Small-Signal Intensity Modulation Response

The preceding subsections have highlighted the importance of parasitics in the overall modulation response. However, the central issue in high-speed modulation is the response of the intrinsic laser itself [5.91, 92]. In this subsection the small-signal intensity modulation characteristics of the intrinsic device are explored using the rate equations as a basis. The small-signal modulation response is of practical interest because it is easy to measure and provides useful information on the modulation dynamics for large-signal modulation.

An analytical expression for the small-signal response is readily obtained by linearizing the rate equations (5.2.71). This is achieved by writing time-dependent quantities as sums of steady-state and sinusoidally varying small-signal components:

$$N = N_0 + n e^{j\Omega t} \quad , \tag{5.6.1a}$$

$$I_a = I_{a0} + i_a e^{j\Omega t} \quad , \tag{5.6.1b}$$

$$S = S_0 + s e^{j\Omega t} \quad , \tag{5.6.1c}$$

where Ω is the radian frequency of the modulation signal. These expressions are then substituted in the rate equations. Terms in $\exp(-j\Omega t)$ on both sides of the rate equations are equated, while products of two or more small-signal terms are neglected. This approach leads directly to an expression for the intensity modulation response $M(\Omega)$, which is given by

$$M(\Omega) = \frac{p(\Omega)}{i_a(\Omega)} \tag{5.6.2}$$

where $p(\Omega)$ is the small-signal optical output power. With small terms neglected, the IM response (normalized to the response at zero frequency) becomes

$$\frac{M(\Omega)}{M(0)} = \frac{\Omega_0^2}{\Omega_0^2 - \Omega^2 + j\Omega\{(\gamma\Gamma I_{\text{th}}/evS_0) + (S_0\varepsilon/\tau_{\text{p}}\}} \tag{5.6.3}$$

where I_{th} is the threshold current, S_0 is the steady-state (average) photon density, $v = whL$ is the volume of the active region, and

$$\Omega_0^2 = \frac{g_0 S_0}{\tau_{\text{p}}} \quad . \tag{5.6.4}$$

The zero-frequency (dc) response $M(0)$ is the slope of the static light-output-power versus current characteristic above threshold. The slope of the output power against the drive current I_{d} is given in (5.2.91b). For our

278

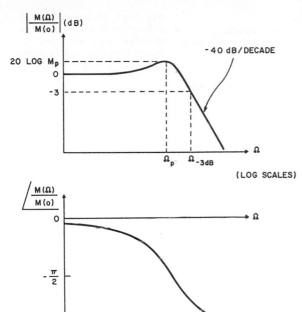

Fig. 5.40. Magnitude and phase of small-signal frequency response

present purposes it is the slope against I_a that is of interest. This slope is

$$M(0) = \frac{\eta_T \hbar\omega}{\eta_I 2e} = \frac{v_g \alpha_m \tau_p \hbar\omega}{2e} . \qquad (5.6.5)$$

Equation (5.6.3) is a conventional second-order, low-pass transfer function with damping characteristics controlled by the magnitude of the jΩ term in the denominator. For practical semiconductor lasers, this term is usually sufficiently small for the transfer function to be underdamped and to exhibit a clearly distinguishable relaxation oscillation resonance peak. The magnitude and phase of the transfer function is illustrated as a function of frequency in Fig. 5.40. The frequency of the resonance peak is Ω_p, the height of the peak is $M_p = |M(\Omega_p)/M(0)|$, and the 3 dB roll-off frequency is $\Omega_{-3\,dB}$. Above the resonance peak, the transfer function asymptotically approaches a slope of −40 dB/decade. The phase of $M(\Omega)/M(0)$ is zero at low frequencies but undergoes a −π radian phase shift as the frequency is increased above the resonance frequency.

The response of a laser with moderate bandwidth is illustrated in Fig. 5.41, which shows the measured and calculated small-signal frequency responses of an InGaAsP ridge laser for three bias currents I_d above threshold [5.90]. For these calculations, the value of the threshold current was obtained by a direct measurement ($I_{th} = 45\,mA$), the parameters γ, Γ, and V_{act} were calculated from known device dimensions and properties, and the

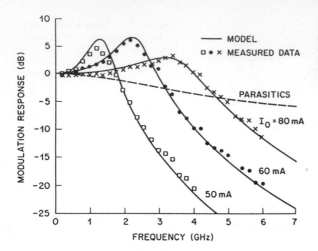

Fig. 5.41. Measured and calculated small-signal response of ridge laser for three different bias currents I_d above threshold [5.90]

Table 5.1. Device parameters and chip parasitics for InGaAsP ridge laser [5.90]

Parameter	Unit	Value	Parasitic	Unit	Value
υ	m^3	4.1×10^{-16}	C_p	pF	0.23
γ	–	2.0×10^{-4}	L_P	nH	0.63
Γ	–	0.3	R_P	Ω	1.0
ε	m^3	6.7×10^{-23}	C_s	pF	8
τ_p	ps	1.0	R_s	Ω	5.5
I_{th}	mA	45	$R_s C_s$	ps	44
I_L	mA	0			
g_0	m^3/s	3.2×10^{-12}			

remaining parameters in (5.6.3) (τ_p and ε) were obtained by fitting the modeled data to the measurements. The package and chip parasitics were determined by a series of microwave scattering parameter measurements and the influence of these parasitics was included in the response calculations. Values for the intrinsic laser parameters and the parasitics [5.90] are given in Table 5.1[3]. The broken curve in Fig. 5.41 shows the high-frequency roll-off of the parasitics acting alone. For this device, the parasitics are small and there is no significant roll-off in the overall response below the resonance peak.

Damping of the relaxation oscillation resonance is important because it affects the height of the resonance peak and the maximum achievable bandwidth. The first and second terms in the damping coefficient in the

---

[3] This method of fitting device parameters leads to some uncertainties in parameter values. A recent estimate of the gain slope, gain compression parameter and photon lifetime for an InGaAsP constricted mesa laser [5.104] gives $g_0 = 1.1 \times 10^{-12}$ m^3/s, $\varepsilon = 8.0 \times 10^{-24}$ m^3, and $\tau_p = 0.9$ ps.

denominator of (5.6.3) arise from spontaneous emission and gain compression, respectively. The spontaneous emission term is inversely proportional to S_0 and is large at low output power levels. This results in strong spontaneous emission damping when the laser is biased close to threshold and the proportion of spontaneous emission in the output is relatively large. On the other hand, the gain compression damping term is directly proportional to S_0 and becomes large at high output power levels where gain compression is largest.

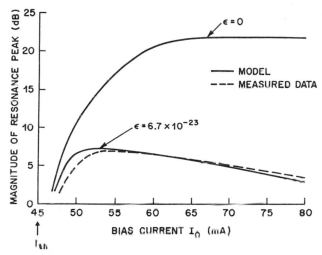

Fig. 5.42. Measured (*dashed*) and calculated (*solid*) resonance peak height [5.90]

The relative significance of the spontaneous emission and gain compression terms is illustrated in Fig. 5.42, which shows the calculated and measured resonance peak height M_p for the ridge laser of Fig. 5.41 as a function of bias current with $\varepsilon = 6.7 \times 10^{23}\,\mathrm{m}^3$ [5.90]. The calculated resonance peak height in Fig. 5.42 is small at bias currents close to threshold due to the spontaneous emission term in the damping coefficient. For bias currents above 53 mA, the peak height M_p decreases due to the increasing gain compression term. Also shown in Fig. 5.42 is the calculated M_p for zero gain compression ($\varepsilon = 0$). In this case the calculated M_p rises to a value of approximately 22 dB at high currents. This is much larger than is observed experimentally and highlights the fact that damping is dominated by gain compression except at bias currents close to threshold. For bias currents well above threshold, (5.6.3) simplifies to

$$\frac{M(\Omega)}{M(0)} = \frac{\Omega_0^2}{\Omega_0^2 - \Omega^2 + \mathrm{j}(\Omega S_0 \varepsilon / \tau_p)} \,. \tag{5.6.6}$$

The corresponding damping time constant is

$$\tau_1 = \frac{\tau_p}{S_0 \varepsilon} \quad . \tag{5.6.7}$$

5.6.6 High-Frequency Limitations

The main parameters of interest in high-frequency response calculations are the $-3\,\mathrm{dB}$ bandwidth $\Omega_{-3\,\mathrm{dB}}$, the resonance frequency Ω_p, and the height M_p of the resonance peak. In high-frequency applications the laser is usually biased well above threshold where the spontaneous damping term is negligible. From (5.6.6), it can be shown that [5.92]

$$\Omega_\mathrm{p} = \Omega_0 + \text{terms of higher order in} \frac{\Omega_0}{\Omega_\mathrm{m}} \tag{5.6.8}$$

$$\Omega_{-3\,\mathrm{dB}} = 1.55\Omega_0 + \text{terms of higher order in } \frac{\Omega_0}{\Omega_\mathrm{m}} \tag{5.6.9}$$

$$M_\mathrm{p} = \frac{\Omega_\mathrm{m}}{\Omega_0} + \text{terms of higher order in } \frac{\Omega_0}{\Omega_\mathrm{m}} \tag{5.6.10}$$

where Ω_0 is given by (5.6.4), and

$$\Omega_\mathrm{m} = g_0/\varepsilon \quad . \tag{5.6.11)}$$

At low and medium power levels the higher-order terms can be neglected. The parameters Ω_p, $\Omega_{-3\,\mathrm{dB}}$, and M_p from (5.6.8–10) are plotted against Ω_0 with solid lines in Fig. 5.43. Since Ω_m is a material parameter which is independent of bias current and Ω_0 is proportional to $\sqrt{S_0}$, the horizontal

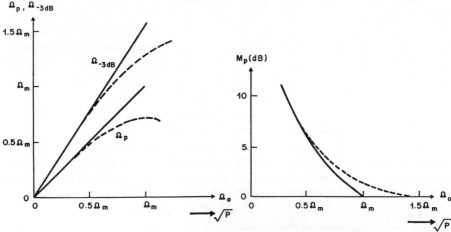

Fig. 5.43. Resonance frequency Ω_p, $-3\,\mathrm{dB}$ frequency $\Omega_{-3\,\mathrm{dB}}$ and resonance peak height M_p against Ω_0. (*Solid lines*) higher order terms in (5.6.8,9,10) neglected. (*Dashed lines*) higher order terms included. The horizontal axis scales as the square root of the optical output power

axis scales as the square root of steady-state optical output power P. The important conclusion here is that the bandwidth rises as $\sqrt{P}$, while the resonance peak height decreases with increasing output power.

If the higher-order terms in (5.6.8–10) are included [5.92], Ω_p, $\Omega_{-3\,dB}$, and M_p follow the broken curves in Fig. 5.43. It can be seen that the rate of increase of bandwidth with increasing $\sqrt{P}$ decreases at high output power and the reduction in resonance peak height is less pronounced.

5.6.7 Design Considerations for Wideband Lasers

In addition to low parasitics, a wide-band laser should have a large resonance frequency and a strongly damped resonance peak to provide a flat response. These objectives are met simultaneously if Ω_0 is large. A number of methods for maximizing Ω_0 [5.91] are immediately obvious from (5.6.4), $\Omega_0^2 = S_0 g_0 / \tau_p$, which can be written as

$$\Omega_0^2 = P \frac{2\Gamma g_0(\alpha + L^{-1} \ln R^{-1})}{(\hbar\omega)(wh)\ln R^{-1}} \tag{5.6.12}$$

using (5.2.27 and 77c).

The first method for maximizing Ω_0 apparent from (5.6.4 or 12) is to maximize the average photon density S_0. For a given output power this can be achieved by decreasing the width of the optical field distribution in the transverse direction parallel to the junction plane. Thus, lasers such as the constricted mesa laser with narrow active regions and tight optical field confinement parallel to the junction plane have wide bandwidth and strong damping. The effect of reducing the width w of the active region is clear from (5.6.12). Note, however, that reducing the thickness h of the active region will have little effect on Ω_0 because Γ is approximately proportional to h. The photon density can be further increased by driving the laser well above threshold. Heating may lead to reliability problems at high current levels and limit the available bandwidth due to a reduction in the value of the gain slope g_0. Well-designed wide-band lasers should therefore have low threshold currents and good heat sinking. Self-sustained pulsations can occur at high-power levels, and catastrophic mirror damage may also place a limit on high-power operation, particularly in AlGaAs devices.

A second method for increasing Ω_0 is to increase the gain slope g_0. This can be achieved by decreasing the temperature [5.93] and by increasing the doping level in the active layer [5.94]. To illustrate the temperature dependence of g_0, Fig. 5.44 shows the $-3\,dB$ frequency of a constricted mesa laser [5.44] plotted against the square root of output power of a range of temperatures decreasing from $15°$ to $-70°$ C. At an output power level of $9\,mW$, the $-3\,dB$ bandwidth increases from $12.5\,GHz$ at $15°$ C to approximately $22\,GHz$ at $-70°$ C. Note that the curves are approximately linear, indicating that Ω_m is large in this device. The bandwidth begins to saturate at an

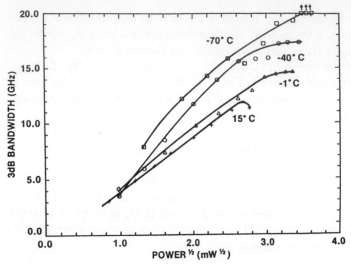

Fig. 5.44. Bandwidth of constricted mesa laser against square root of output power [5.44]

output power of approximately 9 mW. Two other promising approaches to achieving high values of g_0 are quantum-confined lasers [5.96] and DFB or DBR lasers that are detuned to the short wavelength side of the gain peak [5.97] (see Fig. 5.6).

A third method for increasing Ω_0 is to reduce the photon lifetime τ_{p} by increasing the $(\alpha + L^{-1} \ln R^{-1})$ term in (5.6.12). This can be achieved by reducing the cavity length of the device. The effectiveness of this technique has been demonstrated both in AlGaAs [5.93] and InGaAsP [5.95] devices.

5.6.8 Large-Signal Modulation – PCM

The large-signal dynamic response of a semiconductor laser is complex due to the highly nonlinear properties of the intrinsic device. The output waveform depends strongly on the frequency and amplitude of the input signal and distortion can be large. Large-signal behavior has been investigated for a variety of modulation schemes, including conventional pulse code modulation (PCM) for digital commmunications and short pulse generation by gain switching. This subsection describes some of the main speed limitations for the PCM scheme.

In simple two-level PCM, the laser is switched to the "on" state to signal a mark or one and to the "off" state to signal a space or zero. The speed of transmission is therefore limited by the speed at which the laser can be switched from the "off" state to the "on" state and vice versa. To determine the parameters which affect this speed limitation, simple analytical expressions can be obtained for the turn-on time and turn-off time of a laser driven by a current pulse. Figure 5.45 shows the input current pulse I_{a} and

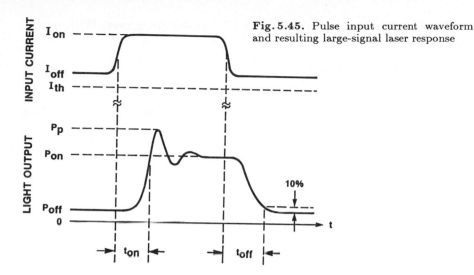

Fig. 5.45. Pulse input current waveform and resulting large-signal laser response

the resulting large-signal response. The off-level drive current and output power are I_{off} and P_{off}, respectively, and the corresponding on-level values are I_{on} and P_{on}. It is common practice in high bit-rate systems to bias the laser slightly above threshold, such that $I_{off} > I_{th}$, in order to improve speed and single-mode behavior. The turn-on time t_{on} is defined here as the time taken for the output power $P(t)$ to first reach the value P_{on}. The turn off time t_{off} is the time taken for the output to fall to within 10 % of its final value. The peak value of $P(t)$, due to overshoot, is P_p.

a) Turn-On. A numerical analysis of the full rate equations [5.98] shows that if the laser is switched on from below threshold, then the stimulated and spontaneous recombination terms in the photon rate equation are small and can be ignored for $t \leq t_{on}$. The same conclusion applies if I_{off} is just above threshold. Therefore for $0 \leq t \leq t_{on}$, (5.2.71a) can be simplified to

$$\frac{dN(t)}{dt} = \frac{I_{on} - I_{off}}{ev} \quad , \tag{5.6.13}$$

where it is assumed that the electrical pulse has a short rise time. The solution to (5.6.13) (for $0 \leq t \leq t_{on}$) is

$$N(t) = N_{off} + \frac{(I_{on} - I_{off})t}{ev} \tag{5.6.14}$$

where N_{off} is the off-level carrier density and $t = 0$ corresponds to the leading edge of the electrical pulse. The output power $P(t)$ during the turn-on transient is obtained by substituting (5.6.14) in (5.2.71b). In order to obtain a simple analytical solution for $P(t)$, the gain compression term $(1 - \varepsilon S)$ in (5.2.71b) is set to unity. The solution therefore ignores bandwidth saturation due to damping at high output power levels. However, the results

285

are reasonably accurate for low and moderate output power. With $\gamma = 0$, the output power becomes

$$P(t) \simeq P_{\text{off}} \exp\left(\frac{g_0(S_{\text{on}} - S_{\text{off}})t^2}{2\tau_{\text{p}}}\right) \quad , \tag{5.6.15}$$

where S_{off} and S_{on} are the off-level and on-level photon densities. For $S_{\text{off}} \ll S_{\text{on}}$ (5.6.15) reduces to

$$P(t) = P_{\text{off}} \exp(0.5\Omega_{0,\text{on}}t^2) \quad \text{where} \tag{5.6.16}$$

$$\Omega_{0,\text{on}} = \left(\frac{g_0 S_{\text{on}}}{\tau_{\text{p}}}\right)^{1/2} \tag{5.6.17}$$

is the *small-signal* resonance frequency Ω_0 when the laser is biased at the on-level drive current I_{on}, see (5.6.4). The turn-on time becomes [5.99]

$$t_{0,\text{on}} = \frac{\sqrt{2}}{\Omega_{0,\text{on}}} \left[\ln\left(\frac{P_{\text{on}}}{P_{\text{off}}}\right)\right]^{1/2} \quad . \tag{5.6.18}$$

This equation highlights the importance of the on-level photon density or output power (which affects $\Omega_{0,\text{on}}$) and the on-off ratio $P_{\text{on}}/P_{\text{off}}$. The turn-on time is decreased by increasing S_{on}, and by decreasing $P_{\text{on}}/P_{\text{off}}$. In other words, rapid turn-on is achieved with wideband lasers using small on/off ratio. A similar analysis [5.99] has shown that the overshoot $P_{\text{p}}/P_{\text{on}}$ also decreases with decreasing $P_{\text{on}}/P_{\text{off}}$. It is worth noting that in communications links a reduced on-off ratio introduces a system power penalty [5.6, Chap. 18], so the optimum on-off ratio may involve a trade off between turn-on time and receiver sensitivity. Figure 5.46 shows the calculated turn-on time and overshoot of a ridge waveguide laser [5.99] against the on-off ratio and the corresponding value of $I_{\text{off}}/I_{\text{th}}$. The turn-on time curve was obtained from (5.6.18) above threshold ($I_{\text{off}}/I_{\text{th}} > 1.0$), while the broken curve below threshold was calculated numerically from the rate equations (5.2.71). Note that the turn-on behavior is improved as the device is biased above threshold.

b) **Turn-Off.** Simple analytical expressions for the turn-off characteristics can be obtained by dividing the turn-off transient into two time periods: period (i) where $P > P_{\text{on}}/2$, and period (ii) where $P < P_{\text{on}}/2$. For the first of these periods (5.2.71a) becomes

$$\frac{dN(t)}{dt} = -\frac{I_{\text{on}} - I_{\text{off}}}{ev} + [r_{\text{st,on}} - r_{\text{st}}(t)] \quad \text{where} \tag{5.6.19}$$

$$r_{\text{st}}(t) = g_0[N(t) - N_{\text{t}}][1 - \varepsilon S(t)]S(t) \tag{5.6.20}$$

is the stimulated emission term in (5.2.71a) and $r_{\text{st,on}}$ is the steady-state

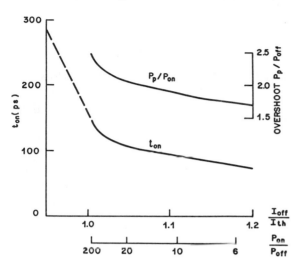

Fig. 5.46. Calculated turn-on time and overshoot against on-off ratio and I_{off}/I_{th}

value of r_{st} at the on-level drive current. The spontaneous recombination term has been neglected in (5.6.19) because N is approximately constant. It is difficult to obtain an exact solution to (5.6.19) because of the time dependence of $r_{st}(t)$. However an approximate solution can be obtained for period (i) by recognizing that the first term on the right of (5.6.19) dominates while the $[r_{st,on} - r_{st}(t)]$ term increases from zero to approximately $0.5(I_{on} - I_{off})/ewhL$. In the interests of simplicity, we assume an average (constant) value of $0.25(I_{on} - I_{off})/ewhL$ for the second term in (5.6.19) during period (i) and substitute this into (5.6.19) to give

$$\frac{dN(t)}{dt} = -\frac{0.75(I_{on} - I_{off})}{ewhL} \quad . \tag{5.6.21}$$

The soluton for $P(t)$ during period (i) becomes

$$P(t) = P_{on}\exp(-0.375\Omega_{0,on}t^2) \quad \text{(period (i))} \tag{5.6.22}$$

where $t = 0$ now corresponds to the trailing edge of the electrical drive pulse.

During period (ii) the second term on the right of (5.6.19) dominates. An approximate solution for $P(t)$ in period (ii) can be obtained by neglecting the first term on the right, replacing the time-varying gain in (5.6.19) by an average (constant) value, and matching the slope of $P(t)$ at the boundary of regions (i) and (ii). This results in

$$P(t) = \frac{P_{on}}{2} \exp[1.018\Omega_{0,on}(t_{1/2} - t)] \quad \text{(period (ii))} \tag{5.6.23}$$

where

$$t_{1/2} = \frac{1.36}{\Omega_{0,on}} \tag{5.6.24}$$

is the time at which $P(t)$ falls to $P_{on}/2$. From (5.6.23 and 24), t_{off} becomes

$$t_{off} = t_{1/2} + \frac{1.58}{\Omega_{0,on}} = \frac{2.94}{\Omega_{0,on}} \quad . \tag{5.6.25}$$

Note that the turn-off time in (5.6.25) is independent of the on-off ratio. For typical values of on-off ratio (~ 10) in (5.6.18), the turn-on time t_{on} is always less than the turn-off time t_{off}. Therefore the turn-off time is the dominant switching-speed limitation in high bit-rate systems [5.99].

Figure 5.47 shows measured eye patterns for the same ridge laser as in Fig. 5.46, at a bit rate of 2 Gbit/s. These patterns were obtained by driving

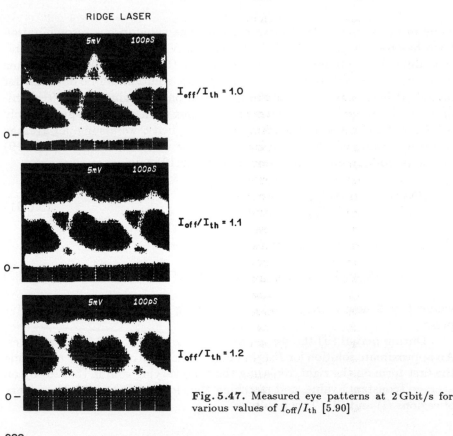

RIDGE LASER

$I_{off}/I_{th} = 1.0$

$I_{off}/I_{th} = 1.1$

$I_{off}/I_{th} = 1.2$

Fig. 5.47. Measured eye patterns at 2 Gbit/s for various values of I_{off}/I_{th} [5.90]

the laser with a non-return-to-zero (NRZ) pseudo-random bit stream and using a sampling oscilloscope to superimpose a large number of on-off and off-on transitions of $P(t)$. As I_{off}/I_{th} is increased, overshoot decreases and the pulse width increases due to the decreasing value of t_{on}. Note that the turn-off time is larger than the turn-on time, as shown by the theory above.

c) Frequency Chirp. Experimental studies of frequency chirp [5.100] have shown that the chirp reaches its maximum value during the turn-on transient. From (5.6.14) it can be seen that N, and therefore the frequency shift $\delta\omega$, increases linearly with time for $0 < t < t_{on}$. An estimate of the chirp $\delta\omega_{on} - \delta\omega_{off}$ between times $t = 0$ and $t = t_{on}$ can be obtained from (5.2.19b), (5.2.39), (5.6.14), and (5.6.18). For $S_{off} \ll S_{on}$,

$$\delta\omega_{on} - \delta\omega_{off} = \frac{a\Omega_{0,on}}{\sqrt{2}} \left[\ln\left(\frac{P_{on}}{P_{off}}\right) \right]^{1/2} . \tag{5.6.26}$$

Note that $\delta\omega_{on} - \delta\omega_{off}$ is proportional to $\Omega_{0,on}$, and is therefore large for wide-band devices. Like the turn-on time t_{on}, the chirp can be made smaller by reducing the on-off ratio P_{on}/P_{off}. The chirp is directly proportional to the linewidth enhancement factor a.

The analysis presented so far has assumed an ideal (zero rise-time) step input. However, in practical applications the rise-time of the input signal may be significant, due to the finite speed of the electronic drive circuit and the slowing-down effect of the parasitics. If the turn-on time is not critical, then it may be feasible to tailor the drive waveform to improve the chirp characteristic. For example, a double step on the leading edge of the drive pulse [5.101] can significantly reduce overshoot in the electron density and thus reduce the chirp. A different approach is to use a pulse-shaping circuit with a band-stop characteristic centered at the laser resonance frequency [5.102], thereby removing components of the drive signal at the resonance frequency. Other techniques for chirp reduction include the use of quantum-confined structures with low a values [5.96], the use of an external grating to reduce the effective value of a [5.84], injection locking [5.103], and the use of external modulators (Chap. 4).

The chirp can be related directly to the photon density S or output power P, without reference to the drive current or electron density [5.105]. This enables a pulse shape to be tailored in order to achieve a prescribed chirp [5.106]. To obtain a relationship between the chirp and P, we take δg in (5.2.19b) to be the difference between the gain g and the (constant) loss term $(\Gamma \tau_p v_g)^{-1}$ which serves as a reference gain and is close to the steady-state gain. Then (5.2.19b) gives

$$\delta\omega = \frac{1}{2} v_g a \Gamma \left(g - \frac{1}{\Gamma \tau_p v_g} \right) . \tag{5.6.27}$$

Substituting (5.6.27) and (5.2.39) in (5.2.71b) with $\gamma = 0$, we obtain

$$\frac{\dot{S}}{S} = \frac{2\delta w}{a} - v_g \Gamma g \varepsilon S \quad . \tag{5.6.28}$$

To solve this equation we note that, above threshold, the gain g in (5.6.28) is approximately clamped to a constant value above threshold, and is equal to the loss $(\Gamma \tau_p v_g)^{-1}$. With this approximation, (5.6.28) simplifies to

$$\delta w = \frac{a}{2}\left(\frac{\dot{S}}{S} + \frac{\varepsilon S}{\tau_p}\right) \quad . \tag{5.6.29}$$

It is useful to express δw in terms of optical output power P rather than the photon density. Thus, using (5.2.23), (5.6.29) becomes [5.105]

$$\delta\omega(t) = \frac{a}{2}\left[\frac{d}{dt}\ln P(t) + \kappa P(t)\right] \quad \text{where} \tag{5.6.30a}$$

$$\kappa = \frac{2\Gamma\varepsilon\eta_{\mathrm{I}}}{v\hbar\omega\eta_{\mathrm{T}}} \quad . \tag{5.6.30b}$$

Note that κ is a coefficient which determines gain compression in terms of optical power rather than photon density. In terms of this parameter, the gain compression characteristic in (5.2.59) becomes

$$g = \bar{g}_0(N - N_{\mathrm{t}})(1 - \kappa\tau_p P) \quad . \tag{5.6.31}$$

The chirp $\delta\omega(t)$ and output power $P(t)$ are written in (5.6.30) explicitly as functions of time to emphasize that if $P(t)$ is known, then the time-dependence of chirp can readily be calculated.

The magnitude of the first term in either (5.6.29) or (5.6.30a) is small at the on and off levels, where $\ln P(t)$ is constant. However, it may be large during high-speed turn-on and turn-off transients and usually oscillates between positive and negative values due to ringing. This term can be controlled to an extent by shaping the optical pulse to limit the rate of change of $\ln P(t)$ after turn-on. The second term in (5.6.30a) gives rise to frequency offset between the on and off levels due to differences in steady-state values of N.

To illustrate the above points, Fig. 5.48 shows measured transient wavelength shift data measured for a number of different laser structures [5.107]. The wavelength chirp $\delta\lambda$ is given for a ridge waveguide cleaved-coupled cavity (C^3) laser and three types of DFB laser; a vapor phase transported (VPT) laser, a double-channel planar buried heterostructure (DCPBH) laser, and a heterostructure ridge overgrown (HRO) laser. Each laser was biased slightly above threshold and turned on at time $t = 0$ by a current step with a rise time of 90 ps. For the ridge and HRO lasers the ringing of the output power is large due to the relatively large active region volume in

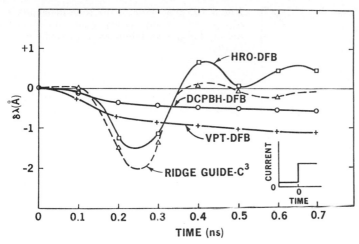

Fig. 5.48. Measured transient wavelength shift for a number of InGaAsP lasers [5.107]

these devices and the correspondingly small photon density. The ringing in $\delta\lambda$ for these lasers is dominated by the first term in (5.6.30a). In contrast to the ridge and HRO devices, the VPT and DCPBH lasers show little ringing, and the VPT laser has a significant wavelength offset between the on and off levels. The reduced ringing in these devices is partially due to reduced active region volume and partially due to the slowing-down effects of the chip parasitics. The frequency offset in the VPT laser is caused by the relatively large value of κ for this device due to its narrow active region ($\sim 1\,\mu$m). Note that the chirping characteristics of all lasers are expected to be the same and independent of structure if the photon densities and photon lifetimes are the same, as shown in (5.6.29).

5.6.9 Large-Signal Modulation − Gain Switching

In the pulse modulation scheme considered in Sect. 5.6.8 above, the laser is driven by electrical pulses which are approximately rectangular in shape. The objective is to obtain an optical output waveform that mimics the electrical drive waveform. Limitations on modulation rate arise primarily from the turn-on and turn-off delays. An alternative approach is for generating short optical pulses is known as gain-switching. In this method, a periodic electrical drive signal is applied to the laser with the bias level and pulse parameters adjusted such that the optical output falls to zero immediately after the first spike of each relaxation oscillation. The following spikes are quenched [5.108].

The drive current waveform in gain switching can either be a pulse train, or a sinusoidal function. Figure 5.49 shows the time dependence of (a) the injected current density, (b) the carrier density or gain, and (c)

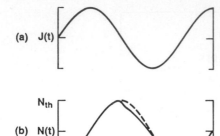

(a) J(t)

Fig. 5.49. Schematic of the time dependence of injected current density, gain, and photon density for gain switching with sinusoidal drive [5.108]

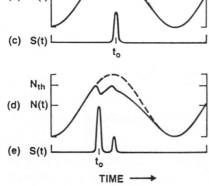

(b) N(t)

(c) S(t)

t_o

(d) N(t)

(e) S(t)

t_o

TIME ⟶

the photon density for sinusoidal drive. The decrease in gain due to stimulated emission is the difference between the dashed and the solid curves. In Fig. 5.49d and e, the gain and photon density at a higher bias current are shown. In this case the gain reaches threshold a second time and a second (smaller) pulse is emitted.

There are a number of inherent advantages to gain-switching. First, if the electrical drive signal is a train of sufficiently short pulses, the amplitude of the pulses can be considerably larger than would be allowable in ordinary PCM. This leads to very short turn-on times. Second, the rate of change of output power on the trailing edge of the single relaxation oscillation spike is large, and slow turn-off times are not as large a problem as in conventional PCM modulation. Finally, the overall pulse width obtainable by gain switching is very short ($\sim 20\,\mathrm{ps}$ or less) which, in principle, suggests the possibility of operation at bit rates in the tens of Gbit/s regime. Although gain-switching is usually repetitive, gain-switching with digital data at bit rates as high as 8 Gbit/s have been demonstrated [5.109]. The main disadvantage of gain-switching for communications applications is the relatively large chirp associated with the rapid change in output power during the gain-switched pulse [5.109].

5.6.10 Active Mode-Locking

Another approach to the generation of repetitive short optical pulses in semiconductor lasers is active mode-locking by gain modulation [5.110, 111].

In this technique the laser is anti-reflection coated on one facet and is placed in an extended cavity with the same basic structure as the extended cavity laser shown in Fig. 5.36. The external cavity can be in air, as indicated in Fig. 5.36, or in a waveguide such as an optical fiber [5.112]. Like gain switching, the active device is driven by a dc bias current and a periodic RF signal. The RF current is adjusted to the Fabry-Perot mode spacing frequency $f_s = \omega_s/2\pi$, which corresponds to the round-trip time T_R in the cavity

$$f_s = 1/T_R \quad \text{and} \tag{5.6.32}$$

$$T_R = 2L_1/v_g \quad . \tag{5.6.33}$$

Mode-locking can also be achieved at higher harmonics of this frequency [5.112].

In essence, the mode-locking process operates as follows. A pulse circulates in the cavity in synchronism with the RF drive signal. The pulse is in the active medium at the instant the periodic optical gain reaches its maximum value and is amplified by a short burst of stimulated emission. The gain rises as the leading edge of the pulse passes through the gain medium and falls on the trailing edge, as depicted in Fig. 5.50. These effects tend to sharpen the pulse by shaving the leading and trailing edges, respectively. In the steady state, this sharpening is balanced by pulse dispersion and the pulse amplification is balanced by losses.

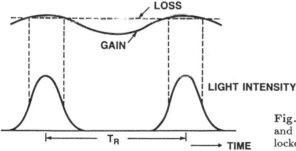

Fig. 5.50. Time dependence of gain and photon density in a mode-locked laser [5.110]

An alternative picture of mode-locking can be obtained using a frequency domain model. The longitudinal modes of the extended cavity are separated by the Fabry-Perot frequency spacing, of the cavity, f_s, which is usually in the microwave range. The modulating signal at this frequency excites a number of these modes in such a way that there is a definite phase relationship between them. (This is the origin of the term "mode-locking"). It follows, by Fourier transform techniques, that the output comprises a train of pulses, each separated by T_R. The amplitudes of all locked cavity modes reach their maxima at the instant corresponding to a pulse peak.

5.7 Luminescent Diodes and Laser Amplifiers

The index-guided laser structures described in Sect. 5.3 can be used below threshold either as incoherent sources or as amplifiers. These incoherent sources are known as edge-emitting light-emitting diodes (ELED) or superluminescent diodes (SLD), depending on the linearity of operation. The below-threshold regime can be extended to higher current and correspondingly higher output by minimizing the feedback from one or both cleaved mirror facets by means of antireflection (AR) coatings or other methods.

The ELED and SLD are sources of amplified spontaneous emission with operating conditions such that the output is a linear function of drive current in the former case and a superlinear function in the latter. The amplifier, on the other hand, amplifies an incident coherent signal, but the amplified signal is accompanied by the same amplified spontaneous emission present in the ELED and SLD. Hence index-guided LEDs and amplifiers are closely related and will be discussed together in the following subsections. The ideal amplifier has no feedback from the facets and amplifies the incident signal in one pass; it is called a traveling-wave amplifier (TWA). Practical amplifiers, even those with good AR coatings, have some residual facet reflectivity; these are Fabry-Perot amplifiers (FPA). A thorough review of semiconductor laser amplifiers has been given elsewhere [5.113] .

5.7.1 Edge-Emitting and Superluminescent Diodes

A conventional surface-emitting light-emitting diode (SELED) emits incoherent spontaneous emission over a wide spectral range into a 2π-radian solid angle [5.114]. The light emerges in one pass from a surface that is parallel to the junction plane. Since the injection current is usually too small to produce appreciable gain, the emitted light originates in a thin portion of the active layer that has an approximate thickness α^{-1}, such that effective attenuation is $1/e$, limited by the material absorption coefficient α. The unpolarized output passes through the transparent substrate material to the surface and power increases linearly with drive current. The modulation bandwidth of the diode is limited by the spontaneous lifetime τ_n, and is typically $\sim 100\,\mathrm{MHz}$. The spectral width is $\sim 125\,\mathrm{nm}$. LEDs have been reviewed in [5.114].

In contrast, the spontaneous emission generated in the active layer of an ELED emerges parallel to the junction from an edge of the active layer and operates at sufficiently low gain that the output power increases linearly with current. Thus, the net gain $(\Gamma g - \alpha)$ is less than zero in the ELED regime but the effective emission path length $(\alpha - \Gamma g)^{-1}$ may be greater than for the surface-emitting SELED.

As $(\Gamma g - \alpha)$ becomes positive, the amplified spontaneous emission increases superlinearly with current and the device enters the SLD regime. As the amplified photon density increases further with path length, gain compression, due to depletion of carriers from the active layer, takes place and the gain increases less rapidly with increasing current. As a consequence the output characteristic becomes more linear. Note that this gain compression is a phenomenon different from the optical field-dependent gain compression described in Sect. 5.2.6 by εS.

The idealized SLD behavior supposes negligible round-trip mirror feedback. In practice, this feedback is minimized by (a) shortening the electrode to leave a substantial portion of one end of the laser structure unpumped [5.115], (b) orienting the gain stripe at a sufficiently large angle from the normal to the cleaved end to prevent the reflected beam from coupling back into the gain region [5.116], or (c) anti-reflection coating one or both facets [5.117]. None of these schemes is effective in reducing the feedback completely to zero, and, unless heating effects dominate, a laser threshold can be reached at sufficiently large drive current. The behavior of an edge emitter in terms of spectral width, radiated beam angle and modulation bandwidth, can be adjusted to fall somewhere between that of an SELED and a laser as the net gain is increased. The spectral width is reduced as the drive current is increased because the exponential gain favors the peak of the spontaneous emission distribution. Similarly, the beam angle is narrowed because the on-axis rays experience gain over a longer path than the off-axis rays, and the bandwidth is increased as the carrier lifetime is shortened by stimulated emission.

Early ELEDs and SLDs were gain guided structures with no lateral index-guiding in the junction plane, but with index-guiding normal to the junction provided by the heterostructure. The electrical-to-optical conversion efficiency of the SLD, especially for coupling to a fiber of given numerical aperture (NA_f), can be improved significantly by providing lateral index guiding as well [5.118]. If the NA of the guide, NA_g, matches the fiber, $NA_g = NA_f$, then all the spontaneous emission generated in the active region within a solid angle that can later be collected by the fiber will be guided over the full length of the device. Components of spontaneous emission outside this solid angle, on the other hand, will not be guided and amplified, and therefore will not needlessly deplete the excited carriers and contribute to gain saturation.

The analytic approach to amplified spontaneous emission in the small-signal linear approximation and in limiting saturated cases will be discussed in the next subsection. However, because of the nonlinearities introduced by gain saturation and the general complexity of the situation outlined above, the description of real devices does not lend itself to analytic solution, and the behavior of an index-guided SLD must be treated by numerical analysis [5.118]. In the computer model, spontaneous emission is generated at source points distributed throughout the active volume. The guided rays from the sources are traced and their amplified outputs summed; the unguided rays are ignored after passing through the guide boundary. The results confirm that for given drive current the maximum power coupled into the fiber occurs when $NA_g = NA_f$. The narrowing of the radiation beam angle with current is also confirmed.

The numerical analysis of photon density $S(z)$ for forward and backward waves, taking gain saturation into account, illustrates the importance of having a low reflectivity output facet and a high reflectivity back facet. This criterion is just the reverse of the situation with the unpumped back region described in (a) above. In this case, there is no reflection at the back end ($z = -L/2$) and the forward wave grows from the spontaneous emission to a value $S_+(L/2)$ at the output end. If a finite reflectivity $R(L/2)$ is present at the cleaved output facet, the reflected backward-wave signal, $R(L/2)S_+(L/2)$ is much larger than the spontaneous emission that initiated the forward wave and the backward wave grows to a much larger value $S_-(-L/2)$ at the back-end than $S_+(L/2)$. The numerical analysis [5.118] shows that ten times more power is lost in the unpumped region than is coupled out the cleaved end. In addition, the large and useless backward photon density depletes the carriers substantially. The off-normal approach (b) is difficult to implement for the index-guided structures and, furthermore, because of the large refractive index of the semiconductor, the output beam approaches the grazing angle with respect to the cleaved facet. Thus, the best arrangement is to have a good AR coating (c) on the output facet and a high reflector on the back facet [5.119]. The situation is then optimized and most of the photon density contributes to the output while very little is lost at the back-end.

An index-guided SLD, operating at 1.3 μm, was fabricated from a ridge waveguide laser with an AR-coated front facet and high-reflection back facet [5.119], as shown in Fig. 5.51. The power versus current curve (Fig. 5.52a)

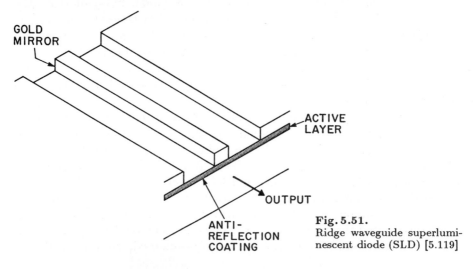

GOLD MIRROR

ACTIVE LAYER

OUTPUT

ANTI-REFLECTION COATING

Fig. 5.51.
Ridge waveguide superluminescent diode (SLD) [5.119]

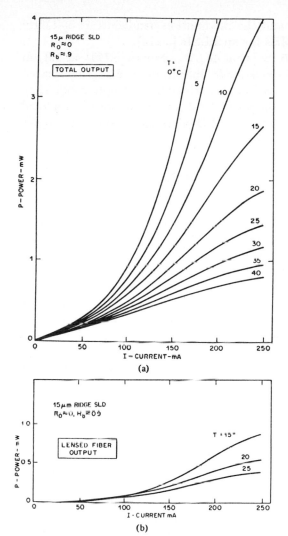

Fig. 5.52. (a) Output power from a 1.3 μm ridge waveguide SLD with ridge width 15 μm, output reflectivity $< 10^{-2}$, back reflectivity 0.9. (b) Power coupled into a lensed 0.23 NA multimode fiber [5.119]

illustrates the knee, which is characteristic of superluminescent behavior, followed by an approach to linearity due to gain saturation and heating. Figure 5.52b shows that a substantial fraction ($\sim 25\%$) of the power is coupled into the 0.23 NA, 50 μm-core, lensed multimode fiber because of the narrowed radiation pattern in the lateral dimension. At 20° C, 500 μW is coupled into the fiber. Still better coupling could have been achieved with a cylindrical lens to match the NA of the diode in the plane normal to the junction. It is clear from Fig. 5.52 that the output is small, linear and insensitive to temperature at low current, in the ELED regime, but is large, nonlinear and temperature sensitive in the SLD regime at high current. The spectral linewidth is 30 nm in the SLD regime [5.119], compared with

$\sim 75\,\text{nm}$ for an ELED [5.114], and the modulation bandwidth is $350\,\text{MHz}$ [5.119], compared with $\sim 100\,\text{MHz}$ for an ELED [5.114].

Recently, it was reported [5.120] that $45\,\mu\text{W}$, or more [5.121], at $1.3\,\mu\text{m}$ could be coupled into a lensed *single-mode* fiber from an ELED fabricated from a $2\,\mu\text{m}$-wide AR-coated CMBH laser. Although the coupling efficiency into a single-mode fiber is small, the power level is adequate for certain local network applications and the use of the ELED offers the prospect of simpler and cheaper operation compared to a laser.

5.7.2 Linear Amplification and Amplified Spontaneous Emission in TWAs and ELEDs

Consider a long single-transverse-mode traveling-wave amplifier (TWA) or traveling-wave ELED, in which the signal and detected spontaneous emission travel in the $+z$-direction only. Assume that only the component of spontaneous emission polarized parallel to the signal is detected. The rate equation in the total photon number representation is given by (5.2.54). The number I of photons in the mode is converted to power P traveling in the forward direction by multiplying by $v_g \hbar \omega \nu$. The quantity ν is the number of forward-traveling modes per unit length for the given polarization, which can be determined as follows: The mode spacing ω_s is given by (5.2.18c) for *standing waves* in an active region of length L; the number of standing-wave modes in length L within an incremental frequency band $\Delta\omega$ is obtained by dividing by ω_s; half this number are forward-traveling modes [5.122],

$$\nu = \Delta\omega/2\pi v_g \quad , \tag{5.7.1}$$

where $L \to \infty$ in such a way that $\Delta\omega \gg \omega_s$ but $\Delta\omega$ is sufficiently small that other parameters in the problems we will be discussing do not vary significantly in that increment.

The time derivative in the rate equation is related to the spatial derivative by $d/dt = v_g \partial/\partial z$, assuming steady-state conditions for an observer traveling with the forward wave so that $\partial/\partial t = 0$. Then the rate equation (5.2.54) becomes for a given mode

$$\frac{\partial P}{\partial z} = (\Gamma g - \alpha)P + \frac{\Delta\omega}{2\pi}\hbar\omega\Gamma g n_{\text{sp}} \quad , \tag{5.7.2}$$

where we have taken $\tau_p^{-1} = v_g \alpha$ from (5.2.27b) with $R_1 R_2 = 0$, $R_{\text{stim}} = v_g \Gamma g I$ from (5.2.45), $R_{\text{spon}} = v_g \Gamma g n_{\text{sp}}$ from (5.2.47a) and ν from (5.7.1). In the absence of gain saturation, the carrier density N and the gain,

$$g = \overline{g}_0(N - N_t) \quad , \tag{5.7.3}$$

are independent of z. The inhomogeneous differential equation (5.7.2) with constant coefficients can be solved by the method of variation of constants yielding for a device of length L

$$P(L) = P(0) \cdot G_0 + \frac{\Delta\omega\hbar\omega\Gamma g n_{sp}}{2\pi(\Gamma g - \alpha)}(G_0 - 1) \quad, \tag{5.7.4a}$$

where the single-pass power gain is

$$G_0 = e^{(\Gamma g - \alpha)L} \quad. \tag{5.7.4b}$$

We find from (5.2.81d) that the total quantum efficiency for no leakage current ($\eta_I = 1$) is $\eta_T = (\Gamma g - \alpha)/\Gamma g$. Thus, with no input power the amplified spontaneous emission power (in a given polarization) P_N is given by [5.122]

$$P_N = \frac{\Delta\omega\hbar\omega n_{sp}}{2\pi\eta_T}(G_0 - 1) \quad, \tag{5.7.5}$$

where $n_{sp} = \eta_T = 1$ in the ideal case and $n_{sp} \approx 2$ and $\eta_T \approx 0.5$ in practical cases, due to incomplete inversion and imperfect quantum efficiency. In the physical absence of the amplifier ($L = 0$, $G_0 = 1$), no spontaneous emission noise is introduced. In the lmit of large gain, $G_0 \gg 1$, the noise power is equivalent to one spontaneous photon per mode per hertz amplified by $G_0(\omega)$ for the ideal case, whereas, for the practical case, about four ($\sim n_{sp}/\eta_T$) spontaneous photons per mode per hertz are subject to gain $G_0(\omega)$,

$$P_N/(\Delta\omega/2\pi) \approx (n_{sp}/\eta_T)\hbar\omega G_0(\omega) \quad, \tag{5.7.6}$$

where G_0 and, to a lesser extent, η_T are functions of ω, and $\Delta\omega$ is a small increment centered at ω. The increment $\Delta\omega$ must be small enough that the right-hand side of (5.7.6) is approximately constant within it. Otherwise, the total noise power is obtained by replacing $\Delta\omega$ with $d\omega$ and integrating over the band. The spectral distribution will be limited chiefly by the exponential gain factor, which narrows with increasing amplification. The total noise power detected with a polarizer at the output of a linear amplifier or ELED is obtained by integrating (5.7.6) over the receiver bandwidth. If G_0 and η_T do not vary appreciably over $\Delta\omega$, then (5.7.6) can be used directly with $\Delta\omega$ as the effective optical bandwidth. In Sect. 5.7.6 it is shown that the noise performance of an amplifier is optimized by employing optical filtering to limit the spontaneous emission noise to a band equal to the signal bandwidth.

Following engineering practice for radio-frequency amplifiers, the noise introduced by the amplifier can be represented by a noise factor F defined [5.122] as the ratio of signal-to-noise ratio (SNR) at the input to that at the output

$$F = \frac{2\pi(P_S)_{IN}}{\hbar\omega\Delta\omega} \bigg/ \left(\frac{P_S}{P_N}\right)_{OUT} \quad, \tag{5.7.7}$$

where P_S and P_N are signal and noise powers in an incremental amplifier bandwidth $(\Delta\omega/2\pi)$. The input noise power $\hbar\omega\Delta\omega/2\pi$ is the quantum-limited minimum value specified by the zero-point fluctuations of the vacuum. Taking $(P_S)_{\text{OUT}}/(P_S)_{\text{IN}} = G_0$ and taking $(P_N)_{\text{OUT}}$ as the sum of P_N from (5.7.5) introduced by the amplifier plus the vacuum contribution, which passes through the amplifier unchanged, gives

$$F = \frac{(n_{\text{sp}}/\eta_{\text{T}})(G_0 - 1) + 1}{G_0} \quad . \tag{5.7.8a}$$

The minimum noise factor is unity, occurring for $n_{\text{sp}}/\eta_{\text{T}} = 1$ or $G_0 = 1$. Thus, the SNR detected at the output can never exceed that at the input. In the limit of large gain $(G_0 \gg 1)$,

$$F \approx n_{\text{sp}}/\eta_{\text{T}} \quad . \tag{5.7.8b}$$

It may be noted from [5.2.47a] and Fig. 5.11 that n_{sp} may be made to approach unity in an amplifier by operating on the low-energy side of the gain peak, where $A(\omega) \to 0$.

To obtain the total noise factor F_{T} of two amplifiers (either TWA or FPA) in tandem with amplifications G_1 and G_2 and noise factors F_1 and F_2, one proceeds as follows: The output noise is

$$(P_N)_{\text{OUT}} = [F_1 G_1 G_2 + (F_2 - 1)G_2]\hbar\omega\Delta\omega/2\pi \quad . \tag{5.7.9}$$

where the vacuum quantum noise is introduced only at the input of the amplifier chain. The output signal is

$$(P_S)_{\text{OUT}} = G_1 G_2 (P_S)_{\text{IN}} \quad \text{and} \tag{5.7.10}$$

$$F_{\text{T}} = F_1 + \frac{F_2 - 1}{G_1} \quad . \tag{5.7.11}$$

Hence the noise factor is determined by the first stage alone for large G_1. As in the case of radio amplifier design, the best strategy for overall low-noise performance is to employ a low noise, high gain amplifier in the first stage and a noisier, high power amplifier in the second stage. The first stage also dominates the noise performance for multi-stage amplifiers.

5.7.3 Fabry-Perot Amplifiers and ELEDs

The ideal traveling-wave devices described in the preceding subsection have not been realized in practice because of the difficulty in eliminating the facet reflectivity to a sufficient degree. Theoretically the reflectivity of a cleaved facet cannot be reduced below 10^{-6} by means of a single "quarter-wave" coating, and then only under critical conditions on thickness, refractive index and wavelength [5.123–126]. The reason is that the modal field does not have a planar wavefront after it exits the guide. The best experimental results are in the range $R \approx 2 \times 10^{-4}$ to 2×10^{-5} [5.125, 127, 128]. Therefore, at present, all ELEDs and amplifiers exhibit Fabry-Perot behavior to some extent. The Fabry-Perot modulated spontaneous emission from an ELED is illustrated in Fig. 5.15. The variation with wavelength of the amplification for a FPA and for a near-TWA is illustrated in Fig. 5.53 [5.129, 130]. The FPA is an ordinary cleaved-facet laser operated just below threshold while the near-TWA has AR-coated facets with $R \sim 3 \times 10^{-3}$.

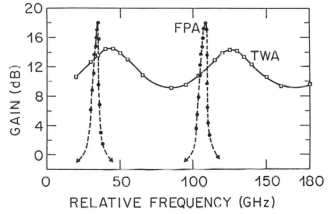

Fig. 5.53. Measured gain spectra for an FPA and near-TWA operating at wavelengths near 1.5 μm [5.129]

The amplification at the peaks of a FPA, at frequencies ω_p with mode spacing $\omega_s = \pi v_g / L$, is given by (5.2.20). The general frequency-dependent expression for the power gain in an FPA, G_{FP} (in contrast to G_0 for the single-pass gain), can be obtained from (5.2.17),

$$G_{FP}(\omega) = \frac{(1 - R_1)(1 - R_2)G_0}{(1 - \sqrt{R_1 R_2}G_0)^2 + 4\sqrt{R_1 R_2}G_0 \sin^2\left[(\omega - \omega_p)(L/v_g)\right]} .$$

(5.7.12)

The ratio of maximum and minimum gains is

$$\varrho = \left(\frac{1 + \sqrt{R_1 R_2}G_0}{1 - \sqrt{R_1 R_2}G_0}\right)^2 .$$

(5.7.13)

Far from threshold, $\sqrt{R_1 R_2}G_0 \rightarrow 0$, the FPA approaches a TWA, and $\varrho \rightarrow 1$; for $\sqrt{R_1 R_2}G_0 \rightarrow 1$, the FPA approaches laser threshold, and $\varrho \rightarrow \infty$, as shown in Fig. 5.53. Measurement of ϱ can be used to determine G_0 if $R_1 R_2$ is known, in analogy with the method in Sect. 5.2.11 for measuring $R_1 R_2$ when G_0 can be deduced. The strong frequency dependence of G_{FP} for large ϱ is unsatisfactory because it requires that the signal frequency be precisely aligned with ω_p, which varies rapidly with v_g and L, and because the operating bandwidth is limited. If $G_{FP}(\omega)$ is required to be flat within 1 dB ($\varrho < 1.26$), then for $R_1 = R_2 = R$, we require $RG_0 < 6 \times 10^{-2}$, and, for $G_0 = 10^2$ (20 dB), R must be less than 6×10^{-4} at the operating wavelength, a result that may be difficult to achieve reproducibly.

The spontaneous emission noise $(P_N)_{FP}$ also contains a resonance factor [5.21]

$$(P_N)_{FP} = \frac{\Delta\omega\hbar\omega n_{sp}}{2\pi\eta_T}(G_0 - 1)$$

$$\times \frac{(R_1 G_0 + 1)(1 - R_2)}{(1 - \sqrt{R_1 R_2}G_0)^2 + 4\sqrt{R_1 R_2}G_0 \sin^2\left[(\omega - \omega_p)(L/v_g)\right]}$$

(5.7.14)

where the output is at the R_2 facet. If G_{FP} is substituted from (5.7.12), Eq. (5.7.14) can be rewritten as

$$(P_N)_{FP} = \frac{\Delta\omega\hbar\omega n_{sp}}{2\pi\eta_T}(1 - G_0^{-1})\frac{(R_1 G_0 + 1)}{(1 - R_1)}G_{FP}(\omega) \quad . \tag{5.7.15}$$

If the FPA is operated near threshold

$$(\sqrt{R_1 R_2}G_0 \approx 1) \quad , \tag{5.7.16a}$$

then

$$(P_N)_{FP} \approx \frac{\Delta\omega\hbar\omega n_{sp}}{2\pi\eta_T}\chi G_{FP}(\omega) \quad , \tag{5.7.16b}$$

where

$$\chi = \frac{(1 - \sqrt{R_1 R_2})(1 + \sqrt{R_1/R_2})}{1 - R_1} \quad . \tag{5.7.16c}$$

Near threshold, ϱ is large and the Fabry-Perot peaks in (5.7.16b) may be narrower than the band $\Delta\omega$. In this case the FPA acts as its own optical bandpass filter with transmission function $G_{FP}(\omega)/G_{FP}(\omega_p)$, normalized to its value at a resonance peak, where

$$G_{FP}(\omega_p) = \frac{(1 - R_1)(1 - R_2)G_0}{(1 - \sqrt{R_1 R_2}G_0)^2} \quad . \tag{5.7.17}$$

Then replacing $\Delta\omega$ by the differential $d\omega$ and integrating over a band that includes only one of the resonance peaks, we find the noise factor at ω_p

$$F \approx \frac{n_{sp}}{\eta_T}\chi \quad , \tag{5.7.18}$$

where the asymmetry factor for various mirror combinations is

$$\chi \approx \begin{cases} 1 & , \quad R_1 \approx 0 \\ 2 & , \quad R_1 = R_2 \\ \infty & , \quad R = 1 \text{ or } R_2 = 0 \end{cases} . \tag{5.7.19}$$

The gain $G_{\mathrm{FP}}(\omega_{\mathrm{p}})$ is the same for all values of χ, but in (5.7.19a) most of the spontaneous emission is reflected back to the source, in (5.7.19b) spontaneous photons emitted in both forward and backward directions in the resonator contribute to the output, and in (5.7.19c) all the spontaneous emission appears at the output.

Thus, an asymmetric FPA near resonance satisfying (5.7.19a) can have a noise factor as small as a TWA (5.7.8) but with a higher resonant gain (5.7.17). The narrow FP bandwidth (FWHM),

$$\Delta\omega_{\mathrm{p}} = (1 - \sqrt{R_1 R_2 G_0})(\omega_{\mathrm{s}}/\pi) \quad , \tag{5.7.20}$$

assures a large signal-to-noise ratio. However, in practice it is difficult to maintain the alignment of the signal frequency and ω_{p}. Furthermore, the high internal photon density for given amplified output power leads to gain compression effects.

5.7.4 Amplifier Gain Compression

In Sects. 5.7.2, 3 we have assumed linear amplification, i.e. that the gain is independent of the input optical power and the output power is directly proportional to the input power. As the input power increases in a practical amplifier, the output power saturates and the gain becomes compressed[4] (i.e., reduced). This gain compression is partly due to the optical field dependence of the optical gain g as described in Sect. 5.2.6, but is dominated by depletion of carriers from the active layer. Thus the mechanism that causes the gain to saturate in SLD's, see Sect. 5.7.1, also causes gain compression and output power saturation in optical amplifiers.

Gain-compression effects have been taken into account in a numerical analysis of the performance of TW and FP amplifiers [5.131]. Among other results, the analysis shows that spontaneous emission can saturate the amplifier, even with weak input signals, if the total spontaneous emission band is not filtered in some way. Similarly, spontaneous emission in the polarization orthogonal to the signal also contributes to saturation. The numerical analysis also shows, as in (5.2.74), that the photon density varies radically along the length of a TWA, being small at the center and large at the ends due to the growing waves. Thus, the ends saturate first.

---

[4] Note that we use the term "compression" to refer to a decrease in the value of a parameter and the term "saturation" to refer to the leveling out or reduced rate of increase of a parameter. Thus the output power of an amplifier saturates and the gain is compressed as the input power increases. In some of the literature, such as [5.132], the term "saturation" is used for both of these phenomena.

Gain-compression effects have been treated by analytic approximations in various limits under the assumption of a homogeneously broadened gain curve; i.e. one that is reduced by a constant scale factor over the spectrum without hole burning [5.132]. An insight to saturation behavior in these limits can be obtained by considering (5.7.3 and 4). As the intensity of the signal and spontaneous emission increases with length, N is depleted and g is reduced. In some range of path length, L_1 to L_2, the condition

$$0 < (\Gamma g - \alpha)(L_2 - L_1) \ll 1 \qquad (5.7.21a)$$

holds, and for that section

$$G_0 \approx 1 + (\Gamma g - \alpha)(L_2 - L_1) \approx 1 \quad . \qquad (5.7.21b)$$

Then (5.7.4a) becomes

$$P(L_2) \approx P(L_1) + \frac{\Delta\omega\hbar\omega n_{sp}\alpha}{2\pi}(L_2 - L_1) \quad . \qquad (5.7.21c)$$

In this regime, the compressed modal gain coefficient Γg is approximately equal to the attenuation α. The signal no longer experiences any amplification and the spontaneous emission noise increases linearly with length rather than exponentially. The noise factor for this segment is

$$F = 1 + \alpha n_{sp}(L_2 - L_1) \quad . \qquad (5.7.22)$$

If it is placed in tandem with an uncompressed high gain segment, then the total noise factor should not be badly degraded according to (5.7.11) but neither will there be any benefit of added gain. However, the SLD output will be enhanced somewhat.

Finally, in the compression regime for which

$$\alpha \gg \Gamma g \quad , \qquad (5.7.23a)$$

(5.7.4) becomes

$$P(L_2) = \frac{\Delta\omega\hbar\omega n_{sp}\Gamma g}{2\pi\alpha} \quad . \qquad (5.7.23b)$$

Here the signal is totally lost and the effective length for generating spontaneous emission is α^{-1}.

Gain compression or, equivalently, output power saturation is clearly detrimental to device performance. For a given output power, gain-compression effects can be reduced most directly by reducing the internal photon density at the output. Therefore, the modal cross-section wh/Γ should be made as large as possible. Facet reflectivity should be minimized to approach TWA performance, since the Fabry-Perot cavity resonance increases the internal photon density for given power output. The photon density for gain compression can be increased by shortening the spontaneous emission time τ_n by increasing the doping in the active region, with n-type doping having a larger effect [5.113]. Since both spontaneous and signal photons contribute to gain compression, the optical bandwidth should be reduced to match that required for signal transmission by introducing an optical filter intrinsic to the amplifier, as in an FPA near threshold, or by introducing optical filters in amplifier chains.

5.7.5 Receiver Noise

The optical spectral density of the output of an amplifier or chain of amplifiers consists of the coherent modulated signal surrounded by a band of spontaneous emission noise, as illustrated in Fig. 5.54a, where, for simplicity, the signal density is represented by the amplified carrier alone, $\frac{1}{2}G_0 P_c \delta(f \pm f_c)$, with $G_0 P_c$ the optical power in the carrier at frequency f_c and $\delta(f)$ is the delta function. The noise density $D(f)$ is assumed to be constant with value $\frac{1}{2}G_0 F h f_c$ in the band B_0 limited by an optical filter. This combined optical signal is then detected by a square-law detector such that the mean photocurrent is proportional to optical power P,

$$\langle i \rangle = aP \quad , \tag{5.7.24}$$

with a the photodetector response coefficient. The receiver power density in this case is well-known in electrical engineering [5.134],

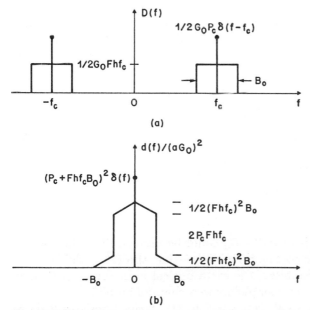

(a)

(b)

Fig. 5.54. (a) Optical power density $D(f)$ for a coherent carrier P_c and spontaneous emission bandwidth B_0 limited by an optical filter. (b) Detector power density $d(f)$ with a square-law detector, ignoring components at $\pm 2f_c$

305

$$d(f) = (aG_0)^2 \left[(P_c + Fhf_cB_o)^2 \delta(f) \right.$$

$$+ \begin{cases} 2P_cFhf_c & , \quad 0 < |f| < B_o/2 \\ 0 & , \qquad \text{otherwise} \end{cases}$$

$$+ \left. \begin{cases} (Fhf_c)^2(B_o - |f|) & , \quad 0 < |f| < B_o \\ 0 & , \qquad \text{otherwise} \end{cases} \right] . \tag{5.7.25}$$

The result is plotted in Fig. 5.54b, where we assume the components near $2f_c$ are filtered out at the receiver. The integrated power response over the baseband $\pm B_d$, where $B_d \geq B_o$, is

$$\langle i^2 \rangle = (aG_0)^2 [(P_c + Fhf_cB_o)^2 + 2P_cFhf_cB_o + (Fhf_cB_o)^2] . \tag{5.7.26}$$

The first term is due to the dc current and can be eliminated by a dc block. The second term is due to current fluctuations produced by the mixing of carrier and spontaneous emission, and the last term is due to fluctuations produced by the mixture of spontaneous emission with itself. The ratio of carrier-spontaneous to spontaneous-spontaneous noise power is

$$\frac{\langle i^2 \rangle_{cs}}{\langle i^2 \rangle_{ss}} = \frac{2P_c}{Fhf_cB_o} = 2(\text{SNR})_{\text{opt}} . \tag{5.7.27}$$

Thus, carrier-spontaneous fluctuations dominate for the usual case of large $(\text{SNR})_{\text{opt}}$.

A rough estimate of the detected signal power for a modulated carrier can be obtained by considering an optical signal of power $P_s = P_c$ and with a signal bandwidth B_s, where $B_d \geq B_s$. Then the detector signal-to-noise ratio is

$$\begin{aligned} (\text{SNR})_{\text{det}} &= \frac{\langle i^2 \rangle_s}{\langle i^2 \rangle_n} = \frac{P_s}{2Fhf_cB_o(1 + Fhf_cB_o/2P_s)} \\ &\approx \tfrac{1}{2}(\text{SNR})_{\text{opt}} , \quad \text{for} \quad (\text{SNR})_{\text{opt}} \gg 1 . \end{aligned} \tag{5.7.28}$$

It is clear from the preceding discussion that, if $B_o \geq B_d \geq B_s$, the $(\text{SNR})_{\text{det}}$ is a maximum when

$$B_o = B_s . \tag{5.7.29}$$

In principle, the receiver bandwidth B_d could be used to filter the spontaneous noise, but reference to Fig. 5.54b shows that the $\langle i^2 \rangle_n$ will be minimum for $B_d = B_o$. Furthermore, the extra optical noise for $B_o > B_s$ may saturate the amplifier, or later amplifiers in a chain of amplifiers.

Shot-noise is usually small compared to the terms considered above [5.113, 135] and can be ignored for present purposes. The limiting result in (5.7.28) indicates that the detector SNR is at least 3 dB worse than the optical SNR so that it might appear that the amplifier has made the sensitivity worse. However, if the optical signal is weak, electronic and/or APD amplification is required and the thermal and/or multiplication noise degrade the sensitivity even more than 3 dB [5.135]. The receiver noise for various modulation formats with the electronic noise sources included are given elsewhere [5.113, 135].

5.8 Tunable and FM Lasers

Optical frequency-division multiplexed (FDM) networks require a comb of optical carrier frequencies with spacings on the order of ten times the modulation bit-rate, which may be in the Gbit/s range. Thus, the channel spacings are about 10 GHz (or 0.8 Å at 1.55 μm and 0.56 Å at 1.3 μm). We have already described two forms of single-frequency laser, the DFB (Sect. 5.4.3) and the DBR (Sect. 5.4.4), that can serve as sources for the carriers in the transmitter. In coherent systems, these lasers can also be used as local oscillators in the receivers. Present-day manufacturing tolerances lead to a wide distribution of operating wavelengths for devices from the same wafer. Thus, one can often select a suitable set of transmitters for a simple FDM network but this is generally an uneconomical solution. Tighter tolerances will allow better control of the laser wavelength but the demanding specifications of future FDM networks will require tunable lasers, just as FDM radio networks require tunable oscillators. Tunability of DFB and DBR lasers can be realized by adding electrodes as shown in Fig. 5.55 and varying the bias current i. Then FDM networks with nominally identical lasers for each transmitter and receiver local oscillator become a possibility. Systems with tens of channels require lasers with tuning ranges of hundreds of gigahertz.

The digital modulation format in these networks may be amplitude-shift keying (ASK) or, more often, frequency-shift keying (FSK). We will see that the same process that allows static tuning of the optical frequency also allows a small-signal FM modulation of the optical frequency f_o at the modulation frequency f_m, where we define a modulation response $\varrho(f_m) = \partial f_o/\partial i$. These devices can be used for wide-deviation FSK provided $\varrho(f_m)$ is sufficiently flat over the FSK Fourier spectrum and has sufficient magnitude to provide FSK without excess ASK.

5.8.1 Tunable DBR

The tunable DBR is illustrated in Fig. 5.55a. It consists of a Bragg reflector section, a phase control section and a gain section driven by separate currents i_b, i_p and i_g, respectively [5.136–139]. Only the gain section contains an active layer; the other sections contain passive waveguides. The operating wavelength is determined by the Fabry-Perot mode nearest the peak of

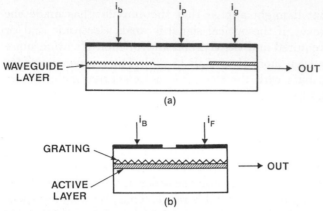

Fig. 5.55a,b. Tunable single-frequency lasers; (a) DBR, (b) DFB

the grating reflection band. The current i_p injects carriers into the passive waveguide, reducing its effective refractive index and continuously tuning the FP mode within the grating band until the next FP approaches the reflection peak, causing a "mode-hop". Continuous tuning ranges, on the order of the FP mode spacing of 380 GHz (31 Å), are possible. The Bragg grating can be tuned by i_b, covering a wide optical frequency range with contiguous FP modes. Then continuous coverage, by overlapping single-longitudinal-mode regions, of 1250 GHz (100 Å) is possible [5.139].

The static tuning response with i_p is as large as $\varrho(0) \approx 13\,\text{GHz/mA}$ for a given mode, but is limited to ∼300 MHz in modulation frequency response by the spontaneous carrier lifetime (∼1 ns) in the phase region.

5.8.2 Tunable DFB

The tunable two-section DFB laser shown in Fig. 5.55b offers the prospect of high-speed FM modulation because the carrier lifetimes are shortened by stimulated emission in both sections. The tunability is due to a relative shift of the Bragg frequencies in the front and back sections due to a difference in bias currents i_F and i_B, respectively [5.140]. If we consider a reference plane between the two sections, assumed to be of equal length, the round-trip reflections between the two gratings must produce a phase shift of 0 or 2π at the DFB resonances, allowing two stop-band modes. With a quarter-wave shift (Sect. 5.4.4) between gratings, only one DFB mode is allowed, although, even without the phase slip, asymmetry in the DFB structure or in the two bias currents usually favors one mode. As noted in (5.4.9c), the phase shift on reflection varies approximately linearly with the offset δ of the operating frequency from the Bragg frequency. The DFB frequency tunes as the Bragg peaks separate, roughly over a range limited by the width of Bragg reflection. Beyond this range the threshold increases and higher-order grating modes are preferred.

Continuous tuning ranges of 260 GHz (21 Å at 1.55 μm) at constant output power have been reported [5.141] for a two-section DFB. A three-electrode DFB with a quarter-wave shifted grating and 1.2 mm length gave a similar tuning range but with a linewidth as low as 500 kHz [5.142]. The two outer sections were connected in parallel and the center current was larger than the side currents in order to compensate for the non-uniform carrier density with length.

The FM responses of one- and two-section DFBs consist of a strong relaxation peak at Ω_p, the same relaxation frequency observed in intensity modulation (Sect. 5.6.5). However, the response function may not be flat in magnitude or phase below ω_p [5.143, 144]. Under high bias current conditions, the relaxation peak and other variations in response are suppressed to give a uniform response out to high frequencies. A flat FM response bandwidth of 8 GHz has been reported [5.145] for a three-section DFB, although with reduced tunability. This device produced good FSK modulation up to 8 Gbit/s. The general analytical FM transfer function [5.143], however, can lead to distorted FSK spectra, as indicated by numerical analysis [5.146] and experiment [5.147], when wide tunability is required.

Appendix 5A

Glossary of Symbols

Notation	Definition	Sect.
a	photodetector response coefficient	5.7.5
$a = \delta n'/\delta n''$	linewidth parameter	5.2.2
$\hat{a}$	Fabry-Perot factor	5.2.11
$A(\omega)$	stimulated absorption constant (nepers/length)	5.2.4
B_{d}	receiver bandwidth	5.7.5
B_{s}	optical signal bandwidth	5.7.5
B_{o}	optical filter bandwidth	5.7.5
$d(f)$	receiver power density	5.7.5
$D(f)$	optical spectral density	5.7.5
e	electron charge	5.2.3
E	optical field	5.2.2
E	electron energy	5.2.3
$E_{\mathrm{Fc}}, F_{\mathrm{Fv}}$	quasi-Fermi levels	5.2.3
E_{g}	bandgap energy	5.2.1
$E_{\mathrm{L}} = \hbar\omega_{\mathrm{L}}$	laser energy	5.2.2, 4
$f = \omega/2\pi$	frequency	5.2.2
F	noise factor	5.7.2
g	gain coefficient (nepers/length)	5.2.2
$\overline{g}_0$	spatial gain constant	5.2.3
$g_0 = v_{\mathrm{g}}\overline{g}_0$	temporal gain constant	5.2.5
$G_0 = \mathrm{e}^{(\Gamma g - \alpha)L}$	single-pass power gain in an ideal TWA	5.2.2, 7.2
G_{FP}	power gain in an FPA	5.7.3
h	active layer thickness	5.2.1
h	Planck constant	5.7.5
$\hbar = h/2\pi$	Planck constant	5.2.3
i	small-signal injected current	5.6.5
I	total number of photons in a cavity mode	5.2.3
$I_{\mathrm{a}} = \eta_{\mathrm{I}}I_{\mathrm{d}}$	injected current	5.2.1
I_{a0}	steady-state injected current	5.6.5
I_{d}	external drive current, including leakage	5.2.9
J	injected current density	5.2.1
$k_0 = 2\pi/\Lambda$	free-space propagation constant	5.2.2
K	wavefront curvature factor	5.2.8
$K = 2\pi/\lambda$	grating wavenumber	5.4.2
L	active layer length	5.2.1
m	modulation index of spontaneous spectrum	5.2.11
M	intensity modulation response	5.6.5
M_{p}	height of resonance peak	5.6.5
n	phase refractive index	5.2.8
n	small-signal carrier density	5.6.5
n_{f}	active layer index	5.2.1
$n_{\mathrm{g}} = c/v_{\mathrm{g}}$	group refractive index	5.2.8
$\hat{n}_p$	modal index for pth mode	5.2.2
n_{s}, n_c	cladding layer indices	5.2.1
n_{sp}	spontaneous emission factor	5.2.4, 5.1, 7.2
N	carrier density	5.2.2, 3
N_0	steady-state carrier density	5.6.5
N_{t}	transparency carrier density	5.2.3
N_{th}	threshold carrier density	5.2.2
p	small-signal optical power	5.6.5

Appendix 5A (cont.)

Notation	Definition	Sect.
P	optical power	5.2.9
P_{off}	off-level power	5.6.8
P_{on}	on-level power	5.6.8
$r(z)$	amplitude reflection coefficient	5.4.2
R	mirror power-reflectivity	5.2.2
R_{spon}	spontaneous emission rate into a cavity mode	5.2.4
R_{stim}	stimulated emission rate into a cavity mode	5.2.3
s	small-signal photon density	5.6.5
$S = \Gamma I / v$	photon density	5.2.5
S_0	steady-state photon density	5.6.5
S_{off}	off-level photon density	5.6.8
S_{on}	on-level photon density	5.6.8
S_m	photon density in the mth longitudinal mode	5.2.8
t_{on}	turn-on time	5.6.8
t_{off}	turn-off time	5.6.8
T	temperature (absolute)	5.2.3
$v = whL$	active volume	5.2.3
v_g	group velocity	5.2.3
V	waveguide V-value	5.2.2
V_a	forward bias voltage	5.2.3
V_g	energy-gap voltage	5.2.1
w	active layer width	5.2.1
W	contact stripe width	5.3.1
α	power attenuation (nepers/length)	5.2.2
$\alpha_{\text{m}} = (\ln R_1 R_2)/2L$	mirror loss	5.2.9
$\overline{\beta}$	complex modal propagation constant	5.2.2
β	real modal propagation constant	5.2.2
$\beta_{\text{B}} = 2\pi/\lambda_{\text{B}}$	Bragg propagation constant	5.4.2
γ	fraction of total spontaneous emission coupled into a cavity mode	5.2.4, 5, 8
γ	exponential decay constant	5.4.2
Γ	mode confinement factor	5.2.2
$\delta = \beta - K/2$	Bragg detuning parameter	5.4.2
δg	incremental gain due to δN	5.2.2
$\delta n = \delta n' - j\delta n''$	incremental index due to δN	5.2.2
$\delta n_g = \delta n'_g - j\delta n''_g$	incremental index referred to zero gain	5.2.2
δN	increment in carrier density	5.2.2, 6.8
$\delta \omega$	frequency chirp	5.2.2, 8
$\delta \lambda$	wavelength chirp	5.6.8
$\Delta\Lambda$	spontaneous emission width (HWHM)	5.2.8
$\Delta\omega$	incremental bandwidth	5.7.2
$\Delta\omega$	laser linewidth	5.5.1
$\Delta\omega$	stop-band width of grating	5.4.2
$\Delta n = n_{\text{f}} - n_{\text{s}}$	index step	5.2.2
ε	gain compression coefficient (photon density)	5.2.6
ζ_0	mode impedance	5.2.2
ζ	slope efficiency	5.2.10
η	external quantum efficiency	5.2.9
η_{I}	current injection efficiency	5.2.9
κ	electron momentum	5.2.3
κ	grating coupling coefficient	5.4.2

Notation	Definition	Sect.
$\kappa = 2\Gamma\varepsilon\eta_{\mathrm{i}}/v\hbar\omega\eta_{\mathrm{T}}$	gain compression coefficient (power)	5.6.8
λ_{g}	energy-gap wavelength	5.2.1
λ_{B}	Bragg wavelength	5.4.2
λ_{s}	longitudinal mode wavelength spacing	5.2.8
$\lambda = c/f$	wavelength	5.2.2
Λ	grating period	5.4.2
$\nu = \Delta\omega/2\pi v_{\mathrm{g}}$	number of modes per unit length	5.7.2
ϱ	ratio of maximum and minimum gains	5.7.3
τ_n	spontaneous emission time	5.2.3
τ_{p}	photon lifetime	5.2.2, 6.5
τ_{i}	intraband relaxation time	5.2.3, 6
χ	mirror asymmetry factor	5.7.3
ϕ	phase of optical field	5.5.1
$\omega = 2\pi f$	optical (radian) frequency	5.2.2
$\omega_{\mathrm{B}} = 2\pi c/\lambda$	Bragg frequency	5.4.2
ω_{L}	laser frequency	5.2.4
ω_{s}	longitudinal mode frequency spacing	5.2.2
Ω	modulation (radian) frequency	5.6.5
$\Omega_{-3\text{-dB}}$	3-dB bandwidth	5.6.6
Ω_p	resonance frequency	5.6.6
$\Omega_m = g_0/\varepsilon$		5.6.6
$\Omega_0 = (S_0 g_0/\tau_{\mathrm{p}})^{1/2}$		5.6.6
$\Omega_{0,\mathrm{on}} = (S_{\mathrm{on}} g_0/\tau_{\mathrm{p}})^{1/2}$		5.6.8

References

5.1 A. Yariv: *Optical Electronics*, 3rd ed. (Holt, Rinehart and Winston, New York 1985)

5.2 H. Kressel, J.E. Butler: *Semiconductor Lasers and Heterojunction LEDs*, (Academic, New York 1977)

5.3 H.C. Casey, Jr., M.B. Panish: *Heterostructure Lasers*, Vols. A and B (Academic, New York 1978)

5.4 G.B.H. Thompson: *Physics of Semiconductor Laser Devices* (Wiley, New York 1980)

5.5 W.T. Tsang (ed.): *Semiconductors and Semimetals: Lightwave Communications Technology*, Vol. 22, Part A *"Material Growth Technologies"*, Part B *"Semiconductor Injection Lasers I"*, Part C *"Semiconductor Injection Lasers II, Light Emitting Diodes"* (Academic, New York 1985)

5.6 S.E. Miller, I.P. Kaminow (ed.): *Optical Fiber Telecommunications II* (Academic, New York 1988)

5.7 R.E. Nahory, M.A. Pollack, W.D. Johnston, R.L. Barns: Appl. Phys. Lett. **33**, 659–661 (1978)

5.8 S. Adachi: J. Appl. Phys. **58**, R1–R29 (1985)

5.9 S. Adachi: J. Appl. Phys. **53**, 5863–5869 (1982)

5.10 S. Adachi: Appl. Phys. **53**, 8775–8792 (1982)

5.11 C.H. Henry, L.F. Johnson, R.A. Logan, D.P. Clarke: IEEE J. **QE-21**, 1887–1892 (1985)

5.12 J. Manning, R. Olshansky, C.B. Su: IEEE J. **QE-19**, 1525–1530 (1983)

5.13 S.E.H. Turley: Electron. Lett. **18**, 590–592 (1982)

5.14 G.P. Agrawal, N.K. Dutta: *Long-wavelength Semiconductor Lasers,* (Van Nostrand Reinhold, New York 1986)

5.15 C.H. Henry: Spectral Properties of Semiconductor Lasers, [Ref. 5.5, Pt. B, Chap. 3]
C.H. Henry, R.H. Logan, K.A. Bertness: J. Appl. Phys. **52**, 4453 (1981)
C.H. Henry, R.A.Logan, H. Temkin, R.F. Merritt: IEEE J. **QE-19**, 941 (1983)

5.16 L.D. Westbrook, B.Eng: IEE Proc. **133**, 135–142 (1986)

5.17 T. Ikegami: **QE-8**, 470–476 (1972)

5.18 J.E. Kardontchik: IEEE J. **QE-18**, 1279–1286 (1982)

5.19 C.H. Henry: IEEE J. **QE-18**, L59–L64 (1982)

5.20 C. Kittel: *Elementary Statistical Physics* (Wiley, New York 1958) p. 141

5.21 C.H. Henry: J. Lightwave Technol. **LT-4**, 288–297 (1986)

5.22 M. Asada, Y. Suematsu: IEEE J. **QE-21**, 434–442 (1985)

5.23 A.P. Bogatov, P.G. Eliseev, B.N. Sverdlov: IEEE J. **QE-11**, 510–515 (1975)

5.24 H. Ishikawa, M. Yano, M. Takasagawa: Appl. Phys. Lett. **40**, 553–555 (1982)

5.25 J. Manning, R. Olshanky, D.M. Fye, W. Powazinik: Electron. Lett. **21**, 496–497 (1985)

5.26 T.P. Lee, C.A. Burrus, J.A. Copeland, A.G. Dentai, D. Marcuse: IEEE J. **QE-18**, 1101–1113 (1982)

5.27 K. Petermann: Opt. Quantum Electron. **10**, 233–242 (1978);
W. Streifer, D.R. Scifres, R.D. Burnham: Electron. Lett. **17**, 933 (1981)

5.28 K.Y. Lau, A. Yariv: High-Frequency Current Modulation of Semiconductor Injection Lasers [Ref. 5.5, Pt. B, Chap. 2]

5.29 P.M. Boers, M.T. Vlaardingerbroek, M. Danielson: Electron. Lett. **11**, 206–208 (1975)

5.30 B.W. Hakki, T.L. Paoli: J. Appl. Phys. **44**, 4113–4119 (1973)

5.31 I.P. Kaminow, G. Eisenstein, L.W. Stulz: IEEE J. **QE-19**, 493–495 (1983)

5.32 J. Wang: Electron. Lett. **21**, 929–931 (1985)

5.33 N.K. Dutta, R.B. Wilson, D.P. Wilt, P. Besomi, R.L. Brown, R.J. Nelson, R.W. Dixon: AT&T Technol. J. **64**, 1857–1884 (1985)

5.34 J.P. van der Ziel, H. Temkin, R.A Logan: Electron. Lett. **19**, 113–115 (1983)

5.35 N. Chinone, M. Nakamura: Mode Stabilized Semiconductor Lasers for 0.7–0.8- and 1.1–1.6 μm Region [Ref. 5.5, Pt. C, Chap. 2]

5.36 T. Hayakawa, N. Miyauchi, S. Yamamoto, H. Hayashi, S. Yano, T. Hijikata: J. Appl. Phys. **53**, 7224–7234 (1982)

5.37 T. Tsukada: J. Appl. Phys. **45**, 4899–4906 (1974)
M. Hirao, A. Doi, S. Tsuji, M. Nakamura, K. Aiki: J. Appl. Phys. **51**, 4539–4540 (1980)
M. Hirao, S. Tsuji, K. Mizuishi, A. Doi, M. Nakamura: J. Opt. Commun. **1**, 10–14 (1980)

5.38 I. Mito, M. Kitaura, K. Kobayashi, S. Murata, M. Seki, Y. Odagiri, H. Nishimoto, M. Yamaguchi, K. Kobayashi: J. Lightwave Technol. **LT-1**, 195–202 (1983)

5.39 K. Kobayashi, I. Mito: J. Lightwave Technol. **LT-3**, 1202–1210 (1985)

5.40 U. Koren, B.I. Miller, R.J. Capik: Electron. Lett. **22**, 947–949 (1986)

5.41 H. Ishikawa, H. Imai, T. Tanahashi, K. Hori, K. Takahei: IEEE J. **QE-18**, 1704–1711 (1982)

5.42 Z.L. Liou, J.N. Walpole: Appl. Phys. Lett. **40**, 568–570 (1982)

5.43 T.L. Koch, L.A. Coldren, T.J. Bridges, E.G. Burkhardt, P.J. Corvini, B.I. Miller, D.P. Wilt: Electron. Lett. **20**, 856–857 (1984);
C.B. Su, V. Lanzisera, R. Olshansky, W. Powazinik, E. Meland, J. Schlafer, R.B. Lauer: Electron. Lett. **21**, 577–578 (1985)

5.44 J.E. Bowers, B.R. Hemenway, A.H. Gnauck, T.J. Bridges, E.G. Burkhardt: Appl. Phys. Lett. **47**, 78–80 (1985)

5.45 G. Eisenstein, U. Koren, R.S. Tucker, B.L. Kasper, A.H. Gnauck, P.K. Tien: Appl. Phys. Lett. **45**, 311–313 (1985)

5.46 H. Kawaguchi, T. Kawakami: IEEE J. **QE-13**, 556–560 (1977)

5.47 I.P. Kaminow, L.W. Stulz, J.-S. Ko, A.G. Dentai, R.E. Nahory, J.C. DeWinter, R.L. Hartman: IEEE J. **QE-19**, 1312–1318 (1983)

5.48 I.P. Kaminow, L.W.Stulz, J.-S. Ko, B.I. Miller, R.D. Feldman, J.C. DeWinter, M.A. Pollack: Electron. Lett. **19**, 877–879 (1983)

5.49 I.P. Kaminow, J.-S. Ko, R.A. Linke, L.W. Stulz: Electron. Lett. **19**, 784–785 (1983)
5.50 I.P. Kaminow, R.E. Nahory, M.A. Pollack, L.W. Stulz, J.C. DeWinter: Electron. Lett. **15**, 763–765 (1979)
5.51 M.-C. Amann, B. Stegmuller: Appl. Phys. Lett. **48**, 1027–1029 (1986);
A.M. Rashid, R. Murison, J. Haynes, G.D. Henshall, T.E. Stockton, A. Janssen: J. Lightwave Technol. **LT-6**, 25–28 (1988)
5.52 C.J. Armistead, S.A. Wheeler, R.G. Plumb, R.W. Musk: Electron. Lett. **22**, 1145–1147 (1986);
C.J. Armistead, B.R. Butler, S.J. Clements, A.J. Collar, D.J. Moule, S.A. Wheeler, M.J. Fice, H. Ahmed: Electron. Lett. **23**, 592–593 (1987)
5.53 W.T. Tsang, R.A. Logan: Appl. Phys. Lett. **45**, 1025–1027 (1984)
5.54 W.T. Tsang, N.A. Olsson, R.A. Logan, C.H. Henry, L.F. Johnson, J.E. Bowers, J. Long: IEEE J. **QE-21**, 519–526 (1985)
5.55 S.E.H. Turley, G.D. Henshall, P.D. Greene, V.P. Knight, D.M. Moule, S.A. Wheeler: Electron. Lett. **17**, 868–870 (1981)
5.56 J.P. van der Ziel, R.A. Logan, W.A. Nordland, R.F. Kazarinov: J. Appl. Phys. **57**, 1759–1762 (1985)
5.57 C. Lin, C.A. Burrus, L.A. Coldren: J. Lightwave Technol. **LT-2**, 544–549 (1984)
5.58 K.Y. Liou, C.A. Burrus, R.A. Linke, I.P. Kaminow, S.W. Granlund, C.B. Swan, P. Besomi: Appl. Phys. Lett. **45**, 729–731 (1984)
5.59 W.T. Tsang: The Cleaved-Coupled-Cavity (C^3) Laser [Ref. 5.5, Pt. B, Chap. 5]
5.60 I.P. Kaminow, H.P. Weber, E.A. Chandross: Appl. Phys. Lett. **18**, 497–499 (1971)
5.61 H. Kogelnik, C.V. Shank: Appl. Phys. Lett. **18**, 152–154 (1971)
5.62 H. Kogelnik, C.V. Shank: J. Appl. Phys. **43**, 2327–2335 (1972)
5.63 H.A. Haus: *Waves and Fields in Optoelectronics* (Prentice-Hall, Englewood Cliffs, NJ 1984) Chap. 8
5.64 W. Streifer, D.R. Scifres, R.D. Burnham: IEEE J. **QE-11**, 867–873 (1975)
5.65 M. Kitamura, M. Yamaguchi, S. Murata, I. Mito, K. Kobayashi: J. Lightwave Technol. **LT-2**, 363–369 (1984)
5.66 L.D. Westbrook, A.W. Nelson, P.J. Fiddyment, J.S. Evans: Electron. Lett. **20**, 225–226 (1984)
5.67 H. Temkin, G.J. Dolan, N.A. Olsson, C.H. Henry, R.A. Logan, R.F. Kazarinov, L.F. Johnson: Appl. Phys. Lett. **45**, 1178–1180 (1984)
5.68 W. Streifer, R.D. Burnham, D.R. Scifres: IEEE **QE-11**, 154–161 (1975);
T. Matsuoka, H. Nagai, Y. Noguchi, Y. Suzuki, Y. Kawaguchi: J. J. Appl. Phys. **23**, L38–L140 (1984)
5.69 S.L. McCall, P.M. Platzman: IEEE J. **QE-21**, 1899–1904 (1985)
5.70 Y. Suematsu, S. Arai, K. Kishino: J. Lightwave Technol. **LT-1**, 161–178 (1983)
5.71 M. Yamaguchi, M. Kitamura, S. Murata, I. Mito, K. Kobayashi: Electron. Lett. **21**, 63–65 (1985)
5.72 H. Soda, H. Imai: IEEE J. **QE-22**, 637–641 (1986)
5.73 C.H. Henry: IEEE J. **QE-19**, 1391–1397 (1983)
5.74 L.D. Westbrook, I.D. Henning, A.W. Nelson, P.J. Fiddyment: IEEE J. **QE-21**, 512–518 (1985)
5.75 R. Lang: IEEE J. **QE-18**, 976–983 (1982)
5.76 S. Saito, O. Nilsson, Y. Yamamoto: Appl. Phys. Lett. **46**, 3–5 (1985)
5.77 M. Ohtsu, S. Kotajima: IEEE J. **QE-21**, 1905–1912 (1985)
5.78 S. Saito, Y. Yamamoto: Electron. Lett. **17**, 325–327 (1981)
5.79 E. Patzak, H. Olesen, A. Sugimura, S. Saito, T. Mukai: Electron. Lett. **19**, 938–940 (1983)
5.80 G.P. Agrawal: IEEE J. **QE-20**, 468–471 (1984)
5.81 R.W. Tkach, A.R. Chraplyvy: J. Lightwave Technol. **LT-4**, 1711–1716 (1986)
5.82 D. Lenstra, B.H. Verbeek, A.J. den Boef: IEEE J. **QE-21**, 674–679 (1985)
5.83 R. Wyatt, W.J. Devlin: Electron. Lett. **19**, 110–112 (1983)
5.84 N.A. Olsson, C.H. Henry, R.F. Kazarinov, H.J. Lee, B.H. Johnson: Appl. Phys. Lett. **51**, 92–93 (1987);
N.A. Olsson, B.H. Johnson, H.J. Lee, C.H. Henry, R.F. Kazarinov, D.A. Ackerman, P.S. Anthony: Electron. Lett. **23**, 688–689 (1987)

5.85 K.-Y. Liou, Y.K. Jhee, G. Eisenstein, R.S. Tucker, R.T. Ku, T.M. Shen, U.K. Chakrabarti, P.J. Anthony: Appl. Phys. Lett. **48**, 1039–1041 (1986)
5.86 E. Patzak, A. Sugimura, S. Saito, T. Mukai, H. Olesen: Electron. Lett. **19**, 1026–1027 (1983)
5.87 H. Sato, J. Ohya: IEEE J. **QE-22**, 1060–1063 (1986)
5.88 J.E. Bowers, B.R. Hemenway, A.H. Gnauck, D.P. Wilt: IEEE J. **QE-22**, 833–844, (1986); see also
R. Olshansky, P. Hill, V. Lanzisera, W. Powazinik: IEEE J. **QE-23**, 1410–1418 (1987)
5.89 R.S. Tucker, D.J. Pope: Trans. **MTT-31**, 289–294 (1983)
5.90 R.S. Tucker, I.P. Kaminow: J. Lightwave Technol. **LT-2**, 385–393 (1984)
5.91 K.Y. Lau, A. Yariv: IEEE J. **QE-21**, 121–137 (1985)
5.92 R.S. Tucker: IEEE J. Lightwave Technol. **LT-3**, 1180–1192 (1985)
5.93 K.Y. Lau, N. Bar-Chaim, I. Ury, Ch. Harder, A. Yariv: Appl. Phys. Lett. **43**, 1–3 (1983)
5.94 C.B. Su, V. Lanziseva: Appl. Phys. Lett. **45**, 1302–1304 (1984)
5.95 R.S. Tucker: C. Lin, C.A. Burrus, P. Besomi, R.J. Nelson: Electron. Lett. **20**, 393–394 (1984)
5.96 Y. Arakawa, Y. Yariv: IEEE J. **QE-22**, 1887–1898 (1986)
5.97 K. Kamite, H. Sudo, M. Yano, H. Ishikawa, H. Imai: IEEE J. **QE-23**, 1054–1058 (1987)
5.98 M.S. Demokan, N. Nacaroglu: IEEE J. **QE-20**, 1016–1022 (1984)
5.99 R.S. Tucker: Electron. Lett. **20**, 802–803 (1984); see also
R.S. Tucker, J.M. Wiesenfeld, P.M. Downey, J.E. Bowers: Appl. Phys. Lett. **48**, 1707–1709 (1986)
5.100 R.A. Linke: IEEE J. **QE-21**, 593–597 (1985)
5.101 R. Olshansky, D. Fye: Electron. Lett. **20**, 928–929 (1984)
5.102 L. Bickers, L.D. Westbrook: Electron. Lett. **21**, 103–104 (1985)
5.103 N.A. Olsson, H. Temkin, R.A. Logan, L.F. Johnson, G.J. Dolan, J.P. van der Ziel, J.C. Campbell: J. Lightwave Technol. **LT-3**, 63–67 (1985)
5.104 P.M. Downey, J.E. Bowers, R.S. Tucker, E. Agyekum: IEEE J. **QE-23**, 1039–1047 (1987)
5.105 T.L. Koch, J.E. Bowers: Electron. Lett. **20**, 1038–1039 (1984). See also T.L. Koch, J.E. Bowers: "Factors Affecting Wavelength Chirping in Directly Modulated Semiconductor Lasers", Proc. CLEO '85 (Baltimore, MD) pp. 72–74
5.106 K. Petermann: Electron. Lett. **21**, 1143–1145 (1985)
5.107 T.L. Koch, R.A. Linke: Appl. Phys. Lett. **48**, 613–615 (1985)
5.108 J.P. Van der Ziel, R.A. Logan: IEEE J. **QE-18**, 1340–1350 (1982)
5.109 R.S. Tucker, J.M. Wiesenfeld, A.H. Gnauck, J.E. Bowers: Electron. Lett. **22**, 1329–1331 (1986);
J.M. Wiesenfeld, R.S. Tucker, P.M. Downey: Picosecond Measurement of Chirp in Gain-Switched Injection Lasers, Technol. Digest, CLEO '86, pp. 264–265
5.110 P.-T. Ho: Picosecond Pulse Generation with Semiconductor Diode Lasers, in *Picosecond Optoelectronic Devices*, ed. by C.H. Lee (Academic, New York 1984)
5.111 H.A. Haus: J. J. App. Phys. **20**, 1007–1020 (1981)
5.112 R.S. Tucker, G. Eisenstein, I.P. Kaminow: Electron. Lett. **19**, 552–553 (1983)
See also G. Eisenstein, R.S. Tucker, U. Koren, S.K. Korotky: IEEE J. **QE-22**, 142–148 (1986)
5.113 T. Mukai, Y. Yamamoto, T. Kimura: Optical Amplification by Semiconductor Lasers [Ref. 5.5, Pt. E, Chap. 3]
5.114 R.H. Saul, T.P. Lee, C.A. Burrus: Light-Emitting-Diode Device Design [Ref. 5.5, Pt. C, Chap. 5 pp. 193–237]
5.115 T.P. Lee, C.A. Burrus, Jr., B.I. Miller: IEEE J. **QE-9**, 820–828 (1973)
5.116 L.N. Kurbatov, S.S. Shakhidzhanov, L.V. Bystrova, V.V. Krapukhin, S.I. Kolonenkova: Sov. Phys. Semicond. **4**, 1739–1744 (1971)
D. Marcuse: J. Lightwave Tech. **LT-7**, 336–339 (1989)
C.E. Zah, J.S. Osinski, C. Caneau, S.G. Menocal, L.A. Reith, J. Salzman, F.K. Shokoohi, T.P. Lee: Electron. Lett. **23**, 990–991 (1987)

5.117 M.C. Amann, J. Boeck: Electron. Lett. **15**, 41–43 (1974)
5.118 D. Marcuse, I.P. Kaminow: IEEE J. **QE-17**, 1234–1244 (1981)
5.119 I.P. Kaminow, G. Eisenstein, L.W. Stulz, A.G. Dentai: IEEE J. **QE-19**, 78–82 (1983)
5.120 R. Olshansky, D. Fye, J. Manning, M. Stern, E. Meland, W. Powazinik, L. Ulbricht, R. Lauer: Electron. Lett. **21**, 730–731 (1985)
5.121 G. Arnold, H. Gottsmann, O. Krumpholz, E. Schlosser, E.-A. Schurr: Electron. Lett. **21**, 993–994 (1985)
5.122 D. Marcuse: *Principles of Quantum Electronics* (Academic, New York 1980) p. 251
5.123 G. Eisenstein: AT&T Bell Laboratories Technol. J. **63**, 357–364 (1984)
5.124 D.R. Kaplan, P.P. Deimel: AT&T Bell Laboratories Technol. J. **69**, 857–877 (1984)
5.125 T. Saitoh, T. Mukai, O. Mikami: J. Lightwave Technol. **LT-3**, 288–293 (1985);
      T. Saitoh, T. Mukai: Electron. Lett. **23**, 218–219 (1987)
5.126 C. Vassallo: Electron. Lett. **21**, 333–334 (1985)
5.127 G. Eisenstein, L.W. Stulz: Appl. Opt. **23**, l161–164 (1984);
      G. Eisenstein, L.W. Stulz, L.G. Van Uitert: J. Lightwave Technol. **LT-4**, 1373–1375 (1986)
5.128 N.A. Olsson, M.G. Oberg, L.D. Tzeng, T. Cella: Electron. Lett. **24**, 569–570 (1988)
5.129 G. Eisenstein, R.M. Jopson, R.A. Linke, C.A. Burrus, U. Koren, M.S. Whalen, K.L. Hall: Electron. Lett. **21**, 1076–1077 (1985)
5.130 G. Eisenstein, R.M. Jopson: Int'l. J. Electron. **60**, 113–121 (1986)
5.131 D. Marcuse: IEEE J. **QE-19**, 63–73 (1983)
5.132 A. Yariv: *Quantum Electronics*, 2nd ed. (Wiley, New York 1975) Chap. 12
5.133 P.S. Henry: IEEE J. **QE-21**, 1862–1879 (1985)
5.134 W.B. Davenport, W.L. Root: *An Introduction to the Theory of Random Signals and Noise* (McGraw Hill, New York 1985), Chap. 12
5.135 J.C. Simon: J. Opt. Commun. **4**, 51–62 (1983)
5.136 K. Kobayashi, I. Mito: J. Lightwave Technol. **LT-6**, 1623–1633 (1988)
5.137 Y. Kotaki, H. Ishikawa: IEEE J. **QE-25**, 1340–1345 (1989)
5.138 X. Pan, H. Olesen, B. Tromborg: IEEE J. **QE-24**, 2423–2432 (1988)
5.139 T.L. Koch, U. Koren: J. Lightwave Technol. **LT-8**, 274 (1990)
5.140 M. Kuznetsov: IEEE J. **QE-24**, 1837–1844 (1988)
5.141 Y. Yoshikuni, K. Oe, G. Motosugi, T. Matsuoka: Electron. Lett. **22**, 1153–1154 (1986)
5.142 Y. Kotaki, S. Ogita, M. Matsuda, Y. Kuwahara, H. Ishikawa: Electron. Lett. **25**, 990–992 (1989)
5.143 M. Kuznetsov, A.E. Willner, I.P. Kaminow: J. Appl. Phys. **55**, 1826–1828 (1989)
5.144 P. Vankwikelberge, F. Buytaert, A. Franchois, R. Baets, P.I. Kuindersma, C.W. Fredriksz: IEEE **QE-25**, 2239–2254 (1989)
5.145 H. Onaka, H. Miyata, Y. Kotaki, H. Kuwahara, T. Takada, Y. Takeuchi: Optical Fiber Communication Conf., San Francisco, Jan. 1990, Paper FD4
5.146 D. Marcuse: J. Lightwave Technol. **8**, 1110 (1990)
5.147 A.E. Willner, M. Kuznetsov, I.P. Kaminov, J. Stone, L.W. Stulz, C.A. Burrus: Photonics Technol. Lett. **1**, 412–415 (1989)

6. Semiconductor Integrated Optic Devices [1]

F.J. Leonberger and J.P. Donnelly

With 66 Figures

This chapter provides an overview of semiconductor integrated optics. The work in this field is motivated in part by the long-term desire to fabricate a variety of components on a common substrate. For example, Figure 6.1 shows a futuristic circuit formed on a direct-gap semiconductor that contains lasers, detectors, waveguides and modulators as well as an electronic circuit. Such a circuit could be useful in a variety of applications ranging from telecommunications to microwave systems to controls.

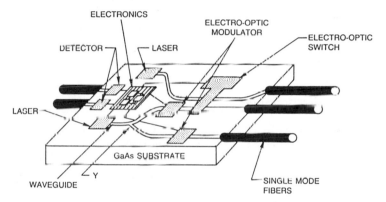

Fig. 6.1. Future GaAs integrated optical circuit

In this chapter the physics and technological status of single-mode guided-wave devices and integrated optoelectronic devices are presented, with primary emphasis on GaAs- and InP-based materials. The theory of semiconductor guides will first be reviewed. Emphasis will be on describing methods of providing index change and single-mode criteria for a variety of channel-guide geometries. A brief overview of epitaxial materials technology will be given to facilitate understanding realizable structures. A review of demonstrated passive guide structures, including a variety of channel guides, directional couplers, Y-splitters and wavelength filters will be presented. The status of guides formed in dielectric films deposited on semiconductors will be included. Next, guided-wave modulators will be reviewed. Opera-

---

[1] The Lincoln Laboratory portion of this work was sponsored by the Department of the Air Force

tional principles and representative performance of many devices including those based on the linear electro-optic effect and electroabsorption will be given. Multiple-quantum-well structures will also be reviewed, including efforts on all-optical devices. Finally, integrated structures will be discussed, including guided-wave structures (e.g., laser/waveguides) and integrated optoelectronic devices (e.g., laser/FET).

6.1 Semiconductor Waveguide Theory

In this section, the theoretical basis of optical waveguiding in semiconductors is discussed. Since optical waveguiding in any material requires a controlled variation of the refractive index in the plane perpendicular to the direction of power flow, methods of controlling the refractive index in semiconductors are discussed first. This is followed by a discussion of slab waveguides and the criteria for single-mode operation. Several methods of lateral confinement are then presented and the criteria for single-mode operation in channel guides discussed. Finally, coupling effects are introduced and loss mechanisms in semiconductor guides briefly discussed. Actual waveguide fabrication and results will be discussed in Sect. 6.3.

6.1.1 Methods of Index Change in Semiconductors

For waveguiding to take place, there must be a region in the semiconductor which has a higher refractive index than its surroundings. In the III-V compound semiconductor material systems, the index can be varied by changing the electrical properties, i.e., the number of free carriers and/or changing the alloy composition of the semiconductor. Strain and/or electric fields can also be used to locally alter the refractive index. Each of these is discussed below.

a) **Free-Carrier Effects.** Optical waveguiding can be obtained in a low-carrier-concentration semiconductor layer on an n^+ (or p^+) substrate of the same material, as illustrated in Fig. 6.2a. The presence of free carriers in the substrate lowers the refractive index from that of pure material [6.1, 2]. The low-carrier-concentration layer therefore has a higher refractive index than the substrate, and if thickness (of the layer) and index criteria, as discussed below, are met, light will be guided in the low-carrier-concentration layer.

The reduction of dielectric constant in n^+ material is due primarily to the negative contribution ε_{fc} associated with the free-carrier plasma to the real part of the dielectric constant. In a semiconductor with N_c free carriers per unit volume, the dielectric constant below the band edge is given approximately by

$$\varepsilon_s = \varepsilon_f - \varepsilon_{fc} = \varepsilon_f - \frac{N_c e^2 \lambda_o^2}{4\pi^2 c^2 m^*} \tag{6.1.1}$$

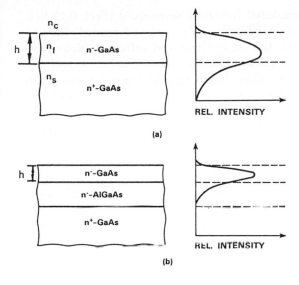

Fig. 6.2. (a) GaAs homojunction and (b) GaAs/AlGaAs heterojunction slab waveguides

where ε_f is the dielectric constant of pure material, e is the electronic charge, λ_o is the wavelength of the optical field in free space, c is the velocity of light, and m^* is the effective mass of carriers in the semiconductor. Since ε_{fc} is small compared to ε_f, the index is given approximately by

$$n - \left(\frac{\varepsilon_s}{\varepsilon_0}\right)^{1/2} = n_f - \Delta n \qquad (6.1.2)$$

where ε_0 is the dielectric constant of free space, n_f is the refractive index of the pure film material at λ_o and

$$\Delta n = \frac{N_c e^2 \lambda_o^2}{8\pi^2 \varepsilon_0 n_f c^2 m^*} \qquad . \qquad (6.1.3)$$

In n^+-GaAs, $n_f = 3.5$, $m^* = 0.067 m_e$, where m_e is the mass of a free electron and

$$\Delta n = 1.8 \times 10^{-21} N_c \lambda_o^2$$

when N_c is in cm^{-3} and λ_o is in μm. For $\lambda_o = 1\,\mu$m and $N_c = 2 \times 10^{18}$ cm^{-3}, $\Delta n = 0.0036$, which is sufficient to permit guiding if the pure layer is sufficiently thick.

The same free-carrier effect that leads to a reduction in the real part of the dielectric constant also leads to free-carrier absorption. As discussed in Sect. 6.1.5, this loss is a limitation on the minimum loss that can be achieved in n/n^+ homojunction waveguides. p^+ substrates can also be used, but the index variation with carrier concentration in the III-V semiconductors is less than in n^+ material and the loss is higher.

b) Effects of Composition Variations. Epitaxial techniques (Sect. 6.2) can be used to grow III-V semiconductor layers with abrupt or graded changes in alloy composition. Since both the energy gap and refractive index are a function of alloy composition, these epitaxial techniques are extremely useful for fabricating optical waveguides, lasers and detectors. To avoid the formation of dislocations and strains at the interface and/or in the grown layer, a good lattice match between the epitaxial layer and the substrate is required. Figure 6.3 shows the band gap vs. lattice constant for the III-V binary and ternary compounds [6.3]. By carefully controlling the composition, it is possible to grow closely lattice-matched ternary and quaternary layers on binary substrates. For example, $Al_x Ga_{1-x} As$ can be grown closely latticed-matched to GaAs over the entire range from GaAs to AlAs, while $Ga_{1-x} In_x As_y P_{1-y}$ $(y = 2.917x)$ can be grown lattice-matched to InP over the range from InP to $Ga_{0.47} In_{0.53} As$. Other quaternary compounds matched to binary substrates such as $Ga_x In_{1-x} As_y Sb_{1-y}$ matched to InAs and $Al_x Ga_{1-x} As_y Sb_{1-y}$ matched to GaSb are also possible. In general, larger index changes can be achieved using compositional changes than by free-carrier effects.

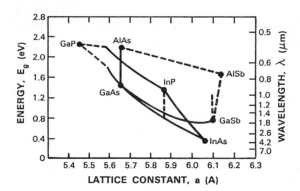

Fig. 6.3. Band-gap energy vs lattice constant for III-V compound semiconductors (after [6.3])

The ability to grow these lattice-matched alloys allows one to change the effective index without substantially affecting the quality of the material. A GaAs/AlGaAs heterojunction slab waveguide is illustrated in Fig. 6.2b. As the Al concentration increases, the band gap of the AlGaAs layer increases, while the refractive index decreases. The band gap of AlGaAs at 295 K as a function of Al composition [6.4, 5] is shown in Fig. 6.4a. For Al compositions greater than 0.45, the band gap is indirect. The refractive index vs. Al concentration for several different wavelengths [6.6–10] is shown in Fig. 6.4b. For $\lambda > 1.0\,\mu$m, Δn relative to GaAs is $\approx 0.45x$.

The band gap [6.11] and the refractive index [6.12, 13] for several different wavelengths vs. composition for quaternary layers grown lattice-matched on InP are shown in Figs. 6.5a, b, respectively. The band gap is direct over the entire range of composition and decreases with increasing Ga (and As).

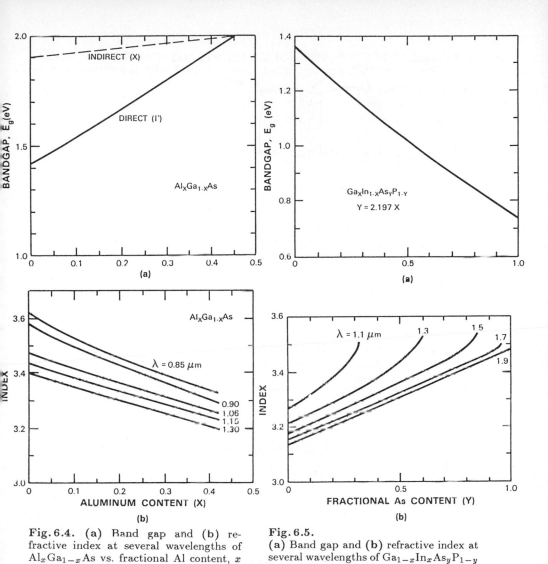

Fig. 6.4. (a) Band gap and (b) refractive index at several wavelengths of $Al_x Ga_{1-x}As$ vs. fractional Al content, x (after [6.4])

Fig. 6.5.
(a) Band gap and (b) refractive index at several wavelengths of $Ga_{1-x}In_xAs_yP_{1-y}$ lattice matched to InP vs. fractional As content, y (after [6.12])

The refractive index at any wavelength increases monotonically from InP to $Ga_{0.47}In_{0.53}As$.

With the development of molecular beam epitaxy (MBE) and met-alorganic chemical vapor deposition (MOCVD), it is also possible to grow structures consisting of alternately thin layers of two different composi-tions, i.e., quantum wells and superlattices [6.14–18]. In a quantum-well structure, the thickness of the low-band-gap, high-refractive-index layers is on the order of the wavelength of the free carrier, i.e., $\lesssim 200\,\text{Å}$. The lower electronic levels are therefore quantized, as illustrated in Fig. 6.6. The

321

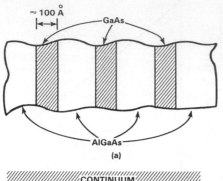

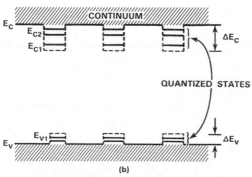

Fig. 6.6. (a) GaAs/AlGaAs multiple quantum well. (b) Quantized states in wells

difference between a quantum-well and superlattice structure is that the quantized states in a quantum well are localized while those of a superlattice are not. A quantum-well structure can consist of a single well while a superlattice must be periodic. Although a quantum-well structure can consist of two or more coupled wells, the thickness of the wide-band-gap layers in a multi-quantum-well structure are usually made large enough so that individual wells (or group of coupled wells) are essentially isolated. In a superlattice, the wide-band-gap layers are thin enough so the waveforms of the states in the wells do overlap resulting in non-localized states. Superlattices and quantum wells are a way of performing "band-gap" engineering and the unique features of these structures should prove useful in waveguides, lasers, detectors and other integrated-optic circuit elements. The layered structure of a superlattice leads to an anisotropy in the index and thus birefringent waveguides [6.19–21] as shown, for example, in Fig. 6.7. Here, light polarized along the plane of the layer interfaces sees an index larger than that for a mixed crystal of the same average composition and larger than that for light polarized perpendicular to the layers. It has also been shown that these structures have a large, easily saturated exciton absorption peak [6.22–23], which should prove useful in nonlinear optical-waveguide devices. Recently there has also been some interest in non-lattice-matched strained superlattices [6.24, 25] and quantum wells [6.26], but it is still too early to tell if they will have use in semiconductor integrated optical circuits.

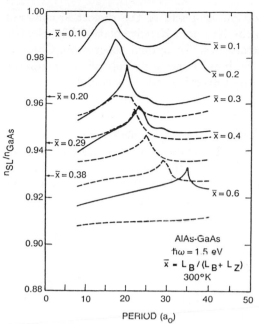

Fig. 6.7.
Calculated normalized anisotropic index of GaAs/GaAlAs superlattice structures. Solid and dashed lines are for light polarized parallel and perpendicular to the layers respectively. Arrows on the left indicate the experimental alloy values for the mole fractions $\bar{x}$ of Al. Mole fractions on the right indicate in descending order the effective values for the curves. L_B and L_Z are the width of the barrier and well respectively [6.19]

c) Electric Field Effects. The index of refraction can also be changed by electric fields via the linear electro-optic effect. Crystals that have a center of symmetry, such as Si or other semiconductors with a diamond lattice, do not exhibit a first-order electro-optic effect, but only second-order effects, i.e. those proportional to E^2, which are similar to the Kerr effect in liquids. The III-V semiconductors, which have a zinc-blende crystal structure ($43m$ point-group symmetry), exhibit a first-order electro-optic effect [6.27]. The only non-vanishing terms in the electro-optic tensor are $r_{41} = r_{52} = r_{63}$. The electro-optic coefficient r_{41} in both GaAs and InP is approximately -1.4×10^{-10} cm/V at $\lambda = 1.3\,\mu$m [6.28, 29]. The electro-optic effect is discussed more fully in Sect. 6.4.

d) Strain Effects. Strain affects the refractive index by changing the lattice constant and therefore the band gap of the semiconductor. For simple compressive and tensile strain, changes in the index are similar to those caused by changes in lattice constant due to temperature and/or pressure. The large photoelastic coefficients of the III-V materials cause significant changes in the refractive index for relatively small strains. Changes in the relative dielectric constant of 10^{-2} are possible with stresses produced by metal and/or oxide layers [6.30]. Changes caused by lattice-matched grown-layer strain are $\sim 10^{-4}$.

Both electric fields and/or strain effects are often inadvertently present in semiconductor waveguides and therefore can have an influence on waveguide results even when they are not the primary cause of waveguiding.

Since changes in the real and imaginary parts of the dielectric constant are related to each other through the Kramers-Kronig relationship, electric field and strain also change the semiconductor optical absorption. Changes in absorption are stronger near the band edge and drop off more rapidly with wavelength than changes in the refractive index (Sect. 6.4).

6.1.2 Slab Waveguides

Some examples of homojunction and heterojunction slab waveguides are illustrated in Figs. 6.2a, b, respectively. For the case of a single higher-index layer on a substrate of lower index (Fig. 6.2a) the guide is a simple asymmetric three layer waveguide with $n_f > n_s > n_c$ (Chap. 2), where n_f, n_s and n_c are the indices of the guide, substrate and superstrate, respectively (usually $n_c = 1$).

As discussed in Chap. 2, both pure TE and TM modes can propagate in a slab waveguide. The number of TE and TM modes that will propagate depends on the guiding layer thickness h and the difference in index between the guiding and confining layers. The cutoff condition for any mode can be determined by setting the reciprocal tail length in the substrate $\gamma_s = 0$ for that mode. This is the point where $\beta = (2\pi/\lambda)n_s$ and light is no longer confined to the high-index layer, but can radiate into the substrate. The resulting equation is

$$\frac{2\pi}{\lambda} h_m (n_f^2 - n_s^2)^{1/2} = \tan^{-1} r_o \left(\frac{n_s^2 - n_c^2}{n_f^2 - n_s^2} \right)^{1/2} + (m - 1)\pi \qquad (6.1.4)$$

where m is the mode number. Here $r_o = 1$ for TE modes and $(n_f^2/n_s^2)^{1/2}$ for TM modes. For the special case where $n_c = n_s$ (the symmetric slab waveguide), there is no minimum thickness needed for the lowest-order mode to propagate, i.e., the lowest order mode of a symmetric slab waveguide has no cutoff thickness. For small h however, most of the optical power will be carried outside the high-index layer.

For a homojunction slab waveguide in GaAs (Fig. 6.2a) in which the substrate has a carrier concentration of $N_c = 2 \times 10^{18}\,\text{cm}^{-3}$, $n_f = 3.5$ and $\Delta n = n_f - n_s = 0.0036$ and $(n_f^2 - n_s^2)^{1/2} = (2n_f\Delta n)^{1/2} \approx 0.247$ for $\lambda = 1\,\mu\text{m}$. For $n_c = 1$, the cutoff thickness [μm] of each mode is given by

$$h_m(\text{TE}) = 1.52779 + (m - 1)3.15051$$

for TE modes and

$$h_m(\text{TM}) = 1.57138 + (m - 1)3.15051$$

for TM modes ($m = 1, 2, 3, \ldots$). For a guide to be single moded (the TE and TM modes are almost degenerate) the thickness of the epitaxial layer must be such that $h_1 < h < h_2$. The cutoff thickness of the lowest and first higher-order modes in a GaAs homojunction slab guide vs. substrate concentration for $\lambda = 1\,\mu\text{m}$ is shown in Fig. 6.8a.

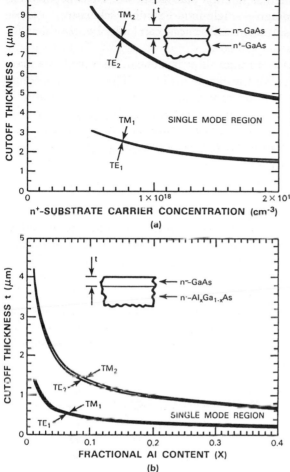

Fig. 6.8. (a) Thicknesses of the n^--layer in an n^-/n^+ GaAs slab waveguide at which the lowest and first higher-order modes for $\lambda = 1\,\mu m$ are cut off. **(b)** Thicknesses of the n^--GaAs layer in a GaAs/AlGaAs heterojunction slab waveguide at which the lowest and first higher-order mode for $\lambda = 1\,\mu m$ are cut off

For a GaAs/AlGaAs heterojunction guide in which the thickness of the AlGaAs layer is much thicker than γ_s, the decay constant in this layer, the solution for a three-layer guide can be used as an approximate solution. Otherwise the solution to the four-layer guide problem must be used. For a thick $Al_xGa_{1-x}As$ layer with $x = 0.3$, $\Delta n = 0.135$ at $\lambda = 1\,\mu m$. The cutoff thickness for each mode is therefore given by

$$h_m(TE) = 0.21156 + (m - 1)0.51938$$

for TE modes and

$$h_m(TM) = 0.25565 + (m - 1)0.51938$$

for TM modes. The cutoff thickness for the lowest- and first-higher-order modes in a GaAs/AlGaAs heterojunction slab guide vs. Al concentration is shown in Fig. 6.8b. Comparable single-mode guide thicknesses to those obtained with carrier concentration differences can only be obtained for layers with small values of Al ($x \lesssim 0.05$).

Slab guides consisting of more than three layers can be solved in a manner similar to the three-layer slab-waveguide problem. The algebra becomes more complex as the number of layers increases because of the number of boundary matching equations involved. In general, slab guides with arbitrary variations in index cannot be analyzed in closed form. A series solution is always possible, however, and in many cases these guides can be approximated by a multi-layer slab guide.

6.1.3 Channel Waveguides

Perpendicular to the direction of propagation, slab waveguides only confine the optical wave in one dimension and the optical wave is free to diffract in the other direction. For integrated-circuit applications, the optical wave must be confined in both directions perpendicular to the direction of propagation. Guides which confine the light in both these directions are called "channel waveguides". To confine the light laterally, an index variation in the y as well as the x direction is required. Compositional and/or carrier concentration variations can be used to provide lateral as well as vertical confinement.

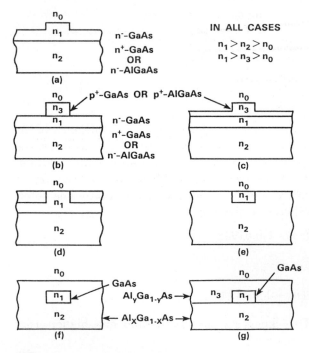

Fig. 6.9a–g. Some examples of channel waveguides: (a) rib guide, (b) and (c) dielectric-loaded strip guide, (d) channel-stop guide, (e) embedded guide and (f) and (g) buried heterojunction guides. With respect to the nomenclature used in Fig. 2, $n_c = n_0$, $n_f = n_1$, and $n_s = n_2$

Some examples of channel guides are shown in Fig. 6.9. In all of these cases, other material combinations, such as different AlGaAs combinations for n^+GaAs or GaInAsP and InP for GaAs and AlGaAs, respectively, can be used to obtain similar waveguide structures. Rib guides (Fig. 6.9a) are commonly used because of their ease of fabrication. They can be either homojunction or heterojunction guides as indicated. Strip loaded guides (Figs. 6.9b and c) and channel-stop guides (Fig. 6.9d) can be considered variations of rib guides. Embedded guides (Fig. 6.9e) have been fabricated using selective proton bombardment of n^+GaAs [6.31]. Buried heterojunction waveguides (Figs. 6.9f, g) are likely to be the most versatile for integrated-optical circuit applications. They should prove the most flexible in design and should be easier to integrate with heterojunction lasers and detectors. It should be stressed that these are only examples. Variations and combinations of these are possible and with rapid advances being made in epitaxial growth, III-V semiconductor waveguide technology should also advance rapidly.

Quantum-well, superlattice or large optical cavities [6.32] may be used for part or all of the guiding region. A p^+-n junction in several guides is illustrated in Fig. 6.9. By reverse biasing the p^+-n junction, an electric field can be applied to the guiding region to either change the refractive index via the electro-optic effect and/or the absorption through electro-absorption [6.33]. The application of reverse bias can also change the index by increasing the guide (depletion) width and/or by reducing the free-carrier concentration in the guide. Forward bias can also be used.

Channel guides can also be made utilizing metal or oxide strips. In metal-gap guides [6.34] in GaAs, as illustrated in Fig. 6.10a, lateral guiding

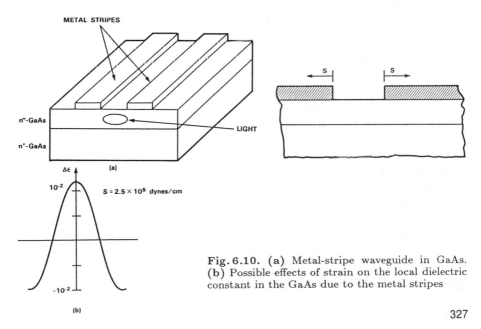

Fig. 6.10. (a) Metal-stripe waveguide in GaAs. (b) Possible effects of strain on the local dielectric constant in the GaAs due to the metal stripes

327

can result from the metal's predominately imaginary index reducing the effective index under the metal or from strain changing the index near the metal edge [6.35]. Strain in the semiconductor produced by stress in the deposited metal can cause substantial variations in the refractive index near the metal's edge (as illustrated in Fig. 6.10b) and in metal-gap guides, this phenomenon can often be the major cause of lateral confinement. Results obtained with metal-gap guides therefore are expected to depend on the metal and deposition procedure used. Strain guiding can produce guiding in several different places and does not necessarily require a higher-index guiding layer [6.36]. Therefore it may be possible to determine which effect is causing the guiding.

In general, closed-form analytic solutions for channel waveguides cannot be obtained. Furthermore, modes in channel waveguides are neither pure TE nor pure TM. Modes in channel waveguides are generally denoted as E^x_{mn} if the transverse electric field is primarily in the x direction and E^y_{mn} if the transverse electric field is primarily along the y axis [6.37]. The m and n subscripts denote the number of maxima in the x and y directions, respectively. The lowest-order mode therefore has $m = 1$ and $n = 1$. For most purposes, the E^x_{mn} modes can be considered quasi-TE modes while the E^y_{mn} modes can be considered quasi-TM modes.

Approximate solutions for channel waveguides can be found using the effective index method [6.38], *Marcatili*'s approximate analysis for rectangular waveguides [6.37], which is useful for buried or embedded guides, or his slab-coupled analysis, which is useful for rib guides [6.39] (which in many respects is similar to the effective-index method). Computer intensive methods such as the variational method [6.40], the mode-matching transverse transmission line technique [6.41] or the beam-propagation method [6.42] can also be used. The effective-index method is the simplest and the results of this method are obtained in a readily useable form which is similar to those obtained for slab waveguides. The results of this method, however, are probably the least accurate and are valid over a smaller range than the more numerically intensive methods.

To provide some insight into the channel waveguide problem, we shall use the effective index method to analyze a semiconductor rib waveguide. Using this method, which was discussed in Chap. 6.2, the rib waveguide is divided into sections I and II, as shown in Fig. 6.11a, and each region is assumed to be a slab waveguide of infinite extent in the x direction. Specifically, region I is assumed to be a three-layer slab guide consisting of an n^+-substrate, and n^--epitaxial layer of thickness h and air ($n_c = 1$), while region II is assumed to be a similar three-guide slab waveguide of thickness t. The effective guide indices N_{gI} and N_{gII} are then obtained by calculating the propagation constant or phase velocity $\beta_{I,II} = 2\pi N_{gI,II}/\lambda_o$ for each mode of the two sections using the usual asymmetric three-layer slab guide eigenvalue equation (Sect. 6.1.2). If both slab regions only support a

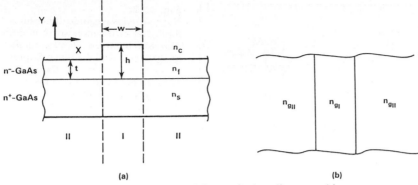

Fig. 6.11a, b. Effective-index method for analyzing rib waveguide

single mode a unique effective index for each region is determined and the rib guide will only support modes which have one maximum in the y direction, i.e., $n = 1$ in $E_{m,n}^{x,y}$. If the slab regions can support more than one mode the problem is more complicated, and the effective index of each mode should be calculated to see if the guide will support higher-order modes in the y direction. This is discussed more fully below. The guide is then analyzed by modeling it as a symmetric slab guide of uniform extent in the y direction with index N_{gI} in the guiding region, as shown in Fig. 6.11b. The second part of the analysis gives the overall propagation constant β (overall effective index N) and transverse propagation constant in the x direction. If more than one mode can be supported in the slab guide of the rib region (region I of Fig. 6.11a) the symmetric slab guide (Fig. 6.11b) must be solved for all combinations of N_{gI} and N_{gII} as long as N_{gI} is greater than the N_{gII} of the lowest order slab mode. Even if the slab guide of the rib region (region I) will support more than one mode, the rib guide itself can still be single moded if the overall effective index of the higher-order mode (taking into account x variation as well as y variation) is less than N_{gII} of the lowest-order mode of the slab guide. In this case the higher-order modes will radiate into the slab.

   Marcatili [6.39] performed an analysis similar to the effective-index analysis outlined here and has shown that the number of modes that the rib guide can support depends on the ratio of the defined ratios T/H and T/W, where T, H and W are the effective dimensions of the guide, obtained from t, h and w by adding the distance it takes the fields to reduce to $1/e$ in the adjacent regions. These distances are essentially the exponential decay constants in the different regions. Figure 6.12 shows *Marcatili*'s curves for the cutoff of the second-order mode in both the x and y directions and the region of single-mode operation. In this analysis, *Marcatili* assumed that the slab guide of the slab region (region II of Fig. 6.11a) was single moded since modes in general that have more than one maximum in this region will usually have an effective index less than the effective index of the lowest

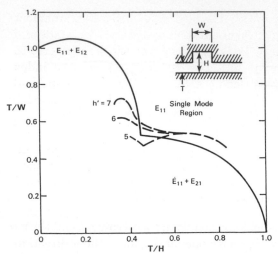

Fig. 6.12.
Normalized waveguide dimensions for higher-order mode at cutoff from *Marcatili's* slab-coupled waveguide analysis [6.39] (*solid line*). Superimposed are results of *Dagli* and *Fonstad* obtained using transverse transmission-line resonance method [6.45] (*dashed lines*)

slab mode. For thick slabs, this assumption is not necessarily valid [6.43]. It should also be noted that in actuality the cutoffs of the higher-order modes are not abrupt and for large guides, the higher-order modes may behave more like leaky-modes and if excited may require a long distance before they become insignificant. For practical purposes therefore, the slab guides of both regions (Fig. 6.11a) are kept single moded to ensure a single maximum in the y direction. *Dagli* and *Fonstad* [6.43–45] have used the mode-matching technique to analyze rib waveguides and compared their results to that obtained with the effective-index method and *Marcatili's* analysis [6.39]. Their cutoff results depend on the actual guide dimensions and not just the ratios and are superimposed on *Marcatili's* results in Fig. 6.12 for several normalized rib heights, h', where $h' = (2\pi/\lambda)h(n_1^2 - n_2^2)^{1/2}$.

The effective-index result is most accurate for cases when both the rib and slab regions (Fig. 6.11a) will only support one mode and the etch depth is small. Using a variational method, *Austin* [6.46–48] has shown that the effective-index method becomes less accurate as the etch depth increases and breaks down completely when the slab region (region II of Fig. 6.11a) becomes so thin that it is cut-off. In general, the effective-index result gives a propagation constant which is too high. As the rib guide is etched deeper, the lateral confinement increases and, in general, a higher number of modes can exist. If the index difference between the substrate and the guiding layer is not too large, however, the effective index of the higher-order mode may be less than that of the substrate and therefore the light will radiate into the substrate. A rib guide may therefore go from single mode to multi-mode back to single mode as it is etched deeper. This type of behavior has also been predicted when either the beam-propagation method [6.49] or mode-matching technique [6.45] is used. Both methods show that the slope of the rib side, and the width at the bottom must be carefully controlled if the higher modes are to be suppressed. The large lateral-mode confinement possible with deeply-etched guides should make it easier to make bends and "Y" branches. Single-mode deeply etched guides, however, are difficult to reproducibly fabricate because they are critically dependent on the index difference between the guiding layer and the substrate, and the shape and width of the bottom of the rib. In addition, scattering losses due to edge roughness can be a significant problem with deeply etched rib guides.

The effective-index or similar methods such as *Marcatili*'s rectangular-waveguide method [6.37] can also be used to give approximate solutions for other channel guides such as embedded or buried heterojunction guides. The effective-index method is much more accurate when the guide is much larger in one dimension than in the other.

6.1.4 Coupling Effects

Several different passive coupling effects are expected to be useful in integrated circuits utilizing semiconductor optical waveguides [6.50,51]. These include coupling from one waveguide to another, coupling from a forward traveling wave to a backward traveling wave and coupling from one mode to another such as coupling from a TE to a TM mode [6.52]. All coupling by its nature is wavelength dependent and in certain situations the coupler can be made to be effective only in a narrow band about a selected wavelength.

To obtain a high percent of power transfer in any coupling phenomenon, a synchronization between input and output waves is usually required. For coupling in or between waveguides this generally means guides and/or modes with equal propagation constants.

a) **Coupling Between Guides.** Coupling between guides can be treated either (i) by solving for the actual normal modes of the coupled system and matching the modes at the input and output to the external system, or (ii) by using coupled-mode theory. A detailed treatment has been presented in Chapter 3.

The modes of a coupled-waveguide system can be obtained in a manner similar to that used for a single waveguide, i.e., by matching the fields at the boundaries. When the coupled-waveguide system is composed of slab waveguides, the analysis is straightforward but becomes algebraically complex as the number of guides in the system increases. As was the case for isolated channel waveguides, an exact analytic solution is not possible for a coupled-waveguide system which has two-dimensional optical confinement. Approximate analytic solutions similar to those discussed for isolated channel waveguides, however, can often be used.

If the individual guides of the coupled-mode system are single moded, the number of modes generally equals the number of waveguides in the system. As with all coupled systems, the propagation constants of the modes

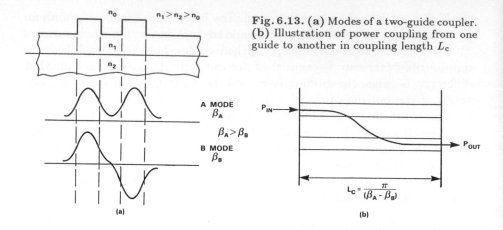

Fig. 6.13. (a) Modes of a two-guide coupler. (b) Illustration of power coupling from one guide to another in coupling length L_c

are split off from the propagation constants of the isolated guides that comprise the system.

The modes of a two-guide coupler comprised of two identical guides are summarized here. As shown in Fig. 6.13a, the symmetric A mode has a propagation constant β_A which is greater than that of an isolated guide while that of the B mode, β_B, is less than that of the isolated guide. If at $Z = 0$, modes A and B are in phase as shown, the fields of the two modes add constructively in the left-hand guide and destructively in the right-hand guide, i.e., the light will be confined primarily in the left-hand guide. Since the velocity of the two guides is different, at a distance $Z = L_C$ down the guide where L_C is the coupling length given by

$$L_C = \frac{\pi}{(\beta_A - \beta_B)} \tag{6.1.5}$$

the modes will be 180° out of phase and will add destructively in the left-hand guide and constructively in the right-hand guide, i.e., the light will have essentially transferred from the left- to the right-hand guide (Fig. 6.13b). In a two-guide coupler comprised of identical guides the sum of the modes at $Z = 0$ and $Z = L_C$ are always mirror images of each other around the mid-point between the guides.

The actual power transfer from one guide to another depends on the input and output optical structure as well as the coupler itself since some power is always lost in the transitions. If the guides are loosely coupled, the modes A and B are approximately linear combinations of the modes of the individual guides and if the input and output structures are individual isolated waveguides identical to the waveguides comprising the structures, little power is lost at the input and output transition. For a two-guide coupler composed of identical guides, we can then talk about "complete" power transfer from one guide to the other. If the guides are not identical, the two modes of the system will generally not cancel each other in the input guide at $Z = L_C$ and "complete" power transfer will generally not occur. Only in the special case where the non-identical guides comprising the coupler have the same propagation constant when isolated is good power transfer obtained. This brings in the concept of "synchronization" which is somewhat easier to comprehend when the problem is analyzed using coupled-mode theory.

Coupled-mode theory [6.53] is a perturbation technique that can be used to analyze loosely coupled waveguide problems. Among the assumptions used is that the modes of the system can be described by linear combinations of the unperturbed modes of the isolated waveguides that comprise the system. As discussed in Sect. 2.6.3, the fields in guides 1 and 2 can be expressed as

$$E_1 = A(z)\xi_1(x, y) \quad \text{and} \tag{6.1.6a}$$

$$E_2 = B(z)\xi_2(x, y) \quad , \tag{6.1.6b}$$

respectively, where ξ_1 and ξ_2 are the normalized transverse field shapes, and

$$A(z) = (A_1 e^{-j\beta_A z} + A_2 e^{-j\beta_B z}) \quad , \tag{6.1.7a}$$

$$B(z) = (B_1 e^{-j\beta_A z} + B_2 e^{-j\beta_D z}) \tag{6.1.7b}$$

where $\exp(j\omega t)$ is understood and

$$\beta_{A,B} = \frac{\beta_1 + \beta_2}{2} \pm \sqrt{\Delta^2 + K^2} \tag{6.1.8}$$

where $\Delta = (\beta_1 - \beta_2)/2$. Here β_1 and β_2 are the propagation constants of an isolated guide 1 and 2, respectively, and K is the coupling coefficient. The plus sign is used for β_A and the minus sign for β_B. From the coupled-mode equation $B_1 = (\beta_A - \beta_1)A_1/K$ and $B_2 = (\beta_B - \beta_1)A_2/K$. A_1 and A_2 are found by matching the input boundary conditions.

If all the power is in guide 1 at $Z = 0$, the power in each guide as a function of distance Z down the guide is given by

$$P_1 = P_T - P_2 \quad , \tag{6.1.9a}$$

$$P_2 = P_T \left(\frac{K^2}{\Delta^2 + K^2} \sin^2 \sqrt{\Delta^2 + K^2} Z \right) \tag{6.1.9b}$$

where P_T is the total power. Complete power transfer can only occur if the guides are synchronous, i.e., $\Delta = (\beta_1 - \beta_2)/2 = 0$. For $\Delta = 0$, power transfer occurs at

$$Z = L = \frac{\pi}{2K} = \frac{\pi}{(\beta_A - \beta_B)} \tag{6.1.10}$$

which is the same as (6.1.5).

333

If the guides are not loosely coupled, first-order coupled-mode theory does not give an accurate picture of the two-guide coupler. The propagation constants of the actual modes are shifted from those given above and complete power transfer is usually not obtained. More accurate propagation constants and power transfer can be obtained using a more precise coupled-mode formalism [6.54, 55].

For a coupler consisting of more than two guides, the situation is more complex [6.56]. When there are more than two modes that take part in a coupling phenomenon, good beats (i.e., power transfer) cannot be obtained unless the propagation constants of the modes are related to each other in a simple manner, i.e., the splitting of the propagation constants must be equal. For example, a three-guide coupler has three modes, as shown in Fig. 6.14. To couple power from one outside guide to another with high efficiency, the difference in propagation constants must be such that

$$\beta_A - \beta_C = \beta_C - \beta_B = \frac{\beta_A - \beta_B}{2}$$

or $2\beta_C - \beta_A - \beta_B$ must equal zero. For a three-guide coupler of identical guides, this is not the case unless the guides are very loosely coupled [6.57]. If the index or width of the center guide is increased a few percent in relation to the outside guides, $2\beta_C - \beta_A - \beta_B$ can be adjusted to zero [6.58]. From these results, it would appear that when more than two modes take part in power transfer in a multi-guide coupler, careful adjustment of guide widths and/or indices will be needed to obtain good power coupling. Since the changes are small, it is probable that these couplers will have to be adjusted using the electro-optic effect.

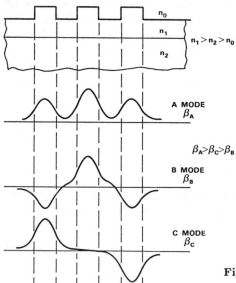

Fig. 6.14. Modes of a three-guide coupler

If first-order coupled-mode theory is used to analyze a three-guide coupler, the solutions for the βs are such that $2\beta_C - \beta_A - \beta_B$ equals zero. A more rigorous coupled-mode formalism which includes higher-order terms [6.54, 55], such as the cross power and detuning terms due to the proximity of the other guides, can give a better approximation for the βs. The inclusion of the higher-order terms, however, makes the solution to the coupled-mode equations much more difficult.

Coupling between tapered waveguides is also possible. Power has been transferred from an AlGaAs waveguide of higher index to one of lower index by tapering the thickness of the higher-index waveguide until it was essentially cut off [6.59]. Tapering of guides or the spacing between coupled guides should be gradual to minimize radiation loss.

b) Coupling Between Modes Due to Perturbations. Perturbations in a waveguide dimension or index can lead to coupling between the waveguide's normal modes, including radiation modes. This type of coupling is often an undesirable side effect of the tolerances and roughness inherent in any waveguide-fabrication procedure. Deliberate periodic perturbations, however, are often useful for coupling from one waveguide mode to another. For example, periodic gratings are used in distributed feedback (DFB) and distributed Bragg reflector (DBR) lasers to couple from a forward to a reverse traveling wave of the same mode and are used in wavelength-selective filters (Sect. 6.3.5). A waveguide with a periodic grating can be analyzed using coupled-mode theory or various eigenmode techniques. When coupled-mode theory is used, a coupling coefficient between the forward and reverse traveling modes is calculated from the perturbation in effective index [6.60]. The resultant coupled-mode equations are similar to those used to determine coupling between guides. The forward traveling wave decays exponentially with distance in the perturbed region with the power being coupled into the

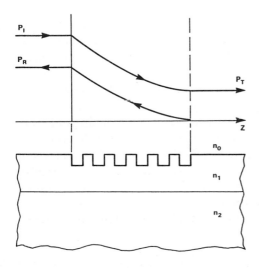

Fig. 6.15. Illustration of effects of periodic perturbation in optical waveguide. P_{I} is incident power, P_{T} transmitted power and P_{R} reflected power

reverse traveling wave, as shown in Fig. 6.15. The length of the perturbation region determines the amount of light transmitted and reflected.

If the grating has vertical side walls, it can be treated as a series of alternating waveguide sections of widths a and b which have slightly different mode shapes and different phase velocities, β_1 and β_2 respectively. A reflection occurs at each boundary and if the reflections constructively interfere a large reverse traveling wave can build up. Constructive interference of the reflected waves occurs if

$$\beta_1 a = \beta_2 b = (2m - 1)\tfrac{\pi}{2} \tag{6.1.11}$$

where $m = 1, 2, 3, \ldots$. For a small perturbation, the period of the grating is related to the average propagation constant $\beta_0 = (\beta_1 + \beta_2)/2$ and is given by

$$\Omega = a + b = \frac{m\pi}{\beta_0} = \frac{m\lambda}{2n} \quad . \tag{6.1.12}$$

Gratings with periods of a multiple of $\frac{\pi}{\beta_0}$ are generally used in AlGaAs/ GaAs DFB and DBR lasers since the fundamental period in the III-V semiconductors is about $1200\,\text{Å}$ for $\lambda = 0.8\,\mu$m.

c) **TE to TM Mode Coupling.** If the index of refraction in the waveguide is anisotropic, coupling between TE and TM modes can occur whenever two principal axes of the index ellipsoid do not lie in the plane of the waveguide. The index of refraction in the III-V and II-VI semiconductors is nominally isotropic but can be made anisotropic by the application of either an electric field or built-in strain. Strain effects have been previously described and electro-optic effects will be discussed in Sect. 6.4.

6.1.5 Optical Loss

Losses in semiconductor waveguides are due primarily to either scattering into radiation modes or absorption. In addition, radiation losses can occur at transitions and in curved sections. Roughness at waveguide boundaries can lead to scattering losses. This could be one reason why guides grown by LPE typically exhibit higher losses than those grown by vapor-phase techniques. These scattering losses are generally higher at boundaries where the refractive index change is large. Deeply etched semiconductor rib guides are especially prone to scattering loss because of the large index difference at the semiconductor-air interface. Smooth side walls are therefore important in these structures. The side-wall shape of these guides can be controlled using ion-beam-etching techniques, but these techniques generally faithfully reproduce roughness in the edges of the etch mask. Chemical etching tends to smooth side-wall roughness, but the side-wall shape tends to be deter-

by crystallographic directions. Scattering can also occur from imperfections, such as precipitates, metal inclusions and other defects in the semiconductor layers.

Various types of absorption can occur in the different semiconductor layers comprising the waveguide. Free-carrier absorption [6.1] is an important loss mechanism in n^+ and p^+ material and can place a lower limit on the losses in n/n^+ homojunction waveguides. Free carriers not only reduce the real part of the dielectric constant, but also increase the absorption of the semiconductor. The free-carrier absorption can be calculated along with changes in the real part of the dielectric constant and is given approximately by

$$\alpha = \frac{ge^3\lambda_o^2 N_c}{4\pi^2 m^{*2}\mu c^3 n\varepsilon_o} \qquad (6.1.13)$$

where μ is the mobility and g is a factor which depends on how the carrier scattering time is related to its energy and is usually slightly larger than one for acoustic lattice scattering and around three for ionized impurity scattering. The other parameters have been previously defined. The free-carrier absorption is proportional to the number of free carriers N_c, to the square of the free-space wavelength and inversely proportional to the carrier mobility. The λ_o^2 dependence of the free-carrier absorption has been found to be only approximately true, probably due to variations in the g factor with wavelength. Losses at $1.3\,\mu\mathrm{m}$ in n^+-GaAs ($n \sim 1\text{--}2 \times 10^{18}\,\mathrm{cm}^{-3}$) are usually $> 10\text{--}20\,\mathrm{cm}^{-1}$ ($43\text{--}86\,\mathrm{dB/cm}$). Minimum losses in a single-mode homojunction slab waveguide due to the n^+-substrate ($10^{18}\,\mathrm{cm}^{-3}$) are expected to be on the order of $0.6\,\mathrm{cm}^{-1}$ ($2.5\,\mathrm{dB/cm}$).

Interband free-carrier absorption, i.e., absorption due to a transition between electrons in a Γ minimum and a higher conduction band, are also possible and have been observed in n^+-InP. In p^+ material, absorption can also occur between the light- and heavy-hole valence bands, adding to the effects of carrier concentrations on absorption coefficient.

Below-bandgap absorption can also occur due to deep levels in the band gap. Figure 6.16 illustrates the absorption coefficient vs. wavelength in n-type InP with several different impurity concentrations [6.61]. As the wavelength approaches the band gap, absorption increases rapidly due to band tailing, exciton absorption, acceptor-donor absorption, etc. The Burstein shift, i.e., band filling, is responsible for the absorption decrease with increasing carrier concentrations for energies greater than the band-gap energy. Electric fields affect this near-band-edge region via the Franz-Keldysh or electroabsorption effect (Sect. 6.4). Below the band-edge, the absorption decreases to a minimum and then increases due to free-carrier absorption. The minimum reached will depend on band tailing, free-carrier interband absorption, deep levels, etc. For InP, the local minimum below the band edge is due to absorption from electrons being excited from the Γ mini-

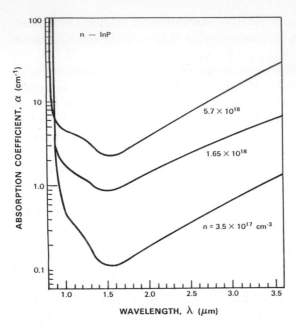

Fig. 6.16. Optical absorption vs. wavelength for n-type InP samples with different impurity concentrations [6.61]

mum to an L minimum. For low loss, waveguides should be fabricated in low-carrier-concentration material and operated below the band edge. In general, this minimum loss occurs about 0.5 μm from the band edge for ~ 1-μm band-gap materials. For devices relying on the electroabsorption or a nonlinear band-edge effect, however, operation near the band edge may be necessary.

In waveguides containing quantum wells or superlattices, additional absorption due to the quantized conduction and valence band levels are possible. An easily saturable exciton absorption, which is electric-field dependent, is also present in these structures and may be useful for some nonlinear optical effects.

A useful way to estimate the absorption loss α in guides is from the expression

$$\alpha = \sum_{i=1}^{3} \alpha_i P_i / P_\mathrm{T} \quad . \tag{6.1.14}$$

Here α_i and P_i are the absorption coefficient and the power carried in each of the guide layers (e.g., core, cladding) and P_T is the total power in the guided mode. For example, one can see that losses in heterojunction guides can be inherently lower than those in the homojunction guides for comparable low-loss core layers, since the wider-band-gap cladding of the heterojunction will have lower absorption than a heavily doped homojunction substrate.

Losses in semiconductor channel waveguides are often determined by measuring the transmission through several different lengths of the same

338

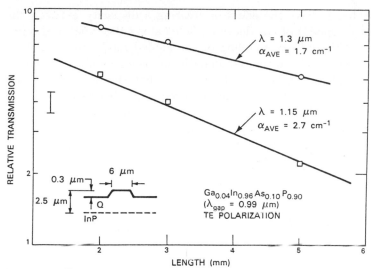

Fig. 6.17. The relative transmission at 1.15 and 1.3 μm vs. sample length for rib GaInAsP/ InP waveguides obtained using end-fire coupling. Waveguide loss is obtained from the slope of the data [6.94]

sample using an end-fire coupling technique. The transmission is plotted as a function of length, as illustrated in Fig. 6.17 for data obtained on some GaInAsP/InP waveguides at 1.15 and 1.3 μm, and the loss obtained from the slope of the data. Although this technique is usually satisfactory when losses are $\geq 0.5 \, \text{cm}^{-1}$ (2 dB/cm), variations in the quality of the cleaved faces as the sample is shortened and the existence of Fabry-Perot modes between the two end faces made it difficult to reproducibly measure very low losses using this technique. It has recently been reported that small changes in sample temperature can be used to change the index and length of the guides and therefore the effective length of the Fabry-Perot cavity [6.62]. By measuring the maxima and minima in the transmission, the losses can be determined without cleaving the sample to different lengths. The same effect can also be obtained by tuning the wavelength of the laser used in the measurements.

Loss has also been determined by measuring the light scattered out of the top surface due to imperfections in the waveguide as a function of distance. The technique is usually only suitable where the losses are high. Raman scattering has also been used to measure loss in GaAs/AlGaAs waveguides [6.63].

6.1.6 Curvature Loss

The ability to change direction on a waveguide chip is limited by mode conversion to radiation modes (assuming single-mode operation). Due to the optical confinement properties in which the optical fields decay outside of the waveguides, bends have the effect of creating a disparity between the travel distances (or speed) of the decaying fields along the inner and outer waveguide sides. This induces coupling of the guided mode into radiation modes. This problem has been treated analytically using coupled-mode theory [6.64, 65]. Qualitatively, the results indicate that the curvature losses are reduced by: (i) increasing the index difference Δn between the guide and its surrounding; (ii) increasing the width of the guide; (iii) choosing a larger height-to-width waveguide aspect ratio.

An expression for the losses per radian α_R can be approximated in terms of the waveguide parameters as follows [6.64]

$$\alpha_R[\text{dB/rad}] = \frac{8.686 \times 2b(1-b)\Delta nR}{w\sqrt{2n\Delta nb + \lambda}}$$

$$\times \exp\left\{-\frac{8\pi\Delta n}{3\lambda}\sqrt{\frac{\Delta n}{n}}R\left[1 - (1-b)\left(1 + \frac{wn}{4\Delta n(1-b)R}\right)^2\right]^{3/2}\right\}$$

$$(6.1.15)$$

where Δn is the index difference, w width, R radius of curvature, n refractive index of the waveguide, and b a normalized factor signifying the difference between the guided mode and the surrounding ($n_{\text{eff}} = n + b \cdot \Delta n$).

The sensitivity of the curvature losses to Δn and w has been estimated for GaAs guides [6.66]. One example is shown in Fig. 6.18 which gives curves

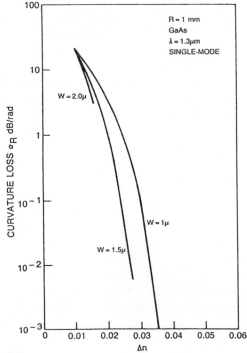

Fig. 6.18. Calculated curvature loss of single-mode GaAs waveguides vs. n with the guide width as parameters (after [6.66])

calculated from (6.1.15) for single mode GaAs waveguides at $1.3\,\mu$m. For small radii of curvature ($R = 1\,$mm in Fig. 6.18) the losses are small only if Δn is sufficiently large and w small enough to conform to single-mode operation. Therefore, for a 1-mm radius of curvature, Δn has to be greater than about 0.022 to yield $\leq 0.5\,$dB/rad of loss. The guide width required to maintain single-mode operation at $1.3\,\mu$m would have to be $1.5\,\mu$m (for a rib guide this can be met by a variety of rib heights and alloy parameters). Notice that as the width is increased to $2\,\mu$m, the range of Δn required to maintain single-mode operation is reduced to < 0.0155, and the loss is $> 3\,$dB/rad.

For wider single-mode guides (several micrometers wide), Δn is further reduced and radii of curvature of $1\,$mm cause excessive loss. For example, a radius of curvature of $10\,$mm is needed to yield losses $< 0.5\,$dB/rad when the guide width is $3\,\mu$m and $\Delta n > 0.005$. For larger single-mode widths, larger radii of curvature are needed to maintain low bending losses.

In order to emphasize the sensitivity of the curvature loss to Δn, the required radius of curvature R which yields 0.3 and $1\,$dB/rad is plotted in Fig. 6.19 as a function of Δn for a 2-μm-wide GaAs waveguide at $1.3\,\mu$m, over a range of Δn extending to 0.1. Naturally, such a range would cover single-mode as well as multi-mode operation. If that is ignored for a moment, one can notice that in order to obtain $0.3\,$dB/rad of loss, a Δn of about 0.019

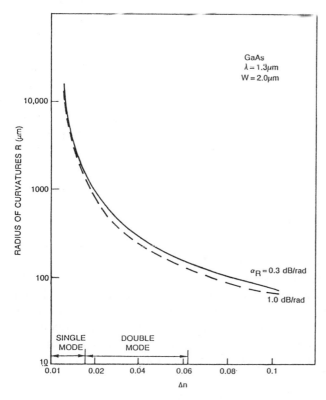

Fig. 6.19. Calculated dependence of curvature radius on n with loss as parameter for GaAs waveguides (after [6.66])

341

is needed with a radius of 1 mm. Similarly, Δn would have to be about 0.039 if the radius of curvature is reduced to 300 μm. Experimentally, a curvature loss of 0.6 dB/rad has been reported for a radius of curvature of 300 μm [6.47]. The reported waveguide was a 2-μm-wide, 1-μm-thick GaAs/Ga$_{0.88}$ Al$_{0.12}$As rib waveguide with a rib etch depth of 0.9 μm. From the reported results, it may be inferred that the Δn of such a waveguide with 12 % Al content in the bottom cladding is in the 0.04 neighborhood, which seems to be consistent with Fig. 6.19. It is difficult to assess the modal properties of such guides since the dimensions are small and the associated parameters are very near the threshold of single- and double-mode operation.

The above calculation indicates that small rib guides are desirable for minimal practical bends in semiconductor guides. Such small dimensions also are advantageous for low-voltage modulators, as discussed in Sect. 6.5. However, these design advantages need, in many cases, to be traded off against increased propagation losses and fiber-coupling losses. In narrow guides, an enhancement of loss due to a given fabrication-induced edge roughness (e.g., 500 Å) occurs as guide dimensions are reduced. With respect to fiber-coupling loss, for the 1.3-μm design case discussed above, for conventional single-mode 1.3-μm fiber (mode diameter $\sim 7\,\mu$m), a 2-μm wideguide can have much more (~ 6 dB) coupling loss than a guide sized to match the fiber mode. It may be possible to minimize these latter effects. For example, combined wet/dry etching can be used to smooth out sidewall roughness. Also, tapered and/or lensed fibers could be used with small guides, or an asymmetry could be created in the cladding index on both sides of a guide so that both the inner and outer tails of the decaying fields appear to travel the same optical path length. Other possibilities include ramping the index of the waveguide to achieve the same effect. This has recently been employed in LiNbO$_3$ waveguides by the superposition of a saw-tooth-like pattern of Ti on the original Ti-waveguide pattern to yield a ramped index within the waveguide [6.67]. These considerations illustrate the tradeoff between fiber insertion loss and curvature sharpness (and, as discussed in Sect. 6.5, modulation efficiency) that must be made for any practical circuit application.

6.2 Material Technology

As discussed in Sect. 6.1, semiconductor optical waveguides are generally fabricated using index differences due either to free-carrier concentration and/or compositional variations. Although some early GaAs waveguides were fabricated using selective proton bombardment of n^+ GaAs [6.68], which reduces the free-carrier concentration in the bombarded region, the bombardment also introduces additional optical loss [6.69], which can only partially be removed by annealing. Low-loss semiconductor waveguides today are therefore made almost exclusively using epitaxial techniques.

The rapid advances of epitaxial techniques over the last ten years have led not only to reduced optical losses ($\lesssim 1\,\mathrm{dB/cm}$) in semiconductor waveguides, but to the ability to conceive and develop integrated optical waveguides and components which are limited to a large extent only by the knowledge and imagination of the researcher. Various epitaxial techniques such as liquid phase epitaxy (LPE), vapor phase epitaxy (VPE), metalorganic chemical vapor deposition (MOCVD), also known as metalorganic vapor phase epitaxy (MOVPE), and molecular beam epitaxy (MBE), can now be used to fabricate homojunction, heterojunction, quantum-well and superlattice waveguides and components suitable for integrated optics in both the AlGaAs and GaInAsP material systems. Each of these epitaxial growth techniques offers various advantages and disadvantages for the two different material systems. A brief description of each is given in this section.

6.2.1 Liquid-Phase Epitaxy (LPE)

Liquid-phase epitaxy is a technology widely used for the fabrication of both AlGaAs/GaAs and GaInAsP/InP heterojunction laser diodes. It was developed to a large extent because of its low cost and the inability of hot wall VPE reactors to handle the growth of layers containing Al. Extensive reviews of LPE technology are available [6.70, 71]. Only the main points of the technique in its relevance to waveguides is reviewed here.

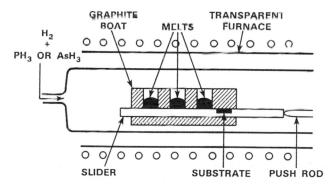

Fig. 6.20. Schematic illustration of a liquid-phase epitaxial (LPE) growth system

An illustration of a typical LPE growth system is shown in Fig. 6.20. The LPE growth system consists of a growth boat which is machined out of high-purity graphite, and a transparent furnace, so the melts can be observed before and during growth. The growth boat has several bins which contain the appropriate melt compositions to grow the required epitaxial layers, and a slider, which allows the substrate to be sequentially slid under each melt. Each melt consists of a saturated solution of Ga, Al and As for the growth of AlGaAs, and In, Ga, As and P for the growth of GaInAsP. Other

343

III-V compounds such as GaAlAsSb and GaAlP have also been grown successfully by LPE using appropriate melt compositions. As required, n- and p-type dopants can be added to the melts. After the substrate is slid under a melt, the temperature is usually reduced at a rate of $0.1-0.5°/\text{min}$, causing the solute in the melt to crystallize out at the substrate-melt interface. The melt can also be supercooled prior to sliding the substrate under the melt, allowing growth to take place under isothermal conditions. In many cases, both a controlled reduction in temperature and a controlled amount of supercooling are used.

The composition of the epitaxial layer depends on the composition of the melt, the growth temperature and, to some extent, the substrate orientation since there is an orientation dependence of the distribution coefficient. Accurate Ga-Al-As and Ga-In-As-P phase diagrams have been empirically determined over most of the range of interest so appropriate melts can be chosen for the desired epilayer compositions. For the growth of GaInAsP on InP substrates, the composition of the melt must be carefully chosen so the epitaxial layer is lattice-matched to the InP. This is less of a problem with AlGaAs since this ternary system is fairly well lattice-matched over its entire range. Since the thermal expansion coefficient of these materials is composition dependent, even when a perfect lattice match at the growth temperature is achieved, a small lattice mismatch occurs as the sample is cooled to room temperature. Growth temperatures for growing GaAs and AlGaAs from a Ga-rich solution on GaAs are typically around 800° C, while those for InP and GaInAsP from an In-rich solution on InP are typically in the range of 600–650° C.

The thickness of the epitaxial layer depends on the size of the melt, the cooling rate, growth time and the degree of supercooling, if any. For near-equilibrium growth, i.e., no supercooling and a slow cooling rate, the thickness depends primarily on the total temperature change during growth.

Although there is no fundamental problem in growing AlGaAs/GaAs/AlGaAs double heterojunction structures over the entire range of Al concentrations, great care must be taken to eliminate oxygen from the system because of the high oxidation rate of Al. On the other hand, there is a serious problem in growing lattice-matched InP/GaInAsP/InP double heterojunction structures if the As concentration in the quaternary layer is too high. GaInAsP layers corresponding to wavelengths beyond 1.4 μm tend to dissolve when brought in contact with an In-P solution because of the large amount of As required in the solution for it to be in equilibrium with the quaternary layer. This problem is often solved in the fabrication of long-wavelength laser diodes by growing an intermediate quaternary layer with a lower As concentration before growing the top InP layer.

Long pre-bake times are usually required to reduce the background concentration of the epitaxial layers. Residual Si doping is often a limiting factor and provisions must be made to prevent the H_2 in the growth stream from reducing the SiO$_2$ walls of the reactor if the lowest possible background carrier concentration is to be achieved. This is more of a problem with InP-based materials than GaAs-based materials because of a difference in the segregation coefficient of Si between the two material systems. Furthermore, if provisions are not made to keep the substrate from thermally etching prior to growth, the growth of a buffer layer is generally required. High-purity GaAs layers with 77 K mobility $> 100,000$ cm^2-V/s and net carrier concentrations in the 10^{14} cm^{-3} range have been grown utilizing a multiple-well graphite boat.

Although LPE has proven extremely useful for laser diode fabrication, it suffers from several severe limitations that tend to limit its usefulness for integrated optical circuits. The thickness of the epitaxial layer and its uniformity across the wafer can be difficult to control, especially when growing thick layers, and the morphology of the top surface is generally poor. These problems tend to increase as sample size increases and therefore LPE has generally been limited to small samples. The optical loss in LPE-grown layers also tends to be higher than in those grown by VPE, MOCVD or MBE. This may be due to roughness at the interfaces. Since dielectric films such as SiO$_2$ or Si$_3$N$_4$ can prevent nucleation, selective growth can be achieved by depositing such a film and opening holes where growth is desired. However, it is difficult to nucleate growth in very small holes and thickness control is also often a problem with small holes.

6.2.2 Vapor-Phase Epitaxy (VPE)

Vapor-phase epitaxial growth utilizing chloride transport of the column III element has been used successfully for the growth of high-purity GaAs and InP epitaxial layers. Since chloride transport requires a hot-wall reactor, the Al compounds cannot be handled and this system is therefore limited to growing GaAs on GaAs. GaInAsP alloys over the entire range from InP to GaInAs, however, can be grown lattice-matched to InP by VPE.

VPE can be performed using either the "trichloride" method, which utilizes AsCl$_3$ and/or PCl$_3$ to introduce both the column V elements and the chlorine needed for transport of the column III element(s), or the "hydride" method, which utilizes AsH$_3$ and/or PH$_3$ and HCl. Otherwise the methods are quite similar. The lowest background impurity concentrations have been obtained with the trichloride process and this process is generally used for growing high-purity GaAs.

A typical AsCl$_3$ Ga-boat VPE reactor is illustrated in Fig. 6.21. Here high-purity H$_2$ is passed through a temperature controlled AsCl$_3$ bath to introduce a controlled amount of AsCl$_3$ into the reactor chamber. The AsCl$_3$ flows over the Ga boat, where it decomposes, forming arsenic vapor and Ga chlorides. The mixture of arsenic and Ga chlorides is then swept downstream into the deposition zone where epitaxial growth takes place on the

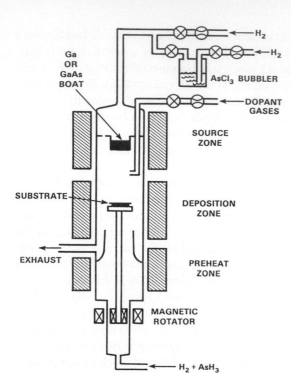

Fig. 6.21. Schematic illustration of $AsCl_3$ GaAs vapor-phase epitaxial (VPE) growth system

substrate. A GaAs skin is usually formed on the Ga in the boat prior to growth. Solid GaAs can be used instead of Ga, but usually an increase in background carrier concentrations results. GaAs layers with net carrier concentrations in the $10^{14}\,\mathrm{cm}^{-3}$ range and 77-K mobilities in the 180,000 to $200,000\,\mathrm{cm}^2/\mathrm{V}$-s range have been grown by this technique. InP layers with mobilities greater than $100,000\,\mathrm{cm}^2/\mathrm{V}$-s have also been grown, but the results are not as extensive or as reproducible as in the case of GaAs.

Both the trichloride and hydride method are used for growing GaInAsP on InP. The trichloride offers the advantage of higher purity, while the latter offers the advantage of a simpler system for growing the entire range of lattice-matched quaternary layers on InP. A hydride system for growing GaInAsP on InP is illustrated in Fig. 6.22. Here, the AsH_3, PH_3 and HCl flows over the Ga and In boats are independently controlled and the ratios

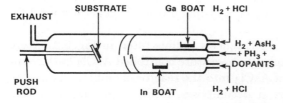

Fig. 6.22. Schematic illustration of GaInAsP HCl-hydride vapor-phase epitaxial (VPE) growth system

are used to control the composition of the epitaxial layer. The PH_3 and AsH_3 decompose and are combined with the In chlorides and Ga chlorides in the mixing zone and are then swept downstream into the deposition zone where they crystallize out on the substrate. Dopant gases can be added either downstream of the metal boats or in the AsH_3-PH_3 stream.

Dual barrel reactors, in which the sample can be rapidly shuttled back and forth between two different growth streams, e.g., an InP growth stream and a GaInAsP growth stream, have been developed to rapidly switch from the growth of one material to another.

VPE provides good epilayer thickness control, good uniformity over large wafers and good surface morphology. Selective growth can be achieved in dielectric film openings even of submicrometer dimensions, and by choosing appropriate orientations, lateral overgrowth of dielectric films can be achieved [6.72, 73]. VPE has been used to fabricate a large number of optical waveguides and components. Optical losses in GaAs n/n^+ waveguides grown by VPE have tended to be $\approx 1\,cm^{-1}$ or $4\,dB/cm$. Lower loss values ($\sim 1\,dB/cm$) can be achieved in heterojunction structures grown by MOCVD and MBE (Sects. 6.1.5 and 6.3.1). For more information on VPE, see [6.74, 75].

6.2.3 Metal Organic Chemical Vapor Deposition (MOCVD)

MOCVD is also a vapor phase epitaxial layer technique, but it differs from the usual VPE techniques in that it utilizes a cold-wall reactor tube and is therefore capable of growing Al compounds. The technique is fairly well developed for growing AlGaAs/GaAs, and GaInAsP/InP growth by this technique is progressing rapidly.

A simple AlGaAs/GaAs MOCVD system is illustrated in Fig. 6.23. Growth is usually achieved by introducing controlled amounts of the group-III elements in the form of metal alkyds, e.g., trimethyl gallium $Ga(CH_3)_3$ for Ga and trimethyl indium $In(CH_3)_3$ or triethyl indium $In(C_2H_5)_3$ for indium, and of the group V element(s) in the form of the hydrides AsH_3 and PH_3 into a cold-wall quartz reaction tube. The chamber can be either at atmospheric pressure or at a pressure of 0.1–0.5 atm for low-pressure MOCVD. Dopants can be introduced either in the form of metalorganics such as $Zn(C_2H_5)_3$ or in the form of hydrides such as SiH_4 or H_2S. The substrate is placed on an RF-heated carbon susceptor.

In the growth of GaAs and AlGaAs, the organometallic compounds and hydrides decompose or crack on the hot substrate and growth therefore takes place primarily on this surface. The organic compounds resulting from the reaction are gaseous and are carried downstream. For GaAs grown from trimethyl gallium and arsine, the chemical reaction at the substrate is

$$Ga(CH_3)_3 + AsH_3 \xrightarrow{\text{heat}} GaAs + 3CH_4 \quad .$$

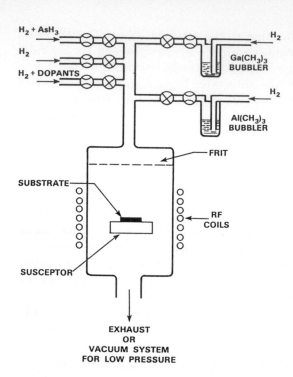

Fig. 6.23.
Schematic illustration of AlGaAs metalorganic chemical vapor deposition (MOCVD) growth system

In the case of both AlGaAs/GaAs growth and GaInAsP/InP growth, the purity of the metalorganic compounds is extremely important not only for enhancing the purity of epitaxial layers but for keeping the growth kinetics under control. In the last several years significant progress has been made in purifying these materials. Both GaAs and InP layers have been grown with net concentrations in the $10^{14}\,\mathrm{cm}^{-3}$ range and 77 K mobilities in the $140,000\,\mathrm{cm^2/V\text{-}s}$ range.

MOCVD is a versatile epitaxial growth method that provides good layer thickness control, good uniformity and surface morphology and good control of heterojunction interfaces. Selective growth is also possible using MOCVD. Some of the lowest optical losses reported for GaAs/AlGaAs waveguides have been grown by MOCVD. More information on MOCVD can be found in [6.76, 77].

6.2.4 Molecular Beam Epitaxy (MBE)

Molecular beam epitaxy is basically an ultra-high-vacuum technique in which the constituent elements are evaporated onto a heated substrate where they are mobile enough to form an epitaxial layer. A simplified MBE system is illustrated in Fig. 6.24. It consists of an ultra-high-vacuum system containing effusion ovens with shutters for each element, Ga, Al, As, In

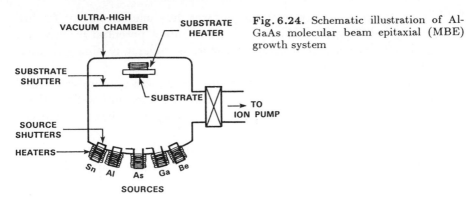

and P, and a heated substrate holder. Effusion ovens containing low vapor pressure dopants such as Be for p-type doping and Si for n-type doping are also included. A variety of analytic instruments such as residual gas analyses and RHEED (Reflection High Energy Electron Diffraction) can also be incorporated to monitor the growth process and the quality of the substrate surface. The composition of the epitaxial layer is controlled by individually controlling the temperature of each oven and opening and/or closing appropriate shutters. The latter permits a rapid change from one composition to another.

Since the sticking coefficients of the group III elements are close to unity, while those of the group V elements are less than unity, it is generally easier to grow alloys consisting of more than one group III element than alloys consisting of more than one group V element. It is also more difficult to work with phosphorus than arsenic because of the multiple forms of solid phosphorus that can occur and the volatile and explosive nature of some of these forms. Recent work has explored the use of gas sources for the group V elements [6.78].

There is a strong interest in using GaAs and AlGaAs/GaAs heterostructures for high-speed digital and microwave applications and the majority of the III-V MBE effort has gone into these materials. Typical growth temperatures for GaAs and AlGaAs are in the range of 575° and 680° C, respectively. High-purity unintentionally doped GaAs grown by MBE with metallic As is p-type with a net acceptor concentration in the low $10^{14}\,cm^{-3}$ range. Low levels of Sn are usually used to provide low n-type carrier concentration. GaAs layers with net electron concentration of a few times $10^{14}\,cm^{-3}$ and 77-K mobility greater than $100,000\,cm^2/V$-s have been grown by this technique.

The growth rate of MBE is low, $\leq 1\,\mu m/h$, so very precise control of layer thickness is possible. In fact, it has been reported that alternate monolayers of GaAs and AlAs can be grown. Heterojunction interfaces between MBE-grown layers are therefore very abrupt. The low growth rates mean, however, that long growth times are required for thick layers. The grown

349

layers are also very uniform and the surfaces very smooth except for low densities of "oval" defects. Since these defects are a potential problem for digital circuits, a considerable research effort is under way aimed at their elimination.

At the present time, MBE is probably the growth system best suited for growing waveguides and components utilizing quantum wells and superlattices. Superlattices represent a form of band-gap engineering and will probably find many uses in future integrated-optic currents. Losses in GaAs waveguides grown by MBE are low and only slightly higher than those grown by MOCVD. Since MBE is a direct line-of-sight evaporation technique, it is possible to position several masks above the substrate during growth to limit growth to selected areas of the wafer or to produce tapers. This should enable the direct fabrication of geometries useful for integrated circuits. Extensive modification of existing equipment, however, will probably be necessary to obtain micrometer-size geometries. Reviews of MBE technology can be found in [6.79 and 80].

6.2.5 Summary

At this point in time, MOCVD and MBE appear to be the most promising techniques for growing III-V epitaxial layers suitable for integrated optics. Recently ultra-low-pressure MOCVD, which can be considered a cross between MBE and MOCVD and is sometimes referred to as metalorganic molecular beam epitaxy (MOMBE), has begun to be developed in the laboratory [6.81] and could prove useful for optical waveguides. Regardless of the growth system, it is generally easier to fabricate devices which require a two-step growth process (i.e., one in which the sample is removed from the growth system for a second growth) in the GaInAsP system than in the AlGaAs system, because of the rapid oxidation of the Al compounds that occurs when they are exposed to air. Development of a completely ultra-high-vacuum processing line should therefore prove extremely useful for fabricating AlGaAs/GaAs integrated-optical circuits.

6.3 Passive Waveguide Devices – Fabrication and Characterization

Many different technologies have been used to fabricate optical waveguides in semiconductors. These include proton bombardment, ion implantation, diffusion, epitaxial growth, and various etching, metallization and dielectric deposition techniques and combinations thereof. In this section, a survey of semiconductor optical-waveguide fabrication techniques and waveguide characteristics is given. Both homojunction and heterojunction guides are discussed. The most promising techniques for future integrated optical circuits are stressed and therefore the text is limited almost exclusively to techniques for fabricating channel waveguides in the III-V semiconductors. The extension of these techniques to multiguide couplers is then discussed and the fabrication of waveguide bends, "Y-branches" and gratings is reviewed. Finally, thin-film-insulator guides on semiconductors are briefly described.

6.3.1 Channel Waveguides

Some of the earliest channel waveguides in GaAs were embedded guides (Fig. 6.9e) which were fabricated using selective proton bombardment of n^+-GaAs [6.31, 68]. It is also possible to make GaAs rib waveguides (Fig. 6.9a) using a combination of proton bombardment and selective etching. As previously discussed, however, proton bombardment, in addition to reducing the carrier concentration and thereby increasing the index, also introduces optical loss. Although the loss can be reduced by appropriate annealing, the loss due to the proton bombardment has made this type of guide impractical for most applications. Proton bombardment can also be used to fabricate waveguides in CdTe [6.82, 83] and ZnTe [6.84].

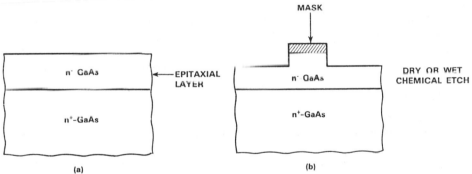

Fig. 6.25a, b. Fabrication of rib waveguide: (a) epitaxial growth of low-carrier-concentration layer on n^+ substrate; (b) etching of rib using suitable etch mask

Homojunction rib waveguides fabricated using a combination of epitaxial growth and selective etching have been the most widely studied channel guides in the III-V semiconductors. As indicated in Fig. 6.25, a low-carrier-concentration epitaxial layer is grown typically on a (001)-oriented n^+ substrate. Initially, VPE or LPE was used to grow the nominally undoped epilayer [6.85]. Later, MOCVD and MBE were used [6.86]. Waveguides made in GaAs/GaAlAs layers have typically shown the lowest optical losses [6.87]. Following the growth of guiding layers, ribs can be etched by means of standard photolithography and wet-chemical, reactive-ion (RIE) or ion-beam-assisted etching (IBAE). Propagation is normal to a cleavage plane (110 or $\bar{1}$10). The (111)A face of GaAs and InP etches more slowly than other orientations in most wet-chemical etches and therefore the side-wall

351

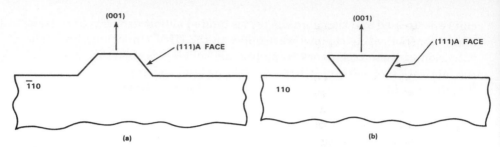

Fig. 6.26a, b. Illustration of the effects of crystal orientation on the slope of ribs wet-chemically etched in III-V semiconductors

slope of the ribs is usually determined by the direction of the (111)A faces for this technique. On (001) samples, ribs formed for the ($\bar{1}$10) propagation direction will generally have sloped side walls, as shown in Fig. 6.26a, while ribs in the (110) direction generally have undercut retrograde side walls, as shown in Fig. 6.26b [6.88]. Several chemical etches have, however, been reported [6.89] that will result in more nearly vertical sidewalls even for the orientations shown in Fig. 6.26. Side-wall shape can be controlled using RIE [6.90] or IBAE [6.91, 92]. Guides with vertical side walls have been fabricated using these techniques. Possible problems associated with RIE and IBAE are damage caused by energetic ions and side-wall roughness. Since RIE and IBAE can faithfully reproduce the etch mask in the semiconductor, any roughness is transferred directly to the semiconductor without smoothing. The roughness can result in scattering loss, especially in deeply etched ribs.

Single-mode rib guides are usually designed using *Marcatili*'s analysis [6.39]. As previously discussed, this analysis gives fairly good results as long as the rib etch depth is not too great and the slab guide outside the rib is single-moded. GaAs single-mode homojunction rib guides with losses in the $1\,\mathrm{cm}^{-1}$ (4 dB/cm) range at $1.06\,\mu\mathrm{m}$ have been formed in VPE material [6.85]. Similar GaAs guides have been fabricated in LPE and VPE InP material with losses in the 6 to 8 dB/cm range [6.93, 94].

Deeply etched single-mode homojunction rib guides in which the slab is cut off have been made in both GaAs and InP [6.46–48, 93]. These guides are usually analyzed by either the variational [6.46–48], the beam-propagation [6.49] or the transverse mode-matching method [6.43–45] (Chap. 2). These guides are difficult to fabricate reproducibly because they depend critically on the concentration in the n^+ substrate and the dimensions and shape of the rib, especially the width at the bottom of the rib. Losses have tended to be higher than in shallow etched rib guides, probably due to scattering from the rib side walls [6.95].

To facilitate the application of electric fields in a rib waveguide, Schottky barrier contacts, MOS-type contacts or a p^+ rib can be used. Schottky barrier contact may perturb the waveguiding and increase losses if placed

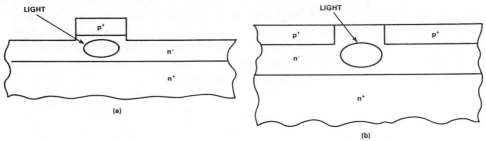

Fig. 6.27a, b. Illustration of channel waveguides with built-in p^+-n junctions

directly on the guiding regions. MOS-type contacts on GaAs show dc drift and can only be used for transient applications. p^+ ribs are applicable to both GaAs and InP waveguides and, as shown in Fig. 6.27a, these guides resemble a dielectric loaded guide. The p^+-layer can be achieved by ion implantation [6.85, 94], diffusion [6.96, 97] or epitaxial growth [6.98] prior to rib etching. In a p^+-n^--n^+ rib guide, an electric field is applied to the waveguide region by reverse biasing the p-n junction. A variation of the rib guide, the channel-stop guide (Fig. 6.27b), has been fabricated using a combination of epitaxial growth and Be-ion implantation and required no etching [6.85, 99]. In this case, an electric field can be applied to one or both of the confining regions by reverse biasing the p-n junction (Sect. 6.5).

As discussed in Sect. 6.1, the optical loss in homojunction rib waveguides is ultimately limited by free-carrier loss in the n^+ substrate (and the p^+ rib layer, if present). Waveguide losses due to the substrate in homojunction structures are typically 2–4 dB/cm at 1.3 μm. To reduce the loss below this level, the n^+ confining substrate must be replaced.

One novel method demonstrated to achieve this was the use of an SiO_2 layer, as shown in Fig. 6.28, to confine the light in the vertical direction. These waveguides were fabricated by epitaxially growing a nominally undoped GaAs layer over an SiO_2 layer using a lateral-epitaxial-growth technique [6.100]. The growth is seeded in the underlying GaAs through slots in the SiO_2. The direction of these SiO_2 strips must be such that the lateral-growth rate is much higher than the vertical-growth rate. It was experimentally determined that, on (100) substrates, the SiO_2 strips had to be at an angle of 10° to a cleavage plane to achieve good GaAs lateral over-

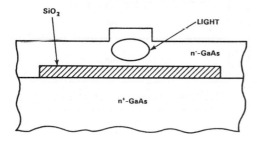

Fig. 6.28. Illustration of GaAs rib waveguide over SiO_2. The n^--GaAs layer was grown using lateral epitaxial growth techniques

353

growth. This places some limits on the direction of waveguides. The SiO_2 has to be thick enough so that the tail of the confined light is negligible at the SiO_2-substrate interface. To provide lateral confinement, ribs can be etched in the overgrown layer, as shown in Fig. 6.28, or a slot can be etched in the SiO_2 prior to overgrowth to form an inverted rib. Losses in these GaAs oxide-confined waveguides were 2 dB/cm at 1.06 μm, which is about half that obtained on similar guides grown on n^+ substrates. In InP, the ratio of the lateral-to-vertical growth rates of VPE material was found to be smaller than in GaAs, and slots in the SiO_2 about 30° to a cleavage plane had to be used. Since this limited the direction of waveguides even more than in GaAs, the effects of waveguides crossing the nucleation slots were investigated. Losses of 6.4 dB/cm at 1.3 μm were measured [6.94] for ribs waveguides formed perpendicular to a cleavage plane on an overgrown sample with 50-μm-wide SiO_2 strips and slots 30° to the cleavage plane.

With the advances made in MBE and MOCVD growth techniques, a more common current technique for reducing the loss is the use of an AlGaAs confining layer which has a lower index than the n^--GaAs guiding layer, as shown in Fig. 6.29. The thickness of the AlGaAs layer is usually made thicker than the decay length of the optical mode in this region, so the field at the AlGaAs/GaAs-substrate interface is negligible. (If the AlGaAs layer is not thick enough, coupling to the GaAs substrate can occur.) Heterojunction rib waveguides of this type with losses less than 1 dB/cm have been fabricated in MOCVD-grown material [6.87, 101, 102].

InGaAsP/InP heterojunction rib guides have also been fabricated. Single-mode InGaAsP/InP rib guides grown by LPE have shown losses of about 7.3 dB/cm at 1.3 μm [6.94, 103]. Guides grown by MOCVD have had losses of 11.5 dB/cm at 1.15 μm [6.104].

Probably the most promising waveguides for integrated optical circuits are buried heterojunction waveguides [6.105, 106]. This type of waveguide should have low loss, have good optical confinement and be compatible with heterojunction lasers and detectors. A buried heterojunction InGaAsP/InP

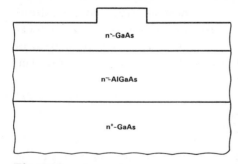

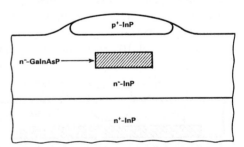

Fig. 6.29. Illustration of n^--GaAs/AlGaAs heterojunction rib waveguide

Fig. 6.30. Illustration of GaInAsP/InP buried heterojunction rib waveguide

waveguide is illustrated in Fig. 6.30. The structure can be fabricated by sequentially growing an undoped InP buffer layer, the InGaAsP guiding layer and a thin InP cap layer. By selective etching, the InP cap layer can be patterned and used as an etch mask to etch the InGaAsP layer. An InP confining layer is then grown to provide lateral confinement [6.105]. By controlling the composition and dimensions of the quaternary layer, the guide can be made single mode. The lateral confinement can also be controlled, which is important in making bends and Y-branches. As illustrated, a p^+ region can be incorporated for applying electric fields across the quarternary guiding layer. The p^+ region can be achieved by epitaxial growth and selective etching or ion implantation. Tradeoffs between optical loss due to free-carrier absorption in the n^+ and p^+ regions and drive voltage requirements are possible by controlling the thicknesses of the top and bottom n^--InP layers. It may also be possible to control the impedance of traveling-wave strip lines by controlling the width of the p^+ region.

Similar guides can be grown in the GaAs-material system. Some problems may occur in trying to grow epitaxial layers on an AlGaAs layer exposed to air, but this problem can probably be circumvented and may not be a problem if the entire fabrication process is carried out in a high-vacuum environment.

When MBE or MOCVD growth is used, quantum wells and superlattices can be used in the waveguides to enhance nonlinear optical effects and electro-optical effects. Such devices are described in Sect. 6.5.

Channel waveguides have also been fabricated by depositing metal or oxide strips on top of a guiding layer. As previously discussed, the metal (oxide) can decrease (increase) the effective index under the strip, leading to lateral confinement [6.34, 107]. Stress and strain effects at the edges of the strips, however, can often be larger, and can dominate the guiding in this type of guide [6.35, 108] (Sect. 6.1.3).

The residual strain in grown heterostructure layers can also lead to a residual birefringence in III-V waveguides [6.52]. This is due to the photoelastic effect. Typical birefringence values are $(N_{TE} - N_{TM}) \sim 10^{-4}$. Further birefringence is present in modulator (e.g., p-n junction) structures due to the effects of built-in electric fields. At wavelengths near the band edge a dichroism of electroabsorption (see 6.4.7) can contribute to the birefringence.

Other semiconductor-based waveguides include those formed on n/n^+ silicon [6.109] and multilayer insulators formed on semiconductor substrates. An example of the latter is a high-index Si_3N_4 guide on a lower-index SiO_2 buffer all grown on an Si substrate, as shown in Fig. 6.31 [6.110]. Losses in planar Si_3N_4 guides of 0.1 dB/cm have been obtained. Dielectric guides have also been formed in grooves in dielectric layers on Si substrates for coupling to detector arrays [6.111, 112]. Recently, multi-mode rib guides have been fabricated from silica layers deposited by flame hydrolysis [6.113].

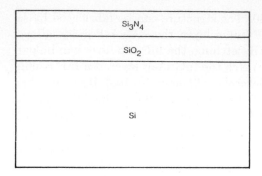

Fig. 6.31. Dual dielectric waveguide on Si substrate

6.3.2 Couplers

Optical couplers are important components of numerous integrated-optical-circuit concepts. Two-guide couplers of various types have been studied fairly extensively [6.99, 104, 114–118]. More recently, interest has been generated in couplers composed of more than two guides [6.56, 119–121]. Schemes for using three-guide couplers for improved sampling and filtering have been proposed [6.120]. Three-guide couplers have also been used as power dividers and combiners, and as the input and output sections of an integrated GaAs interferometer [6.122, 123]. Coupled laser arrays consisting of multiple parallel stripe lasers have also been investigated [6.124], and multi-guide couplers are being investigated for possible lens applications [6.125].

Optical couplers can be made by placing two or more waveguides in close proximity so their fields overlap. Most couplers fabricated to this time have been made by placing two rib-type single-mode waveguides in proximity. With buried heterojunction guides, the dimensions into the crystal can also be used and couplers can be fabricated by placing waveguides either next to each other or on top of each other, as shown in Fig. 6.32.

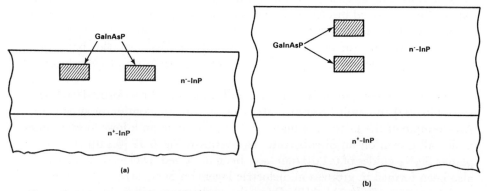

Fig. 6.32a, b. Possible ways of making a two-guide coupler using buried heterojunction waveguides

One of the earliest semiconductor optical couplers made was a multi-guide coupler consisting of embedded waveguides fabricated by selectively proton bombarding n^+-GaAs [6.82, 126]. For an input into a single guide at $x = 0$ (guide 0) and an infinite linear array, the power in each guide is given by [6.127]

$$P_n(z) = J_n^2(2Kz) \tag{6.3.1}$$

where J_n is the Bessel function of nth order, and K is the coupling coefficient. Two-guide directional couplers were also fabricated by this technique [6.82, 126].

Two-guide GaAs and InP directional couplers utilizing rib-type waveguides fabricated in nominally undoped epitaxial layers have been studied fairly extensively [6.86, 115, 128]. n^--n^+ homojunction rib couplers generally have coupling lengths in the 4- to 10-μm range at $1.3\,\mu$m. Photomicrographs of a GaAs p^+-n-n^+ rib and of a channel-stop [6.99] directional coupler are shown in Fig. 6.33. These guides were made using a combination of VPE and Be-ion implantation. As discussed below, these guides can also be used as switches by reverse biasing the p-n junctions. These devices had coupling lengths on the order of $8\,$mm at $1.06\,\mu$m. The wavelength dependence of the coupling length of the channel-stop guides (which are

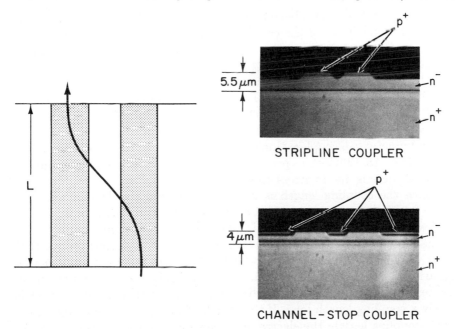

STRIPLINE COUPLER

CHANNEL–STOP COUPLER

GaAs DIRECTIONAL COUPLER

Fig. 6.33. Photographs of a p^+-n^--n^+ GaAs rib waveguide coupler and a p^+-n^--n^+ GaAs channel-stop waveguide coupler [6.99]. The devices were intentionally etched so the various layers could be seen

357

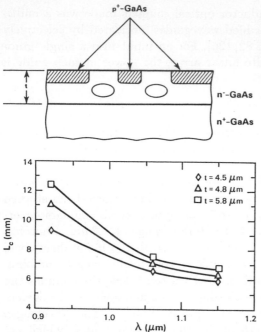

Fig. 6.34.
Variation of coupling length L_c vs. wavelength for GaAs channel-stop directional couplers having an 8-μm guide width, a 4-μm guide spacing and an epitaxial thickness t. The p^+ layer is 2 μm thick on all devices [6.114]

fairly tightly coupled) have been investigated [6.114]. Figure 6.34 shows the wavelength dependence of the coupling length for couplers made in epitaxial layers of several different thicknesses. In all cases, the coupler length decreases with increasing wavelength. Similar wavelength dependence has been observed for rib-guide couplers [6.97, 115].

Rib-type directional couplers have also been fabricated in InP and In-GaAsP epitaxial layers grown on n^+-InP [6.94, 97, 104]. p^+-n^--n^+ rib couplers were fabricated in InP by means of Zn diffusion [6.104].

GaAs metal gap couplers (including a $\Delta\beta$ coupler) have also been fabricated [6.117, 118, 129] but, as previously discussed, it is possible that guiding in these devices is due to stress effects.

Once the coupling length is known, directional couplers in principle can be fabricated with any desired coupling (e.g., 3 dB, 10 dB or 20 dB) at the desired wavelength. In general, however, coupling is not perfect, and the maximum extinction ratio between the input and output guides that can be obtained is usually less than 20 dB. Although there are fundamental limitations, experimental limitations are most likely due to small asymmetries between the guides and the difficulty in fabricating a coupler that is exactly a coupling length long. If the length of the coupler is slightly longer than a coupling length, the electro-optic effect can be used to provide a $\Delta\beta$ configuration [6.130], as shown in Fig. 6.35, to adjust the effective coupling length to the actual interaction length. If the length is slightly shorter than a coupling length, it might be possible to adjust the length by applying

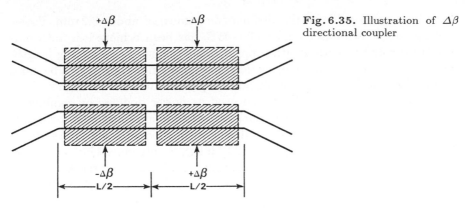

Fig. 6.35. Illustration of $\Delta\beta$ directional coupler

the same bias to all the electrodes in the $\Delta\beta$ configuration (as long as the orientation of the waveguide is chosen correctly, so that the effective-index difference between the guiding and nonguiding region is decreased by the electric field). A detailed discussion of switches formed from such couplers is given in Sect. 6.4.

Another problem is coupling into and out of the coupler. This is generally not a problem for loosely coupled guides but can become a problem for closely spaced guides, since the modes in the coupler become more perturbed from the sum and difference of the modes of the individual isolated guides, as the guides become more tightly coupled. This can affect the extinction ratio, which becomes more dependent on the input and output coupling structure. In actual integrated-optical circuits, it is highly likely that bends will be used at the input and output of couplers, and these must be considered in determining coupling length and input/output coupling. As shown above, the coupling length is wavelength dependent, so couplers have some inherent wavelength selectivity. This selectivity can be increased by using nonidentical guides which are synchronous only at the desired wavelength [6.131, 132].

Three-guide couplers have been fabricated in GaAs [6.56] and InP [6.94]. A cross section of a GaAs rib-waveguide three-guide coupler is shown in Fig. 6.36. Guides of this type had a coupling length for power division

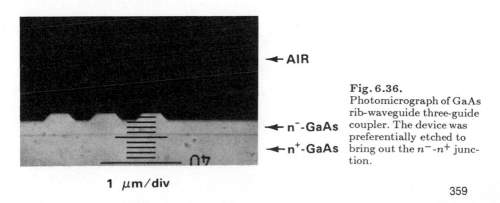

← **AIR**

← **n⁻-GaAs**

← **n⁺-GaAs**

Fig. 6.36. Photomicrograph of GaAs rib-waveguide three-guide coupler. The device was preferentially etched to bring out the n^--n^+ junction.

1 μm/div

(from the center guide to the two outside guides) of about 3.2 mm. Power shifting to the two outer guides of $> 95\%$ has been achieved in both materials for couplers with lengths approximately equal to a coupling length. As previously discussed, the operation of a multi-guide coupler can be more complicated than that of a two-guide coupler because of the higher number of modes that can exist. When used as power dividers or combiners, the three-guide coupler behaves much like a two-guide coupler. Three-guide couplers have been used as the input and output sections of a GaAs Mach-Zehnder-type interferometer [6.122, 123]. If light is introduced into one of the outside guides and extracted from the other outside guide, a "track changer" can be formed. As an approximate rule, if a particular guide geometry and spacing leads to a two-guide coupling length of L_c, then a three-guide coupler of identical geometry will have a coupling length of $L_c/\sqrt{2}$ as a 3-dB splitter and $\sqrt{2}L_c$ as a track changer. Coupler devices in which one or more of the coupled guides terminate on-chip are also under investigation. This approach is appealing if fabrication conditions can be controlled to insure the interaction length is equal to a coupling length. Radiation effects in such structures due to the start/stop of one or more guides have been investigated and found at least for the three-guide power divider to be relatively small and comparable to a Y-junction loss (≤ 1 dB) [6.95, 133].

6.3.3 Bends and Branches

Although couplers of various types can be used to translate light perpendicular to the direction of propagation and to act as power dividers, combiners and splitters, in many applications it would be advantageous to have the ability to make waveguide bends and branches.

As with all optical-waveguide transitions, a certain amount of power is lost into radiation modes at a bend (Fig. 6.37a), or at a transition to a curved waveguide (Fig. 6.37b). There is also some reflection of power, but this is usually negligible unless there is a fairly large discontinuity in effective index at the transition. For a curved waveguide, losses, in addition

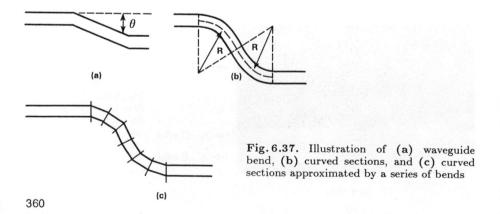

Fig. 6.37. Illustration of (a) waveguide bend, (b) curved sections, and (c) curved sections approximated by a series of bends

to any loss at the transition, depend on the radius of curvature and the lateral mode confinement, as discussed in Sect. 6.1.6. Instead of a smooth curve, a series of bend sections can be used, as indicated in Fig. 6.37c. In some instances, coherent coupling back into the waveguide from radiation modes is possible if the length and angle of each section is correct [6.134].

To achieve reasonable bend angles and radii of curvature with minimum loss, the lateral confinement of the waveguide must be carefully controlled, as described in Sect. 6.1.6. The lateral confinement in rib waveguides, in which the slab is not cut off is generally not high enough for most applications. Deeply etched rib guides, however, have better lateral confinement, and bends and Y-branches have been made using this type of guide. The additional losses in Y-branches compared to straight sections for deeply etched GaAs rib guides in which the ribs were formed by ion-beam-assisted etching was about 1.4 dB at a total "Y" angle as high as 2° [6.95]. In the GaAs Y-branches, the losses did not decrease significantly as the branch angle decreased, indicating that there may be an inherent mode-matching loss at the tip of the Y in Y-branches made out of deeply etched guides. Similar Y-branch losses were obtained for InP rib guides with a Y angle of 1.8° [6.93, 133]. As previously mentioned, however, critical issues for these structures are scattering loss due to side-wall roughness and the strong dependence on etch depth, side-wall shape and index difference between the rib and confining layer. The problems are exacerbated because of the relatively narrow width (2–3 μm) generally required to maintain single-mode operation.

A more practical solution to making bends and branches in semiconductor optical circuits may be to utilize buried heterojunction waveguides. In this case, the waveguide parameters are more flexible and scattering losses are expected to be lower. GaInAsP/InP buried heterojunction waveguide bends and branches have been reported [6.105]. The measured Y-branch transmission as a function of Y-branch angle θ, defined as the ratio of the sum of output powers of the two output branches to that of a straight waveguide, is shown in Fig. 6.38. The $Ga_{0.17}In_{0.83}As_{0.4}P_{0.6}$ guiding region was ≈ 2000 Å thick and 2 μm wide. This data indicated that Y-branches with transmission greater than 90 % (losses ≈ 1 dB) should be possible using buried heterojunction waveguides.

In all of the above considerations, the lowest bend loss occurs for mode sizes small compared to that of a conventional single-mode fiber. Thus a trade-off exists between fiber/chip coupling loss and on-chip propagation loss. This issue is further discussed relative to modulator efficiency in Sect. 6.5.

An interesting approach to making right-angle bends in semiconductor guides is to utilize etching techniques to form vertical side walls at a 45° angle to the guide to produce a total reflection. These bends can have losses < 2 dB for rib guides but are quite sensitive to mirror placement and thus require very precise fabrication control [6.135].

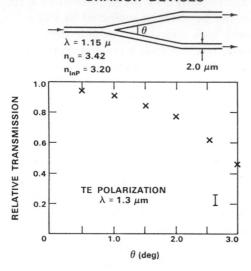

GaInAsP BURIED HETEROSTRUCTURE WAVEGUIDE BRANCH DEVICES

λ = 1.15 μ
n_Q = 3.42
n_InP = 3.20

2.0 μm

RELATIVE TRANSMISSION

TE POLARIZATION
λ = 1.3 μm

θ (deg)

Fig. 6.38. Schematic illustration (a) of a waveguide "*Y*"-branch and (b) measured relative transmission through a GaInAsP/InP buried heterojunction waveguide "*Y*"-branch as a function of total angle θ [6.105]

6.3.4 Grating Filter

Another passive device of interest is a waveguide grating filter. Efforts have followed the theory of distributed gratings presented in Sect. 2.6.4. Results to date are available for InGaAsP [6.136]. Here a passband of 13 Å with side-lobe levels 17 dB down was obtained. In this device, there was a 30 Å shift in the center wavelength between TE and TM modes due to birefringence. One could conceive of tuning such a structure electro-optically by means of a buried grating and a p-n junction. Application of a voltage would adjust the effective index of the guide and thus alter the center wavelength of the filter.

6.4 Electro-Optic Guided-Wave Modulators – Theory

6.4.1 Electro-Optic Effect in III-V Semiconductors

This section summarizes the theory of electro-optical modulators that is particularly relevant to III-V semiconductors. The theory of operation of specific structures (e.g., directional coupler switches, Mach-Zehnder modulators) is discussed in Chap. 4. Phase modulation in semiconductor waveguide devices is achieved primarily via the linear electro-optic (eo) effect. The III-V materials of interest have $\overline{4}3\,m$ symmetry and thus a single electro-optic coefficient r_{41}. It is useful to first review the index ellipsoid for the $\overline{4}3\,m$ group for an electric field applied along the principal crystallographic axes [6.137].

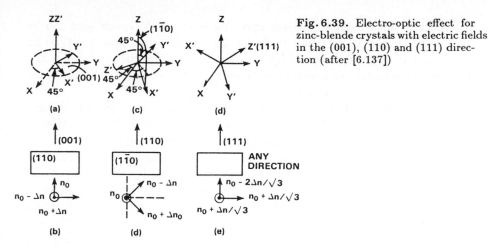

Fig. 6.39. Electro-optic effect for zinc-blende crystals with electric fields in the (001), (110) and (111) direction (after [6.137])

$$\Delta n = \frac{1}{2} \, r_{41} \, n_0^3 \, E$$

$$r_{41} < 0$$

Following conventional techniques, it is easily shown that, for electric fields along a (001) direction (i.e., $E = E_z$), the principal axes of the indicatrix are (001), (110) and ($\bar{1}10$) with

$$n_{x'} = n_0 - \Delta n \quad (110) \quad ,$$
$$n_{y'} = n_0 + \Delta n \quad (\bar{1}10) \quad ,$$
$$n_{z'} = n_0 \quad\quad\quad (001) \quad , \tag{6.4.1}$$

as shown in Figs. 6.39a,b.

For fields applied along (110), (i.e., $E_x = E_y = E/\sqrt{2}$) one finds that the principal axes are

$$n_{x'} = n_0 + \Delta n \quad (11\sqrt{2}) \quad ,$$
$$n_{y'} = n_0 - \Delta n \quad (11\sqrt{2}) \quad ,$$
$$n_{z'} = n_0 \quad\quad\quad (1\bar{1}0) \quad , \tag{6.4.2}$$

as shown in Figs. 6.39c,d.

Similarly, for the electric field along a (111) direction (i.e., $E_x = E_y = E_z = E/\sqrt{3}$) one finds that the three principal axes are along the (111) direction and any pair of directions in the (111) plane that satisfy the right hand rule. One example is

$$n_{x'} = n_0 + \Delta n/\sqrt{3} \quad (1\bar{1}0) \quad ,$$
$$n_{y'} = n_0 + \Delta n/\sqrt{3} \quad (11\bar{2}) \quad ,$$
$$n_{z'} = n_0 - 2\Delta n/\sqrt{3} \quad (111) \quad , \tag{6.4.3}$$

363

as shown in Figs. 6.39e,f. In each case, n_0 is the bulk index and $\Delta n = n_0^3 r_{41} E/2$. It should be noted that, for an n-type guide and a p-type or Schottky top contact, reverse bias results in a positive value of E for the defined crystallographic orientations. Since r_{41} is negative in III-V materials, Δn is generally a negative quantity.

These solutions have a number of implications for modulator design; all of the above orientations will be considered. The (100) direction is a preferred eptiaxial growth direction and is a good orientation for the electro-optic effect. For (001)-oriented fields, modulation only occurs for TE-polarized inputs with no TM modulation for samples with cleaved end faces (i.e., (110) or ($\bar{1}$10) faces). In particular, for TE light incident on a ($\bar{1}$10) cleavage face (i.e., optical fields along (110)), the phase modulation increases the confinement for reverse bias across an n-type guide. There is much confusion in the literature on the difference between the (110) and ($\bar{1}$10) directions. A definitive method of insuring increased confinement for n-type guides is to orient the wafer so that propagation is along the direction that yields, for rib guides, a trapezoidal guide cross section when they are etched with a preferential wet chemical etch, as shown in Fig. 6.26a [6.88].

For (110)-oriented fields and cleaved ($1\bar{1}0$) end faces, the principal axes are at 45° to the TE and TM polarization directions, resulting in polarization rotation for linearly polarized inputs. For (111)-oriented substrates and (111) electric fields, two of the principal axes of the index ellipsoid are coplanar with the waveguide and pure phase modulation is again possible. Now, however, both the TE and TM modes are affected by the electric field. The (111) direction is also relevant to electrodes placed along etched or grown (111) facets on (100)-oriented surfaces or due to fringing field effects.

Representative values reported in the literature for the r_{41} coefficient for $\bar{4}3\,m$ semiconductors at about $1\,\mu$m are given in Table 6.1.

Table 6.1. Electro-optic coefficients

Material	r_{41} ($\lambda \sim 1\,\mu$m) 10^{-10} cm/V	n
GaAs	-1.4	3.43
InP	-1.4	3.3
CdTe	-4.5	2.84

6.4.2 Modulator Design

Electro-optic modulators reported to date have almost always been fabricated on (100) material. For a variety of guide geometries, pn junctions, Schottky barriers, and MOS structures have been used to apply bias. It is useful to first discuss the trade-off and performance criteria of a modulator before discussing specific geometries. The important criteria of a modulator include the required drive power, bandwidth and insertion loss. Obviously different applications will arrive at different figures of merit. One figure of merit that can be carried over from bulk modulators, where insertion loss is much less of a design issue, is power per unit bandwidth $(P/\Delta f)$. We will consider the case of a lumped modulator that is entirely capacitive driven from a 50-Ω system with a 50-Ω termination. It is prudent to include in the equivalent circuit the effect of series resistance and inductance due to packaging and electrode effects, as has been done in the modeling of the frequency response of GHz LiNbO3 modulators [6.138]. The equivalent circuit is shown in Fig. 6.40. At low frequencies where the inductance impedance is negligible it is found that $P/\Delta f = \pi C V^2$, C being the modulator capacitance. The choice of not using a 50-Ω termination $(R_{\text{term}} = \infty)$ reduces P by a factor of 4 but also decreases the 3 dB bandwidth by a factor of 2 (i.e. $P/\Delta f$ decreases by a factor of 2).

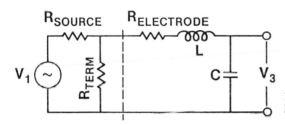

Fig. 6.40. Equivalent circuit of lumped modulator

If we consider the p^+-n-n^+ rib guide shown in Fig. 6.27a, the capacitance is simply $\varepsilon A/d$, where A is the electrode area, d is the depletion width and ε is the dielectric constant. For a punch-through bias condition, d is simply the thickness of the n layer. The value of C can be calculated in general from the well-known expressions of semiconductor physics [6.139] in which partial depletion, heterojunction, graded junction, etc. can be accounted for.

The voltage required for a given degree of modulation can be determined by calculating the effective change in index produced from the overlap of the optical and applied elelctric fields,

$$\Delta n = \frac{1}{2}n_0^3 r_{41}\frac{\iint dx\,dy\,E|\mathcal{E}|^2}{\iint dx\,dy|\mathcal{E}|^2} \qquad (6.4.4)$$

where $\mathcal{E}$ is the optical field and E is the electric field. For the simple case of punch through and tight optical confinement and the geometry shown in Fig. 6.27a, we have $E = V/d$ and $\Delta n = n_0^3 r_{41} V/2d$. The maximum voltage is generally set by the junction breakdown field strength; thus operating at this field, with an appropriate safety margin, suggests that the minimum power is required when d is as small as possible. Of course, d cannot be made arbitrarily small for reasons of mode confinement and insertion loss, as will be discussed later.

Another effect that introduces phase modulation is the depletion of free carriers from the guide region under reverse bias, due to the free-carrier contribution to the index discussed earlier and an effective widening of the guide. This effect occurs in heterojunctions as well as in homostructures with heavily doped confining layers. Thus for a guide layer that is not fully depleted at zero bias, the effective phase modulation will be a combination of this effect and the linear electro-optic effect and leads to a nonlinear dependence of phase on voltage. The phase modulation vs. voltage transfer characteristic for any layer structure can be calculated from the well-known equations of semiconductor junction physics, the mode of the waveguide and the overlap integrals given above.

Modulator insertion loss can be measured as the simple transmission loss of the device or it can be cast in the context of total loss when inserted in a single-mode fiber system. The latter concept is more meaningful for most applications and points to a desire to have a modulator optical mode cross section that provides a good overlap with the fiber mode profile. One could construct a long heterojunction phase modulator with a guide width a few tenths of a micrometer that would have an excellent deg/V modulation characteristic; however, this device would have poor fiber coupling as well as perhaps a high capacitance.

The above discussion does not fully account for waveguide bend requirements. Structures that are needed for tight confinement for highest modulator efficiency can also provide low loss bends of < 1 cm as discussed in Sect. 6.2.6. Again, the resulting guide sizes and confinement are smaller than mode size for optimium fiber coupling.

Results for modulator/fiber coupling have not yet been reported. However, coupling of passive guides to fibers indicates efficiencies of 70 % (1.5 dB loss) for guides typically $4\,\mu$m thick $\times$ $7\,\mu$m wide can be obtained [6.87, 140].

As one example of calculated design curves, Fig. 6.41 illustrates switching $P/\Delta f$ vs. strip width for a p^+-n-n^+ AlGaAs/GaAs directional coupler switch [6.116]. In this type of device, as discussed in detail in Sect. 4.3.1 and in (6.1.9) the induced index change between the guides induces switching. The plots indicate the effect of guide thickness on the required power.

6.4.3 Modulation Frequency Analysis

It is useful to be aware of frequency domain expressions for waveguide modulators because of the many analog applications that exist (e.g., microwave optical-frequency shifting). In this section, expressions for both phase and intensity modulators will be reviewed. For a phase modulator, a sinusoidal

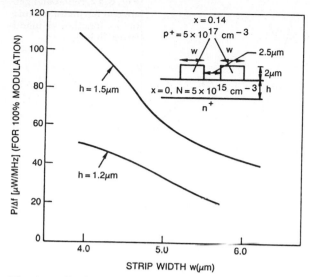

Fig. 6.41. Design curves for modulator drive voltage [6.116]

voltage that produces an RF phase modulation of $\phi_r \sin \omega_r t$ leads to an expression in terms of Bessel functions J_k of the first kind for the output signal of

$$
\begin{aligned}
\mathcal{E} = {} & \frac{1}{\sqrt{2}} J_0(\phi_r) \cos \omega l \\
& + \sum_{k=1}^{\infty} (-1)^k J_{2k}(\phi_r)[\cos(\omega - 2k\omega_r)t + \cos(\omega + 2k\omega_r)t] \\
& + \sum_{k=1}^{\infty} (-1)^k J_{2k-1}(\phi_r)\{\sin[\omega - (2k-1)\omega_r]t \\
& + \sin[\omega + (2k-1)\omega_r]t\} \quad .
\end{aligned}
\tag{6.4.5}
$$

Here ω is the optical angular frequency.

An analysis of this expression yields pairs of upper and lower side bands and, as indicated in Fig. 6.42, it implies that for $\phi_r < 1$ the higher-order side bands are well suppressed with a continuing reduction of the fundamental with modulation depth. It is seen that 10 % conversion efficiency can be achieved with negligible side-band generation, other than the desired upper and lower fundamental side bands.

In a similar manner, an analysis of a Mach-Zehnder interferometric modulator driven push-pull with a sinusoidal modulation signal, viewed as an ideal device with a dc phase shift of ϕ_0 in one arm, leads to an output optical electric field of

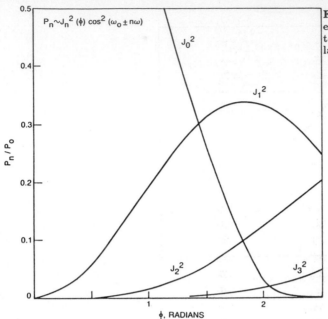

Fig. 6.42. Side-band generation for phase modulator as function of modulation index

$$\mathcal{E} = \frac{1}{\sqrt{2}}\left(\sin(wt + \phi_o)\left[J_0(\phi_r/2) + 2\sum_{k=1}^{\infty}(-1)^k J_{2k}(\phi_r/2)\cos 2k\omega_r t\right]\right.$$

$$+ \cos(\omega t + \phi_o)\left\{2\sum_{k=0}^{\infty}(-1)^k J_{2k+1}(\phi_r/2)\cos\left[(2k+1)\omega_r t\right]\right\}\right)$$

$$+ \frac{1}{\sqrt{2}}\left(\sin\omega t\left[J_0(\phi_r/2) + 2\sum_{k=1}^{\infty}(-1)^k J_{2k}(\phi_r/2)\cos 2k\omega_r t\right]\right.$$

$$\left. - \cos\omega t\left[2\sum_{k=0}^{\infty}(-1)^k J_{2k+1}(\phi_r/2)\cos(2k+1)\omega_r t\right]\right) \quad . \qquad (6.4.6)$$

This expression has been well analyzed for the case of LiNbO$_3$ [6.141] and can be cast in a form similar to (6.4.5) to indicate side-band pairs. For the case of single-polarization modulation (as in (100) GaAs), Fig. 6.43 shows

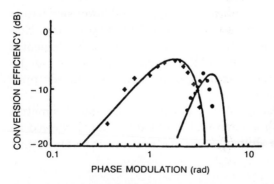

THEORETICAL ———

EXPERIMENTAL $\omega_o \pm \omega_n$ ◆
$\omega_o \pm 3\omega_n$ ●

Fig. 6.43. Side-band generation for LiNbO$_3$ Mach-Zehnder interferometric modulator as a function of modulation index [6.143]

typical side-band generation versus modulation depth. It is apparent that for $\phi = 1.8$ rad, maximum upper and lower first-order side bands are obtained with a suppression of the carrier and all even-order side bands. By combining a pair of these modulators with appropriate phase bias it is possible to generate a single side-band suppressed carrier (SSSC) modulator. Such a structure, while reported in $LiNbO_3$ [6.142, 143], has not yet been reported in semiconductors.

Another method of generating SSSC modulation in a $43m$ crystal is to propagate a microwave signal along a trigonal (111) direction [6.144]. If such a traveling-wave structure could be built in waveguide form (e.g., propagation along (111) using a (110)-oriented wafer) it would have numerous applications because it inherently has 100 % conversion efficiency. Devices have been demonstrated in bulk form in CdTe [6.145], but there are difficulties in generating the required circularly polarized fields in waveguide form and in obtaining the requisite growth orientations. A possible geometry for such a structure, which utilizes a buried heterostructure configuration and coplanar strip lines, is shown in Fig. 6.44. In such waveguide structures, the maximum conversion efficiency is limited by residual birefringence [6.146].

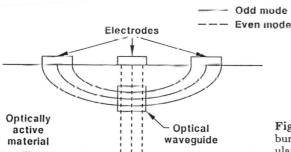

Fig. 6.44. Proposed structure of buried heterostructure SSSC modulator

6.4.4 Traveling-Wave Phase Modulators

Most of the discussion in Sect. 6.4.2 is for conventional types of single-mode phase modulators that function as lumped-element modulators. Traveling-wave structures with these geometries are conceptually appealing because of the good match between the optical and microwave dielectric constant in III-V semiconductors and the implied large bandwidth. That is, the effective bandwidth ν of a traveling-wave modulator of length l is given by the well-known expression

$$\nu = \frac{c}{4nl(1 - n_m/n)} \ . \tag{6.4.7}$$

Here n is the index of the semiconductor at the optical wavelength and n_m is the effective microwave dielectric constant. One can see the advantage of a good index match. For coplanar lines, $n_m^2 = (1 + \varepsilon_s)/2$, which is a combination of the dielectric constant of air and that of the semiconductors. It can be seen that the presence of air or an intervening dielectric ($\varepsilon < 3.5$) severely reduces the index match, resulting in an effective velocity mismatch. This mismatch is still a factor of three better than achievable with LiNbO$_3$. However, there are numerous design approaches that involve effectively burying the electrodes in III-V material to maintain the good velocity match. Quasi-strip-line traveling-wave geometries, i.e. those utilizing p^+-n^--n^+ or Schottky barrier n^--n^+ configurations, could potentially provide a better velocity match. Slow-wave effects [6.147] due to microwave loss in a n^+ substrate, however, must be minimized by making the n^+ (and p^+) layers as thin as possible, or the achieveable bandwidth can be less than in LiNbO$_3$. Recently, a 20-GHz rib-guide modulator utilizing coplanar lines on nominally undoped GaAs/GaAlAs waveguides has been reported [6.148]. Traveling-wave slab-guide modulators for 10.6 μm frequency shifting have also been reported [6.149] (see Sect. 6.5 for both devices).

6.4.5 TE-TM Coupling Analysis

For waveguides made on (110) material with propagation in a (1$\bar{1}$0) direction as shown in Fig. 6.39d, the two major axes of the index ellipsoid are at 45° to the plane of the waveguide when a (110) electric field is present. In this case, the electro-optic effect not only changes the phases of waveguide modes (which are no longer TE and TM modes even in a slab waveguide), but also rotates their plane of polarization. The net result is an effective coupling between the TE and TM modes of the unperturbed waveguide.

   With an applied (110) electric field, the normal orthogonal modes of even a slab waveguide are no longer pure TE or TM [6.52]. The primary electric fields of the new modes are at an angle θ to the electric field of the TE and TM modes of the unperturbed waveguide as indicated in Fig. 6.45.

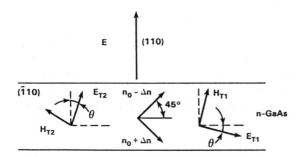

Fig. 6.45. Effects of a (110) electric field on the polarization of the modes in a (110) slab waveguide

370

The angle θ is always less than $45°$, which is the angle of the principal axis of the index ellipsoid to the plane of waveguide. One mode with amplitude $\mathcal{E}_1$ has a propagation constant β_1 which is generally greater than that of the unperturbed TE mode, β_{TE}, while the other mode has amplitude $\mathcal{E}_2$ and a propagation constant β_2 less than that of the unperturbed TM mode, β_{TM}. An input TE wave will excite the two normal modes of the (110) perturbed waveguide with amplitudes in a ratio of $\mathcal{E}_1 - \mathcal{E}_2$ of approximately $\cot\theta$. At a coupling distance L_c down the guide when $L_c = \pi/(\beta_1 - \beta_2)$, the two normal modes will be out of phase and will excite both a TE and TM mode in an output unperturbed waveguide. The amount of TE to TM (or vice versa) mode conversion is approximately equal to $(\sin 2\theta)^2$. For large electric fields the angle θ approaches $45°$, $\beta_1 - \beta_2$ becomes much greater than $\beta_{TE} - \beta_{TM}$ and behavior approaching that obtained in a bulk crystal is obtained.

Because of the difficulty of calculating the normal modes of anisotropic waveguides, coupled-mode theory is generally used to analyze mode conversion problems of this type. For a (110) waveguide with an applied electric field, a coupling coefficient between the TE and TM mode is calculated from the polarization perturbation caused by the electric field. Since the TE and TM modes are non-synchronous, i.e., $\beta_{TE} > \beta_{TM}$, the coupling coefficient K must be greater than $\beta_{TE} - \beta_{TM}$ in order to obtain a large percentage of conversions from the TE to TM and vice versa. This is equivalent to θ approaching $45°$ when the problem is treated using the normal modes of the perturbed system.

6.4.6 Infrared Waveguide Modulators – Wavelength Scaling

Results described above are for III-V semiconductors used in the 1-μm wavelength region. Another wavelength range of interest is the mid-IR (3 to 12 μm) where materials such as undoped GaAs and CdTe are adequately transparent. One can, in principle, fabricate the types of electro-optic modulators previously described, using scaled dimensions (i.e., approximate guide size scales with wavelength). Such modulators require considerably more drive voltage than 1-μm structures. The scaling of guide dimensions, plus the wavelength dependence of the linear electro-optic effect, implies that $V \sim \lambda^2$ for comparable electrode length. Thus, a 2-cm-long modulator that requires $V_\pi \sim 2\,\mathrm{V}$ at 1 μm would require 200 V at 10 μm. For some IR applications, it may not be necessary to shift the wave by π radians, so that the above estimate can be reduced.

In the infrared region, one must account for free-carrier absorption, which varies as $N\lambda^3$, where N is the doping level. For 10.6 μm, it is necessary to use high resistivity material ($N < 10^{12}\,\mathrm{cm}^3$) to keep α in the 1-dB/cm range.

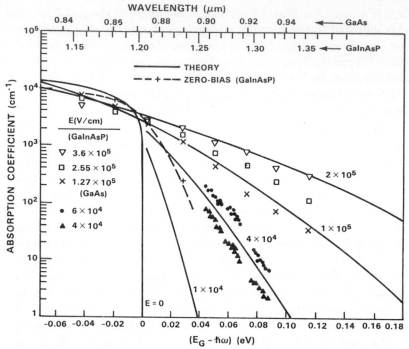

Fig. 6.46. Electroabsorption in GaAs and GaInAsP compounds [6.150]

6.4.7 Electroabsorption Modulation

Electroabsorption, or the Franz-Keldysh effect, is an electric-field-enhanced absorption near the semiconductor band edge. A set of absorption curves with electric field as parameter are shown in Fig. 6.45 [6.150]. The absorption is a function of effective mass m^* and energy difference from the band gap. As shown it is possible to plot the data in wavelength-difference format. As can be seen, the curves are quite similar for GaAs, InP and GaInAsP.

In evaluating optical modulator alternatives, it is worth noting the differences between electroabsorption and electro-optic modulation. First, electroabsorption modulators draw photocurrent and thus can represent a heat dissipation problem not present in electro-optic modulators. Such considerations become important especially for applications requiring high throughput power. To achieve efficient modulation, it is necessary to operate near the semiconductor band edge. This can lead to high insertion loss, unless short devices are used as is discussed in Sect. 6.5.5.

Electroabsorption leads to several other waveguide effects. First, with the enhanced absorption, there is, by Kramers-Kronig, a change in index. This manifests itself as a quadratic electro-optic effect and also leads to chirping effects in modulation. Electroabsorption has also been observed to be dichroic, i.e., it attenuates TM modes more than TE. These effects are also described in Sect. 6.5.5 where experimental work is summarized.

6.4.8 Carrier-Injection Modulator

Another type of modulator under development is the carrier injection modulator. This device is generally based on an X-switch (Sect. 4.3.3) in which the interference of propagating modes is altered by injecting carriers in the crossing region. These injected free carriers can reduce the index more than is achievable by the electro-optic effect if injection currents in the 10's of mA are used. This type of device has been demonstrated in a PIN junction format [6.151] and proposed in a bipolar transistor configuration to reduce switching time to subnanosecond levels [6.152].

6.4.9 Nonlinear Waveguide Modulator

An exciting possibility in semiconductor waveguides is to use light to control light and thus make all-optical modulators and switches. The physics of these devices has been discussed in detail elsewhere [6.153]. Here, device applications are described.

Figure 6.47 shows schematically several generic modulation approaches. In Fig. 6.47a, a light pulse is used to generate an electron-hole plasma. This reduces the effective index via the free-carrier contribution leading to phase modulation of the propagating light. While this device has the attractive feature of normal incident control light, the power required is greater than that generally obtainable from diode lasers for significant modulation (π radians) of a directional coupler or an interferometer. Such an approach may be useful, however, in less confined geometries, such as a cutoff modulator.

Use of the X^3 (or n_2) effect (i.e., $\Delta n = n_2 I$ where I is the light intensity) is another effect being studied for switching. Figure 6.47b shows schematically how light of two different polarizations can be used to drive a 3-input interferometer as an XOR gate. This device was demonstrated in LiNbO$_3$ [6.154]. For such applications, it is desirable to have a very fast nonlinearity. However, for most materials, a rule of thumb is that a nonlinear figure of merit $n_2/\alpha\tau$ is constant within a factor of 10. Here, α is the absorption coefficient and τ is the response time (turn off) of the nonlinearity. This suggests that both high speed and the necessary low absorption are accompanied by a relatively weak n_2. This is indeed the case for LiNbO$_3$. For semiconductor materials, such as some quantum-well structures, high n_2 is achieved in wavelength bands where, by Kramers-Kronig, there also exists high absorption. This can limit transmission lengths, which is a drawback in waveguide structures that are many mm long. (For such long structures, it is desirable to have $\alpha \leq 4\,\mathrm{dB/cm}$ to keep total insertion loss to tolerable levels). Absorption also tends to produce heating, so that

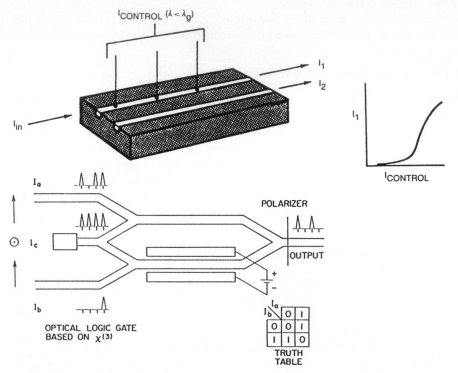

Fig. 6.47a, b. Approaches to waveguide all-optic modulation including **(a)** photo-induced electron-hole plasma and **(b)** X^3-induced

relatively slow thermal (relative to electronic) nonlinearities become significant and can limit response speed and high-speed switch extinction ratios. These generic arguments have led to a search for improved materials. Some of the most promising semiconducting materials in which waveguides have been demonstrated include semiconductor-doped glasses and GaAs/GaAlAs quantum-well structures.

6.5 Electro-Optic Guided-Wave Modulators – Characteristics

6.5.1 Phase Modulators

Here the results reported on phase modulators are briefly reviewed. Devices in GaAs/GaAlAs are first discussed, and recent results in the GaAsInP system are described.

The earliest reported guided-wave modulation was for a reverse biased GaP diode [6.155]. This was subsequently shown to be due to the linear electro-optic effect [6.156] in 1964. Phase modulation via the lin-

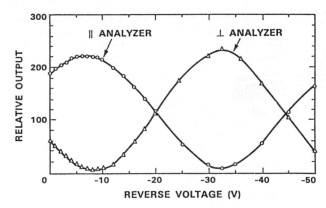

Fig. 6.48. Transfer characteristics of InP waveguide phase modulator [6.94]

ear electro-optic effect as large as $18°/V mm$ has been achieved in a planar GaAs/GaAlAs structure.

Many devices have been reported in the literature and a few of the more recent results are summarized here. A p^+-n-n^+ InP structure similar to that shown in Fig. 6.27a with guides that were $3.5\,\mu m$ high by $5\,\mu m$ wide has been reported [6.94]. For the voltages and doping levels used, full depletion was not achieved and the device was characterized by $1.5°/V mm$ at $1.3\,\mu m$. The full transfer characteristic is shown in Fig. 6.48. For a similar quarternary waveguide, a characteristic of $8°/V mm$ was achieved [6.157]. (As previously noted, the description of these modulators in terms of phase modulation/volt is not necessarily a figure of merit but is more a measure of the effective depletion width.) For both of these devices, the nonperfect sinusoidal transfer characteristics are a result of the varying depletion width with bias.

Recently, heterostructure phase modulators having submicrometer-thick guides which were not fully depleted at zero bias have been demonstrated. These devices had very high modulation characteristics ($56°/V mm$) and are well suited for on-chip coupling to integrated lasers [6.158]. Traveling-wave rib-guide modulators have been reported. Initial devices utilized a GaAlAs/ GaAs heterostructure arrangement with an n^+ substrate and exhibited a bandwidth of $\sim 4\,GHz$ due to slow-wave effects [6.159]. Similar devices formed on a semi-insulating substrate and utilizing a coplanar electrode configuration exhibited a 20-GHz bandwidth at $1.3\,\mu m$ [6.148].

Slab-guide geometries have also been reported in GaAs for traveling-wave phase modulation [6.149]. In this case, the application was for frequency shifting a 10.6-μm laser. The use of a slab geometry permitted prism coupling for large optical power handling capability and the use of wide microstrip lines to minimize series resistance. The device shown schematically in Fig. 6.49 had a center frequency of $13\,GHz$ and a 3-dB bandwidth of $10\,GHz$. The structure was a thinned ($25\,\mu m$) slab of high resistivity GaAs

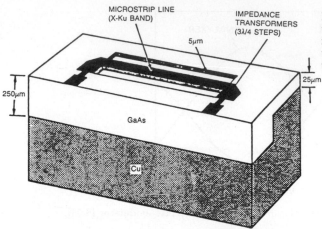

Fig. 6.49. Schematic illustration of GaAs traveling wave phase modulator for 10.6 μm applications [6.149]

that could support a few optical modes. Only a single mode was excited by prism coupling. This structure exhibited output power conversion efficiency of up to about 2 % and optical power handling capability of ∼ 10 W. As described in [6.149], the device can be optimized and efficiency improved by using thinner guide layers with appropriate high-resistivity cladding layers (e.g., ZnSe or high resistivity AlGaAs).

6.5.2 Directional-Coupler Switches

As described earlier, a directional coupler can be formed by placing two guides in close proximity. Switching can then be obtained by spoiling the synchronism of the guides by modulating one propagation constant relative to the other. Theoretical expressions for these switches are given in (6.1.9) and further discussed in Sects. 3.2.6 and 4.3.1.

An initial demonstration of partial switching was obtained for a pair of coupled planar guides (6.160). The first-reported GaAs channel guide switch utilized metal-gap guides and had a 13-dB extinction ratio and required 35-V drive [6.117]. Other types of switches that have been reported include those with rib structures [6.128], ion-implanted channel stops [6.99] and photoelastic confinement [6.108]. The first high-extinction-ratio device [6.129] was a metal-gap structure using a $\Delta\beta$-reversal, electrode configuration as shown in Fig. 6.50. Here each electrode was individually biased to obtain an extinction ratio of 25 dB.

An interesting effect is obtained in these switches due to crystallographic orientation. Transfer characteristics of channel-confined devices formed with guide propagation in a (110) and (1$\bar{1}$0) direction yield different insertion losses [6.99]. For a (001) electric field there is a decrease in confinement in one direction, which results in overall guide attenuation; in the other direction, the mode is better confined with bias and no attenuation occurs (Sect. 6.4.1).

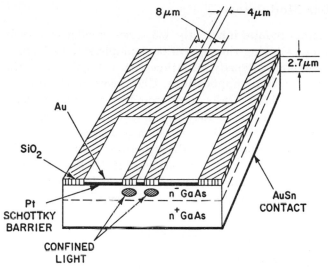

Fig. 6.50. GaAs $\Delta\beta$-reversal directional coupler switch [6.129]

GaAs rib guides switches using MOS [6.161] and p^+n heterojunction [6.162] structures to apply electric fields have also been reported. These rib structures are difficult to fabricate with low leakage current and high breakdown voltage because of the long exposed junction perimeter. The best results have been obtained with a heterojunction geometry as shown in Fig. 6.51. These devices had a demonstrated bandwidth of 250 MHz. Homojunction p^+n switches in InP have also been reported [6.97].

The wavelength dependence of GaAs directional-coupler switches has been studied [6.114]. Figure 6.34 shows the variation in coupling length for a fixed geometry. The voltage required for switching increases as one would expect from changes in coupling length and the variation of electro-optic efficiency with wavelength. Similar results have been obtained for rib guide geometries [6.115].

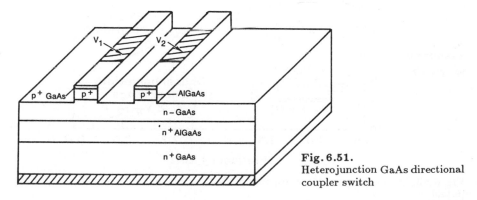

Fig. 6.51.
Heterojunction GaAs directional coupler switch

6.5.3 Interferometric Modulators

Interferometric modulators require inherently $\sqrt{3}$ times smaller change in propagation constant for switching than directional couplers. This feature, in conjunction with a periodic intensity variation with voltage, makes the interferometer attractive for many applications. The theory of these modulators is described in Sect. 4.3.2. GaAs interferometric modulators have been made with Y-junction couplers [6.133] as well as three-guide couplers [6.122] to provide the power splitting. The performance of these passive couplers was described above. Data obtained for the structure shown in Fig. 6.52 indicated $V_\pi \simeq 22$ V at $1.3\,\mu$m. The extinction ratio was 18 dB. Such lumped-element devices have also demonstrated a large flat bandwidth. Figure 6.53 shows results [6.123] of small-signal modulation. The small-signal electri-

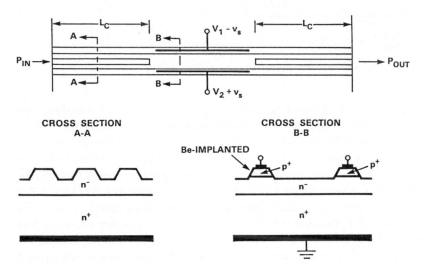

Fig. 6.52. Schematic illustration of a GaAs interferometric modulator [6.122]

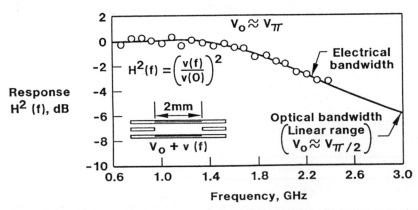

Fig. 6.53. Small-signal frequency response of GaAs interferometric modulator (Fig. 6.52) [6.123]

378

cal bandwidth is 2.2 GHz with a corresponding linear optical bandwidth of 3 GHz. The insertion loss of such structures relative to a straight guide is about 2 dB. Recently, an optical bandwidth of 5 GHz, with some resonances, has been reported [6.133]. Bandwidths approaching 10 GHz should be possible with lumped-element devices.

A method proposed to minimize drive power is to back-bias the junctions to the mid-point of the transmission characteristic and use a push-pull bias arrangement [6.122 and 123]. This reduces the required power by a factor of 4.

6.5.4 Integrated Waveguides/Optoelectronics/Electronics

It is appealing to integrate lasers and/or detectors with optical switches on a single semiconductor substrate. Figure 6.54 shows the only reported coupler/detector [6.163]. This device contains a conventional coupler of Schottky barrier n-n^+ rib design which had a 17-dB extinction ratio. The detector is a reverse-biased junction that was proton implanted (no anneal) to enhance its near-IR sensitivity. This detector had a potential response speed of 1 GHz and a 23 % quantum efficiency at 1.06 μm. The overall structure was evaluated as a bistable switch by feeding the detector output through a resistive load and amplifier back to the switch electrodes. In the initial experiments, switching speed and energy were 1 μs and 1 nJ, respectively. Detector/slab waveguide integration has also been demonstrated in a GaAs structure. The first demonstration [6.164] was a GaInAs Schottky detector formed in a GaAs waveguide structure by selective epitaxy, as shown in Fig. 6.55. This device had good performance but higher than ideal dark current because of the strain due to the lattice mismatch. Other work has been done on integrated photoconductive detectors. Work reported on the integration of silicon detector arrays (photodiode or CCD) and waveguides [6.111] was described at the end of Sect. 6.3.1.

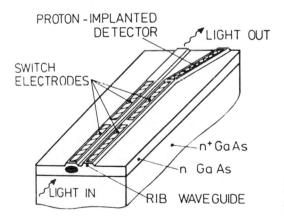

Fig. 6.54.
GaAs integrated detector/directional coupler [6.163]

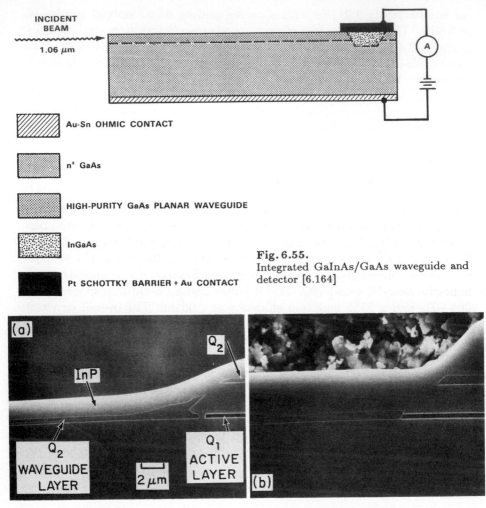

INCIDENT BEAM

1.06 μm

Au-Sn OHMIC CONTACT

n⁺ GaAs

HIGH-PURITY GaAs PLANAR WAVEGUIDE

InGaAs

Pt SCHOTTKY BARRIER + Au CONTACT

Fig. 6.55.
Integrated GaInAs/GaAs waveguide and detector [6.164]

Q₂

InP

Q₂ WAVEGUIDE LAYER

2 μm

Q₁ ACTIVE LAYER

(a)

(b)

Fig. 6.56. GaInAsP laser waveguide integration with regrown guide [6.165]

In laser/waveguide integration, some of the early work was for a GaAlAs/GaAs device with a chemically etched facet and a regrown waveguide of high purity GaAs, which has a low absorption coefficient at the lasing wavelength [6.165]. A more recent version of this approach in GaInAsP is shown in Fig. 6.56 [6.166]. These lasers had threshold current densities of $2-3\,\mathrm{kA/cm^2}$. Here, the two cleaved ends of the chip form the laser cavity. Another technique demonstrated to form a laser/waveguide is to use a double heterostructure and a tapered region [6.167], as shown in Fig. 6.57, to place light in a low-loss guide layer. This structure has the advantage of potentially forming a buried large-optical-cavity-type waveguide. In separate work, lasers have been integrated with rib waveguides [6.168], and switch

380

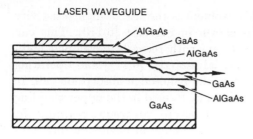

LASER WAVEGUIDE

Fig. 6.57. Laser waveguide integration with tapered transition

TAPERED GUIDE

integration should soon occur. Integration of multiple lasers and waveguides is further discussed in Sect. 6.6.

An exciting class of devices are those that involve the integration of electronic devices with waveguides and modulators. GaAs is the primary material of choice, as it is for other optoelectronic integrated structures. A difficulty here is to obtain adequate drive voltages from available FETs to fully modulate the optical signal. For situations where moderate depth of modulation is adequate, as for some digital and microwave applications, such integrated structures are promising. An initial device incorporating a GaAs directional coupler and a MESFET driver has been reported [6.169]. Because of the need for heterojunction structures and a total epitaxial thickness of 3–5 μm for low-loss fiber coupling, it is apparent that chips with steps or wells, similar to those developed for detectors and lasers could be needed. These techniques will be described in Sect. 6.6 on optoelectronic integration.

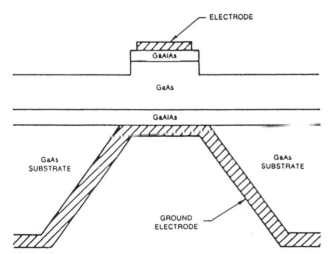

Fig. 6.58. Use of high-resistivity epitaxial growth to form integrated waveguide structures. Shown is a raised rib modulator that utilizes MMIC technology for fabricating the ground electrode (after [6.170])

One interesting approach to this problem is the use of high resistivity material for all the guide layers, as shown in Fig. 6.58 [6.170]. This can be achieved by doping the epitaxial layers with recently developed high-resistivity impurities (e.g., V) for epitaxial material [6.171] and operating at a wavelength somewhat removed from the band-edge absorption tail. In principle, this allows a high-quality FET to be formed in the upper cladding layer so that a significantly more planar structure can result. Forming the FET in the AlGaAs should not cause much change in device performance since the layer contains only a small percent of Al.

6.5.5 Electroabsorption Modulators

The majority of reported waveguide electroabsorption (EA) modulators are short ($\sim 1\,\mathrm{mm}$) slab-guide homojunctions [6.33] and heterojunctions [6.172] with Schottky barrier electrodes. The short length of these structures was dictated by operating near the band edge where absorption is high. Recent results [6.173] for such devices in GaAsInP have shown a 20-dB extinction ratio and bandwidth of $\sim 1.6\,\mathrm{GHz}$. Devices in GaAlAs/GaAs have exhibited a 30-dB extinction ratio and a 3-GHz bandwidth [6.174]. For both devices, drive voltages were $< 10\,\mathrm{V}$ and insertion losses were $\sim 10\,\mathrm{dB}$.

It should be noted that electroabsorption also gives rise to a quadratic electro-optic effect. This has been documented in the GaInAsP materials system [6.175]. Typical plots [6.175] of phase difference vs. voltage are shown in Fig. 6.59. The deviation from linearity is believed to be due to

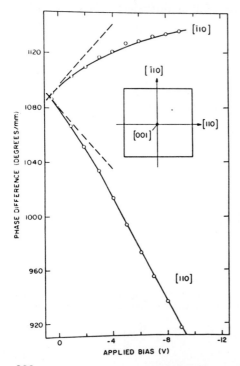

Fig. 6.59. Combination of linear and quadratic phase modulation in GaInAsP waveguide modulators [6.175]

382

index changes near the band-edge induced by electroabsorption and can be referred to as electro-refraction [6.176]. This provides a means of enhancing phase modulation efficiency for some applications. This nonlinear (or Kerr) coefficient is wavelength dependent and can be expressed for the quaternaries as $R_k \sim -1.5 \times 10^{-15} \exp(-8.85\Delta E)\,\mathrm{cm}^2/\mathrm{V}^2$ where ΔE is the difference in eV between the band gap and operating wavelength. In practical cases, at fields below breakdown, this nonlinear effect produces $\sim 10\,\%$ deviation from linearity. However, this dynamic index change can lead to deleterious chirp effects for single-frequency-source modulation [6.177].

It has also been shown that electroabsorption is dichroic [6.175] in GaInAP. That is, a given applied (001) electric field attenuates TM modes more than TE modes. This effect has also been observed in GaAs and has been attributed to a lifting of the valence-band degeneracy by the applied electric field so that, in the electric-field-enhanced tunneling description of Keldysh tunneling, the tunneling probability is enhanced for polarization along the applied field relative to that for polarization perpendicular to the field. Results are independent of whether light propagates in a (110) or (1$\bar{1}$0) direction.

Arrays of EA modulators have also been proposed for signal-processing applications. In one case, a CCD array formed over a planar guide allows the light to be spatially modulated. With a suitable combination of waveguide lenses and modulated lasers, such structures could function as Fourier transformation and correlation devices with predicted GHz bandwidths and attractive dynamic ranges. Figure 6.60 shows an approach for space- and

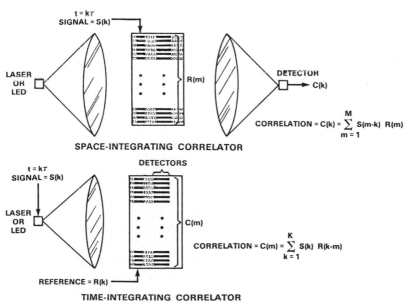

Fig. 6.60. Proposed GaAs CCD waveguide correlators [6.178]

383

time-integrating correlators using a pair of CCDs [6.178]. Similar structures have been analyzed for Fourier transformation [6.179].

6.5.6 Multiple-Quantum-Well Modulators

The birefringent nature of multiple-quantum-well (MQW) waveguides was discussed in Sect. 6.2. MQW devices also have a strong electric-field-dependent absorption near the exciton resonance energy [6.180], as shown in Fig. 6.61. This enhanced absorption had led to the demonstration of non-waveguide EA modulators that are more efficient than those of conventional bulk III-V materials. Initial enhanced EA modulators used transmission normal to the quantum-well plane and thus had a path length of $\sim 1\,\mu$m and small contrast. For fields parallel to the optical path [6.181], the contrast was 3. For electric fields perpendicular to the light path [6.182], i.e., parallel to the plane of the quantum wells, lower contrast was obtained. The difference in orientation effects is related to exciton deformation [6.180].

By utilizing the enhanced absorption in a waveguide geometry, significant modulation can be obtained for electric fields normal to the waveguide (QW) plane. Quantum-well waveguide modulators reported to date have been formed in leaky waveguides having a few quantum wells in the guide center [6.183]. In this device, two 94-Å quantum wells were embedded in the center of a 3.6-μm-thick superlattice which formed the waveguide core. This

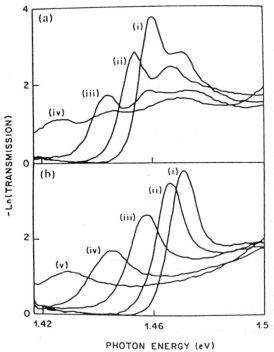

Fig. 6.61.
Electric field dependence of absorption near the band edge of a MQW waveguide structure.
(**a**) Incident polarization parallel to plane of layers for fields of
(i) 1.6×10^4 V/cm,
(ii) 10^5 V/cm,
(iii) 1.3×10^5 V/cm,
(iv) 1.8×10^5 V/cm.
(**b**) Incident polarization perpendicular to plane of layers for fields of
(i) 1.6×10^4 V/cm,
(ii) 10^5 V/cm,
(iii) 1.4×10^5 V/cm,
(iv) 1.8×10^5 V/cm and
(v) 2.2×10^5 V/cm [6.180]

-Ln(TRANSMISSION)

PHOTON ENERGY (eV)

superlattice was surrounded by GaAs cladding layers. The device was doped as a PIN diode and was reverse biased to apply an electric field perpendicular to the quantum wells. This device guides both TE and TM polarizations, but with different modulation characteristics (i.e., selection rules) [6.184]. Here the overlap optical mode with the QW is small. The device reported was $\sim 150\,\mu\text{m}$ long and functioned as a mini-slab-guide modulator. Even for this short length, radiation from the capillary guide led to 3 dB loss. This device demonstrated a 10 : 1 on/off ratio and a 100 ps response speed ($\lambda = 0.85\,\mu\text{m}$). Based on the absorption data for QW GaAs structures, it should be possible to design an EA modulator with enhanced efficiency relative to one formed on bulk material with usefully low insertion loss [6.185].

The electro-optic coefficient of a MQW far from the exciton edge has been determined to be $-1.47 \times 10^{-10}\,\text{cm/V}$ [6.186], a value comparable to that of bulk materials. At wavelengths near the exciton edge (888 μm), there is some evidence for an enhanced r_{41}, as well as a strong quadratic EO effect ($R_{11} - R_{12} = 6 \times 10^{-16}\,\text{cm}^2/\text{V}^2$) [6.187].

The work in the MQW waveguide area is in its initial stages. One clear goal in developing future devices is to design quantum-well structures that maximize modulation efficiency while keeping insertion losses as low as possible. It is also reasonable to expect results to be extended to other MQW materials systems such as InGaAs/GaAs and HgCdTe/CdTe

0.5.7 Nonlinear Waveguide Modulators

Experimental results to date on nonlinear optical waveguide modulators are somewhat limited. Representative results for semiconductor-doped glasses and quantum-well structures are given below. Semiconductor-doped glasses are a class of materials that contain microcrystallites (100–1000 Å diameter) of a semiconductor in a borosilicate glass host. Conventional color glass cut-off filters that contain varying amounts of $\text{CdS}_x\text{Se}_{1-x}$ are one example of such material. A plot of the measured spectral dependence of n_2 and induced absorption for one glass is shown in Fig. 6.62 [6.188]. Here 5-ns low-duty-cycle pulses were used. The change of sign of the index at the band edge is what one expects if the nonlinearity is due to band-filling effects. Response times of ~ 30 ps have been reported, but an underlying slow (~ 10 ns) process, probably thermal in nature was also observed [6.189]. If the glass also contains Na, it is possible to utilize ion-exchange technology [6.190] (e.g., Ag-Na exchange) to form optical waveguides. In this technique, regions of Ag dopant have a higher index of refraction. Work to date has demonstrated slab and channel guides [6.191], but nonlinear measurements have not yet been reported.

In GaAs/GaAlAs quantum-well material, work on optical waveguides is also in its early stages. A fair amount of data exists on the spectral depen-

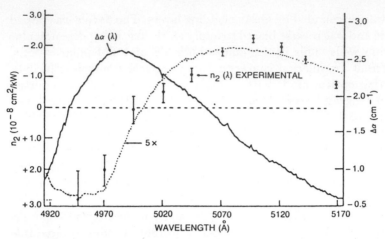

Fig. 6.62. Spectral dependence of nonlinear coefficient in semiconductor-doped glass [6.188]

dence of n_2 and values near the band edge of $n_2 \sim 2 \times 10^{-4}\,\mathrm{cm}^2/\mathrm{W}$ (considerably enhanced over bulk material) have been reported [6.192]. Much of this work has been motivated by interest in bistable thin-film etalon-type devices. These results, as well as enhanced electroabsorption QW-waveguide modulator experiments, suggest that multiple-quantum-well structures, operated at wavelengths of $\leq 300\,\text{Å}$ from the gap may be too lossy to be practical for waveguide use. The peak value of n_2 quoted above corresponds to band-edge absorption ($\sim 10^4\,\mathrm{cm}^{-1}$), and the work of *Wa* et al. [6.193] on nonlinear couplers had $\alpha \sim 15\,\mathrm{cm}^{-1}$ ($64\,\mathrm{dB/cm}$) at a wavelength where $n_2 \sim 10^{-7}\,\mathrm{cm}^2/\mathrm{W}$. This work resulted in switching most of the energy from the unexcited guide at lower power to the excited guide for a higher incident power of $1.5\,\mathrm{mW}$ in an MQW rib guide structure. However, the carrier relaxation time in such material is $\sim 20\,\mathrm{ns}$, which suggests that operation may be limited to $50\,\mathrm{Mbs}$. To increase speed, proton bombardment [6.194] could be used to reduce carrier lifetime, but this could result in increased power requirements and higher absorption. This suggests that operation in specialized structures, such as quantum-well pockets in passive guides or one- or two-well structures, or operation far from the band edge may be required for practical nonlinear guided-wave devices.

6.6 Optoelectronic Integrated Circuits (OEIC)

An area closely related to semiconductor integrated optics is the integration of a number of optoelectronic components on a single substrate along with associated electronics. This area is usually referred to as integrated optoelectronics. Efforts to date are primarily in the GaAs and InP-based material systems, with substantial emphasis on the former. The long-term goal of this work is to achieve fairly complex circuits. Efforts to date have focused on integrated transmitters, receivers, and repeaters, addressable LED and laser arrays, detector arrays, and, as previously discussed, laser/waveguide and detector/waveguide structures.

Emphasis in this section will be on some of the key fabrication and performance issues associated with the basic elements, integrated transmitters and receivers, and on the implications of fabricating these devices with waveguides. For a review of the demonstrated performance through 1985 of a number of these devices, the reader is referred to the paper by *Forrest* [6.195]. Laterally coupled arrays of diode lasers are not treated here.

The basic problem in optoelectronic integration is to solve in some way the conflicting requirements of the optical and electronic components for material composition, layer thickness and heat extraction. These conflicting requirements can lead to large step-height differences on the wafer that make high-resolution photolithography difficult. They also require the development of selective epitaxial techniques and the ability to grow layers and fabricate devices in wells etched in a substrate. For each application, the merits of integration versus sophisticated hybridization must be critically evaluated. Issues such as chip performance, yield, and projected cost must be weighed. While much progress has been made in the research laboratory, monolithic OEICs have not yet performed as well as the best hybrid devices in terms of bandwidth, laser threshold, lifetimes or receiver sensitivity.

The basic problem in an integrated photoreceiver is to maximize the sensitivity of a PINFET, i.e., one must minimize the noise equivalent dark current i_d. This current can be expressed as $\langle i_d^2 \rangle = C_T^2 B^2 / g_m$. One wishes to maximize the transconductance g_m of the FET while minimizing the total capacitance C_T, which is a combination of that of the photodiode, FET, and the interconnection line. This is a difficult task because the optimum conductivity and heterojunction layer thickness for the photodiode conflict with the needs for the FET whether it is formed on a semi-insulating substrate or on a thin doped layer. Figure 6.63 shows one example of an integrated PINFET [6.196].

An alternative to the PINFET is to form a photoconductive detector and preamplifier as shown in Fig. 6.64 [6.197]. This device has the advantage of being inherently more compatible with FET fabrication. However, the device generally has a slower speed of response due to minority carrier lifetime issues and may have less sensitivity due to increased dark current. Nevertheless its ease of fabrication makes it attractive when ultimate sensitivity is not required. A 4-channel photoreceiver utilizing photoconductive detectors has been reported [6.198]. Detector devices which have a fully depleted channel should result in reduced dark current.

In the laser area, one difficulty arises in forming mirror facets in a manner that allows chip design flexibility. Initial devices were laser/FETs which

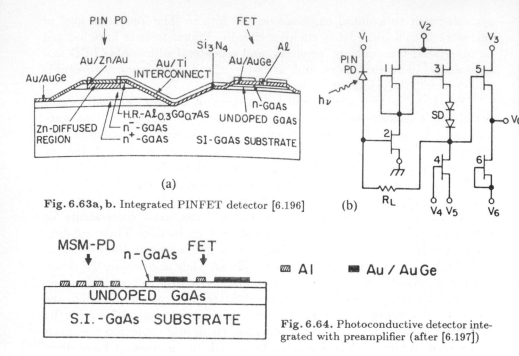

Fig. 6.63a, b. Integrated PINFET detector [6.196]

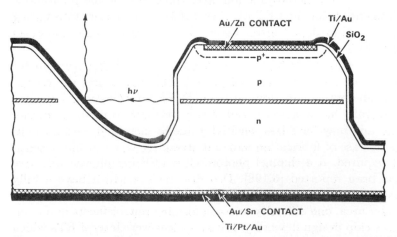

Fig. 6.64. Photoconductive detector integrated with preamplifier (after [6.197])

had chip lengths equal to cleaved laser cavity lengths ($\sim 300\,\mu$m) [6.199]. More sophisticated circuitry is possible with noncleaved facets, i.e. both mirrors of the laser cavity are not at the edge of the chip. Here, both dry etching [6.200] and microcleaving techniques [6.201] have been demonstrated. Another approach is to form a 45° etched mirror for vertical emission [6.202]. As shown in Fig. 6.65, this latter technique allows lasers to be placed anywhere on a chip surface, and laser arrays utilizing this technique have been

Fig. 6.65. Integrated GaInAsP/InP laser and 45° deflection mirror [6.202]

388

demonstrated in GaInAsP [6.203]. In this method, selective chemical etching followed by a vapor-phase mass transport of InP is used to form the laser and deflection mirrors. As yet, no integration of electronics with this laser structure has been reported. Recently, a similar discrete laser structure was reported in GaAs using dry etching [6.204]. Surface emission through a thin epilayer cavity is also under development [6.205].

Another technique that lends itself to flexibility of placement of the laser is a DFB structure coupled to an on-chip waveguide. To date, no integration of DFB lasers with electronic circuits has been reported. However, 6 GaAlAs/GaAs DFB lasers of different wavelengths have been fabricated on the same wafer, and multiplexed using a multi-mode waveguide Y-junction circuit [6.206]. In this case, the threshold current of the lasers was $3\text{--}6\,\mathrm{KA/cm^2}$, wavelength spacing was $20\,\text{Å}$ as determined by the variation in the grating period and the coupled output power was $3\,\mathrm{mW}$.

Two representative examples of laser integration illustrate fabrication techniques being pursued. The first [6.207] is an advanced quantum-well structure with a low threshold and multi-transistor driving circuit, as shown in Fig. 6.66. Here different chip levels were used to define the electronic and

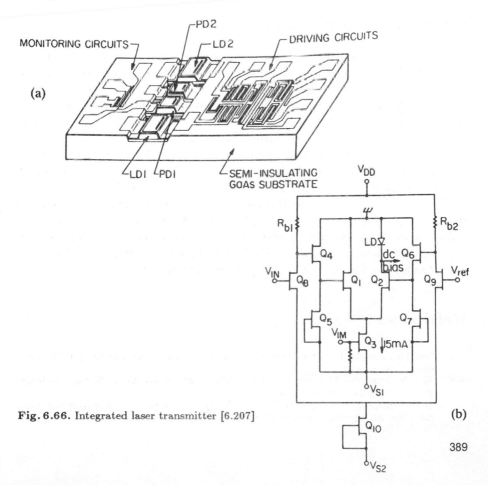

Fig. 6.66. Integrated laser transmitter [6.207]

389

optical portions of the chip. Growth was by MBE. The lasers had dry-etched facets and threshold currents of 40 mA and would be useful in repeater applications. The second example [6.208] is one intended for use in multiplexing operations. Here, a sophisticated digital circuit, a 4 : 1 1-GHz multiplexer, was integrated with a laser driver. In this case, the laser was formed by growth in a well and subsequent wafer resurfacing. The resulting planar structure is needed for the photolithographic processing of the multiplexer circuit. Such a device has exhibited operation to 150 MHz.

An important issue in the development of such laser circuits is the maintenance of low dislocation densities so that long laser lifetimes can be obtained. It is well known that long-life GaAs lasers require substrate dislocations $< 10^3 \, \mathrm{cm}^{-2}$. However, some of the best transistor threshold uniformity results on GaAs digital circuits have been achieved on substrates with high dislocation counts. In recent years, substantial improvements have been made and now good quality electronic circuit yields and uniformities are obtainable on high-resistivity, low-dislocation substrates, so one can be optimistic about the future of opto-electronic integration.

6.7 Concluding Remarks

In this chapter the theory, design considerations, fabrication and experimental results for semiconductor integrated optical devices have been reviewed. Emphasis was placed on the basic passive and active devices in the GaAs and InP materials systems, with limited discussion of growth and fabrication technology, laser and detector integration with waveguides, and optoelectronic integrated circuits. The current state-of-the-art and future prospects for these devices were given. An effort was made to give references that are representative of the field. If all the work in this field had been referenced, the bibliography would easily have been several times its length.

The needs for optical devices especially in telecommunications and information processing are increasing rapidly. At the same time III-V technology is evolving quickly. It will, therefore, probably not be long before complex chips which incorporate guided-wave, optoelectronic and electronic components are demonstrated and put to practical use.

References

6.1   See, for example: R.A. Smith: *Semiconductors*, (Cambridge University Press, London, 1959) p. 222

6.2   See, for example: E. Garmire: In *Integrated Optics*, ed. by T. Tamir (Springer, Berlin, Heidelberg 1979) p. 244;
        R.G. Hunsperger: *Integrated Optics: Theory and Technology*, Springer Ser. Opt. Sci., Vol. 33 (Springer, Berlin, Heidelberg 1982) p. 55

6.3   See, for example: G.A. Antypas: In *GaInAsP Alloy Semiconductor*, ed. by T.P. Pearsall (Wiley, New york 1982) p. 4;
       H.C. Casey, M.B. Parish: *Heterostructure Lasers Pt. B: Material and Operating Characteristics* (Academic, New York 1978) p. 25
6.4   H.C. Casey, M.B. Parish: ibid, p. 17
6.5   A.A. Berg, P.T. Dean: Proc. IEEE **160**, 156 (1972)
6.6   M.A. Afromowitz: Solid State Commun. **15**, 59 (1974)
6.7   H.C. Casey, D.D. Sell, M.B. Parish: Appl. Phys. Lett. **24**, 63 (1974)
6.8   V. Eutuhov, A. Yariv: IEEE Trans. **MTT-23**, 44 (1975)
6.9   S. Adachi: J. Appl. Phys. **58**, 81 (1985)
6.10   B. Jensen, A. Torabi: IEEE J. **QE-19**, 877 (1983)
6.11   R.E. Nahory, M.A. Pollock, W.D. Johnston, R.L. Barns: Appl. Phys. Lett. **33**, 314 (1978)
6.12   Y. Suematsu, K. Iga, K. Kishinu: In *GaInAsP Alloy Semiconductors*, ed. by T.P. Pearsall (Wiley, New York 1982) p. 362
6.13   P. Chandra, L.A. Coldren, K.E. Strege: Electron. Lett. **17**, 6 (1981)
6.14   L. Esaki: IEEE J. **QE-22**, 1611 (1986)
6.15   G. Bastand, J.A. Brum: IEEE J. **QE-22**, 1625 (1986)
6.16   A.C. Grossard: IEEE J. **QE-22**, 1649 (1986)
6.17   J.E. Fouget, R.D. Durnham: IEEE J. **QE-22**, 1799 (1986)
6.18   G.H. Dohler: IEEE J. **QE-27**, 1682 (1986)
6.19   K.B. Kahen, J.P. Leburton: Appl. Phys. Lett. **47**, 508 (1985)
6.20   G.J. Sonek, J.M. Ballantyne, Y.J. Chen, G.M. Carter, S.W. Brown, E.S. Koteles, J.P. Salerno: IEEE J. **QE-22**, 1015 (1986)
6.21   Y. Suzuki, H.O. Kamato: J. Electron. Matl. **12**, 397 (1983)
6.22   J.S. Weiner, D.S. Chemla, D.A.B. Miller, H.A. Haus, A.C. Gossard, W. Wiegmann, C.A. Burrus: Appl. Phys. Lett. **47**, 664 (1985)
6.23   D.S. Chemla, D.A.B. Miller: J. Opt. Soc. **B2**, 1155 (1985)
6.24   G.C. Osbourn: IEEE J. **QE-22**, 1677 (1986)
6.25   I.J. Fritz, L.R. Dawson, T E. Zysperean: Appl. Phys. Lett. **43**, 846 (1983)
6.26   N.G. Anderson, W.D.Lardy, R M. Kolbas, Y.C. Lu: J. Appl. Phys. **60**, 2361 (1986)
6.27   See, for example, A. Yariv: *Quantum Electronics* (Wiley, New York 1967) p. 293
6.28   L. Ho, C.F. Buhrer: Appl. Opt. **2**, 647 (1963)
6.29   K. Tada, N. Suzuki. Jpn. J. Appl. Phys. **50**, 4567 (1979)
6.30   P.A. Kirkby, P.R. Selway, L.D. Westbrook: J. Appl. Phys. **50**, 4567 (1979)
6.31   S. Somekh, E. Garmire, A. Yariv, H.L. Garvin, R.G. Hunsperger: Appl. Phys. Lett. **22**, 46 (1973)
6.32   H.C. Casey, Jr., M.B. Parish: *Heterojunction Lasers* (Academic, New York 1978) Pt. A, p. 83, Pt. B, p. 199
6.33   G.E. Stillman, C.M. Wolfe, C.O. Bozler, J.A. Rossi: Appl. Phys. Lett. **28**, 544 (1976)
6.34   F.A. Blum, D.W. Shaw, H.C. Holton: Appl. Phys. Lett. **25**, 116 (1974)
6.35   L.D. Westbrook, P.J. Fiddyment, P.N. Robson: Electron. Lett. **16**, 169 (1980)
6.36   D.J. Jackson, D.L. Persechine: Electron. Lett. **21**, 44 (1984)
6.37   E.A.J. Marcatili: Bell Syst. Tech. J. **48**, 2071, (1969)
6.38   R. Ramaswamy: Bell Syst. Tech. J. **53**, 697 (1974)
6.39   E.A.J. Marcatili: Bell Syst. Tech. J. **53**, 645 (1974)
6.40   H.F. Taylor: IEEE J. **QE-12**, 748 (1976)
6.41   S.T. Peng, A.A. Oliner: IEEE Trans. **MTT-29**, 843 (1981)
6.42   M.D. Feit, J.A. Fleck, Jr.: J. Opt. Soc. **73**, 1296 (1983)
6.43   N. Dagli, C.G. Fonstad: Appl. Phys. Lett. **49**, 308 (1986)
6.44   N. Dagli, C.G. Fonstad: Appl. Phys. Lett. **46**, 529 (1985)
6.45   N. Dagli, C.G. Fonstad: IEEE J. **QE-21**, 315 (1985)
6.46   M.W. Austin: IEEE J. **QE-8**, 795 (1982)
6.47   M.W. Austin, P.G. Flavin: J. Lightwave Tech. **LT-1**, 236 (1983)
6.48   M.W. Austin: J. Lightwave Tech. **LT-2**, 688 (1984)
6.49   N.A. Wittaker: private communication
6.50   H.F. Taylor, A. Yariv: Proc. IEEE **62**, 1044 (1974)
6.51   H. Kogelnik: IEEE Trans. **MTT-23**, 2, (1975)

6.52   J.M. McKenna, F.K. Reinhart: J. Appl. Phys. **47**, 2069 (1976)
6.53   A. Yariv: IEEE J. **QE-9**, 919 (1973)
6.54   H.A. Haus, W.P. Huang, S. Kawakomi, N. Whitaker: J. Lightwave Tech. **LT-5**, 16 (1987)
6.55   W. Streifer, M. Osinski, A. Hardy: J. Lightwave Tech. **LT-5**, 1 (1987)
6.56   J.P. Donnelly, N.L. DeMeo, G.A. Ferrante: J. Lightwave Tech. **LT-1**, 417 (1983)
6.57   J.P. Donnelly: IEEE J. **QE-22**, 610 (1986)
6.58   J.P. Donnelly, H.H. Haus, N.A. Whitaker: IEEE J. **QE-23**, 401 (1987)
6.59   J.L. Merz, R.A. Logan, W. Wiegmann: Appl. Phys. Lett. **26**, 337 (1975)
6.60   A. Yariv: In *Introduction to Integrated Optics,* ed. by M.K. Barnoski (Plenum, New York 1973) p. 253
6.61   W.P. Dumke, N.M.R. Lorenz, G.B. Pettit, Phys. Rev. **B-1**, 4668 (1970)
6.62   R.G. Walker: Electron. Lett. **21**, 581 (1985)
6.63   A.M. Sergent, J.L. Merz, R.A. Logan: J. Appl. Phys. **50**, 2552 (1979)
6.64   E.A.J. Marcatili: Bell Syst. Tech. J. **48**, 2103 (1969)
6.65   D. Marcuse: Bell Syst. Tech. J. **50**, 2551 (1971)
6.66   T. Findakly, J. Farina, F.J. Leonberger: (unpublished)
6.67   S.K. Korotky, E.A.J. Marcatili, J.J. Veselka, R.H. Basworth: In *Integrated Optics,* ed. by E.P. Nolting and R. Ulrich, Springer Ser. Opt. Sci., Vol. 48 (Springer, Berlin, Heidelberg 1985) p. 207
6.68   E. Garmire, H. Stoll, A. Yariv, R.G. Hunsperger: Appl. Phys. Lett. **21**, 87 (1972)
6.69   J.C. Dyment, J.C. North, L.A. D'Asaro: J. Appl. Phys. **44**, 207 (1973)
6.70   H.C. Casey, Jr., M.B. Parish: *Heterostructure Lasers. Pt. B: Materials and Operating Characteristics* (Academic, New York 1978) p. 109
6.71   K. Nakajima: In *Semiconductor and Semimetals,* Vol. 22, *Lightwave Communication Technology, Part A: Material Growth and Technology,* ed. by W.T. Tsang (Academic, Orlando 1985) p. 1
6.72   R.W. McClelland, C.O. Bozler, J.C.C. Fan: Appl. Phys. Lett. **37**, 560 (1980)
6.73   P. Vohl, C.O. Bozler, R.W. McClelland, A. Chu, A.J. Strauss: J. Crystal Growth **56**, 410 (1982)
6.74   H.C. Casey, Jr., M.B. Parish: In *Heterostructure Lasers, Pt. B: Materials and Operating Characteristics* (Academic, New York 1978) p. 144
6.75   G.H. Olsen: In *GaInAsP Alloy Semiconductors,* ed. by T.P. Pearsall (Wiley, New York 1982) p. 11
6.76   G.B. Stringfellow: In *Semiconductor and Semimetals,* Vol. 22, *Lightwave Communications Technology, Pt. A: Material Growth Technologies,* ed. by W.T. Tsang (Academic, Orlando 1985) p. 209
6.77   J.P. Hertz, M. Razeghi, M. Bonnet, J.P. Duchemin: In *GaInAsP Alloy Semiconductors,* ed. by T.P. Pearsall (Wiley, New York 1982) p. 61
6.78   M.B. Panish, S. Sumski: J. Appl. Phys. **55**, 3571 (1984)
6.79   W.T. Tsang: In *Semiconductors and Semimetals,* Vol. 22, *Lightwave Communication Technology, Pt. A: Material Growth Technologies,* ed. by W.T. Tsang (Academic, Orlando 1985) p. 96;
       C.E. Wood: In *GaInAsP Alloy Semiconductors,* ed. by T.P. Pearsall (Wiley, New York 1982) p. 87
6.80   M.A. Herman, H. Sitter: *Molecular Beam Epitaxy,* Springer Ser. Mat. Sci. (Springer, Berlin, Heidelberg 1988)
6.81   W.T. Tsang: Appl. Phys. Lett. **45**, 1234 (1984)
6.82   D.L. Spears, A.J. Strauss, S.R. Chinn, I. Melngailis, P. Vohl: Tech. Digest Integrated Optics Conf. 1976, IOSA 75CH1039-7 (QEC), paper TuD3
6.83   T. Nishimura, H. Antome, K. Masud, S. Namba: Jpn. J. Appl. Phys. **16**, Suppl. 16-1, 317 (1977)
6.84   S. Valette, G. Labrunie, J.C. Deutsch, J. Lizet: Appl. Opt. **16**, 1289 (1977)
6.85   F.J. Leonberger, J.P. Donnelly, C.O. Bozler: Appl. Phys. Lett. **28**, 616 (1976)
6.86   A. Carenco, L. Menigaux, F. Alexander, M. Abdalla, A. Brenor: Appl. Phys. Lett. **34**, 755 (1979)
6.87   H. Inoue, K. Hiruma, K. Ishida, T. Asai, H. Matsumura: J. Lightwave Tech. **LT-3**, 1270 (1985)
6.88   Y. Tarui, Y. Kamiya, Y. Harada: J. Electrochem. Soc. **118**, 118 (1971)

6.89 L.A. Coldren, K. Furuya, B.I. Miller: J. Electrochem. Soc. **130**, 1918 (1983)
6.90 V.M. Donnelly, D.L. Flamm, C.W. Tu, D.E. Ibbotson: J. Electrochem. Soc. **129**, 2533 (1982)
6.91 M.W. Geis, G.A. Lincoln, N. Efremow, W.J. Piacentini: J. Vac. Sci. Tech. **19**, 1390 (1981)
6.92 N.L. DeMeo, J.P. Donnelly, F.J. O'Donnell, M.W. Geis, K.J. O'Connor: Nuclear Instr. Methods B**7/8**, 814 (1984)
6.93 P. Bachmann, A.J.N. Houghton: Electron. Lett. **18**, 850 (1982)
6.94 J.P. Donnelly, N.L. DeMeo, F.J. Leonberger, S.H. Groves, P.H. Vohl, F.J. O'Donnell: IEEE J. **QE-21**, 1147 (1985)
6.95 N.A. Wittaker: *All-Optical Signal Processing in Semiconductor Waveguides*, Ph.D. Thesis, Massachusetts Institute of Technology (1986)
6.96 T.Tada, T. Suzuki, K. Yamane: Thin Solid Films **83**, 289 (1981)
6.97 A. Carenco, L. Menigaux, N.T. Linl: Appl. Phys. Lett. **40**, 653 (1982)
6.98 K. Okamoto, S. Matsuoka, Y. Niskuwaki, K. Yoshida: J. Appl. Phys. **56**, 2595 (1984)
6.99 F.J. Leonberger, J.P. Donnelly, C.O. Bozler: Appl. Phys. Lett. **29**, 652 (1976)
6.100 F.J. Leonberger, C.O. Bozler, R.W. McClelland, I. Melngailis: Appl. Phys. Lett. **38**, 313 (1981)
6.101 A.J.N. Houghton, P.M. Rodgers, D.A. Andrews: Electron. Lett. **20**, 479 (1984)
6.102 R.G. Walker, R.G. Goodfellow: Electron. Lett. **19**, 590 (1983)
6.103 C. Bornholdt, W. Doldessen, D. Franke, N. Grote, J. Krause, V. Niggerbügge, H.P. Nutberg, M. Schlak, I. Tredle: Electron. Lett. **19**, 81 (1983)
6.104 P. Buchmann, H. Kaufmann, H. Melchior, G. Guekos: Electron. Lett. **20**, 295 (1984)
6.105 L.M. Johnson, Z.L. Liau, S.H. Groves: Appl. Phys. Lett. **44**, 278 (1984)
6.106 T.M. Benson, A.V. Syrbu, N. Chand, P.A. Houston: Electron. Lett. **18**, 812 (1982)
6.107 H. Futura, H. Noda, A. Ihaya: Appl. Opt. **13**, 322 (1974)
6.108 L.D. Westbrook, P.N. Robson, A. Majerfeld: Electron. Lett, **15**, 99 (1070), T.M. Benson, T. Muratoni, P.N. Robson, P.A. Houston: IEEE Trans. ED-**19**, 1477 (1982)
6.109 R.A. Soref, J.P. Lorenzo: Tech. Digest, Integ. Guided Wave Optics (OSA, Washington DC, 1986) paper WDD5
6.110 D.E. Zelmon, J.T. Boyd, H.E. Jackson: Appl. Phys. Lett. **47**, 353 (1985)
6.111 D.A. Ramey, J.T. Boyd: IEEE Trans. **CAS-26**, 1041 (1979)
6.112 J.T. Boyd, R.W. Wu, D.E. Zelmon, A. Naumaan, H.A. Timlin: Opt. Engrg. **24**, 230 (1985)
6.113 Y. Yamada, M. Kawachi, M. Yasu, M. Kobayashi: J. Lightwave Tech. **LT-4**, 277 (1986)
6.114 F.J. Leonberger, J.P. Donnelly, C.O. Bozler: Appl. Opt. **17**, 2250 (1978)
6.115 A. Carenco, L. Menigaux, Ph. Delpech: J. Appl. Phys. **50**, 5139 (1979)
6.116 K. Tada, H. Yanagawa, K. Hirose: Trans. IECE of Japan **E61**, 1 (1978)
6.117 J.C. Campbell, F.A. Blum, D.W. Shaw, K.L. Lawley: Appl. Phys. Lett. **27**, 202 (1975)
6.118 A.R. Reisinger, D.W. Bellavance, K.L. Lawley: Appl. Phys. Lett. **31**, 836 (1977)
6.119 K. Iwoski, S. Kurazomo, K. Itakura: Electron. Commun. Japan **58-C**, 100 (1975)
6.120 H.A. Haus, C.G. Fonstad: IEEE J. **QE-17**, 2321 (1981)
6.121 H.A. Haus, I. Molten-Orr: IEEE J. **QE-19**, 840 (1983)
6.122 J.P. Donnelly, N.L. DeMeo, G.A. Ferrante, K.B. Nichols, F.J. O'Donnell: Appl. Phys. Lett. **45**, 360 (1984)
6.123 J.P. Donnelly, N.L. DeMeo, G.A. Ferrante, K.B. Nichols: IEEE J. **QE-21**, 18 (1985)
6.124 See, for example: J.K. Butler, D.E. Achley, D. Botez: Appl. Phys. Lett. **44**, 293 (1984)
6.125 H.A. Haus, L. Molter-Orr, F.J. Leonberger: Appl. Phys. Lett. **45**, 19 (1984)
6.126 S. Somekh, E. Garmire, A. Yariv, H.L. Garvin, R.G. Hunsperger: Appl. Opt. **13**, 327 (1974)
6.127 S. Somekh, E. Garmire, A. Yariv, A. Garvin, R. Hunsperger: Appl. Opt. **12**, 455 (1973)
6.128 A. Carenco, L. Menigaux: J. Appl. Phys. **51**, 1325 (1980)

6.129 F.J. Leonberger, C.O. Bozler: Appl. Phys. Lett. **31**, 223 (1977)
6.130 H. Kogelnik, R.V. Schmidt: IEEE J. **QE-12**, 396 (1976)
6.131 H.F. Taylor: Opt. Commun. **8**, 421 (1973)
6.132 B. Borlex, B.J. Landgren, M.G. Obey, H. Jiang: J. Lightwave Tech. **LT-4**, 196 (1986)
6.133 P. Bachmann, H. Kaufmann, H. Melchior, G. Guekos: Appl. Phys. Lett. **46**, 462 (1985)
6.134 L.M. Johnson, F.J. Leonberger: Opt. Lett. **8**, 111 (1983)
6.135 P. Buchmann, H. Kaufmann: J. Lightwave Tech. **LT-3**, 785 (1985)
6.136 R.C. Alferness, C.H. Joyner, M.D. Devino, L.L. Buhl: Appl. Phys. Lett. **45**, 1278 (1984)
6.137 S. Namba: J. Opt. Soc. **51**, 76 (1961)
6.138 R.A. Becker, C.E. Woodwand, F.J. Leonberger, R.C. Williamson: Proc. IEEE **72**, 802 (1984)
6.139 See, for example: S.M. Sze: *Physics of Semiconductor Devices,* (Wiley, New York 1979) Chap. 2
6.140 S.H. Lin, S.Y. Wang, S.A. Neuton, Y.M. Huong: Electron. Lett. **21**, 597 (1985)
6.141 F. Auracher, R. Keil: Appl. Phys. Lett. **36**, 626 (1980)
6.142 M. Izutsu, S. Shikama, T. Sueta: IEEE J. **QE-17**, 2275 (1981)
6.143 W.A. Stallard, B.E. Daymond-John, R.C. Booth: In *Integrated Optics,* ed. by H.E. Nolting, R. Ulrich, Springer Ser. Opt. Sci., Vol. 48 (Springer, Berlin, Heidelberg 1985) p. 164
6.144 C.F. Buhrer, D. Baird, E.M. Conwell: Appl. Phys. Lett. **1**, 46 (1962)
6.145 G.M. Carter, H.A. Haus: IEEE J. **QE-15**, 217 (1979)
6.146 J.F. Lotspeich: J. Lightwave Tech. **LT-13**, 746 (1985)
6.147 H. Hasegawa, M. Furukawa, H. Yanai: IEEE Trans. **MTT-19**, 869 (1971)
6.148 S.Y. Wang, S.H. Lin, Y.M. Huong: Proc. 5th Int'l. Conf. Integ. Optics and Optical Fiber Comm. (IEEE 87CH2392-9) paper WK3
6.149 P.K. Cheo: IEEE J. **QE-20**, 700 (1984)
6.150 R.H. Kingston: Appl. Phys. Lett. **34**, 744 (1979)
6.151 K. Ishida, H. Nakamura, H. Inoue, S. Tsuji, H. Matsumura: Tech. Digest 4th Int'l. Conf. Integ. Optics and Optical Fiber Commun. (1985) p. 357
6.152 K. Tada, Y. Okada: IEEE **EDL-7**, 605 (1986)
6.153 H.A. Haus, E.P. Ippen, F.J. Leonberger: In *Optical Signal Processing,* ed. by J.L. Horner (Academic, Orlando 1987)
6.154 A. Lattes, H.A. Haus, F.J. Leonberger, E.P. Ippen: IEEE J. **QE-19**, 1718 (1983)
6.155 A. Ashkin, M. Gershenzon: J. Appl. Phys. **34**, 2116 (1963)
6.156 D.F. Nelson, F.K. Reinhart: Appl. Phys. Lett. **5**, 148 (1964)
6.157 P.M. Rodgers, M.J. Robertson, A.K. Chaterjee, S.Y. Wong: Tech. Digest Topical Meeting on Integrated and Guided Wave Optics (1986) Atlanta, GA, USA, paper THAA4
6.158 A. Alpeing, Y.S. Yar, T.R. Hausha, L.A. Coldren: Appl. Phys. Lett. **48**, 1243 (1986)
6.159 S.H. Lin, S.Y Wang, Y.M. Huong: Electron. Lett. **22**, 934 (1986)
6.160 K. Tada, K. Hirose: Appl. Phys. Lett. **25**, 56 (1974)
6.161 J.C. Shelton, F.K. Reinhart, R.A. Logan: Appl. Opt. **17**, 2548 (1978)
6.162 K. Tada, S. Kawaniski, M.H. Wang, M. Tsuchiya: Tech. Digest 7th Topical Meeting on Integrated and Guided Wave Optics (1984) Kissime, FL, USA, paper WB-4
6.163 A. Carenco, L. Menigaux: Appl. Phys. Lett. **37**, 880 (1980)
6.164 G.E. Stillman, C.M. Wolfe, I. Melngailis: Appl. Phys. Lett. **25**, 36 (1974)
6.165 C.E. Hurwitz, J.A. Rossi, J.J. Hsieh, C.M. Wolfe: Appl. Phys. Lett. **27**, 241 (1975)
6.166 Z.L. Liau, J.N. Walpole: Technical Digest, 1982 Top. Mtg. on Integ. and Guided Wave Optics (OSA, Washington DC) paper WB3
6.167 J.L. Merz, R.A. Logan: Appl. Phys. Lett. **26**, 337 (1976)
6.168 H. Inoue, H. Nakamara, S. Sakaro, T. Katsuyama, S. Tsuji, H. Matsumara, K. Morosawa, Proc. 5th Int'l. Conf. Integ. Optics and Optical Fiber Comm. (1987) Paper ThC3
6.169 J.H. Abeles, W.K. Chen, F. Shokoohi, R. Bhat, M.A. Koza: Tech. Digest CLEO 87 (OSA, Washington DC, 1987) paper MB1

6.170 J. Farina, F.J. Leonberger, A. Shuskus: (unpublished)
6.171 S. Akiyama, Y. Kawarada, K. Kaminishi: J. Crystl. Growth **68**, 39 (1984)
6.172 J.C. Campbell, J.C. DeWinter, M.A. Pollock, R.E. Nahory: Appl. Phys. Lett. **32**, 471 (1978)
6.173 Y. Noda, M. Suzuki, Y. Kushiro, A. Akiba: Electron. Lett. **21**, 1182 (1985)
6.174 C.M. Gee, G.D. Thurmond, H.A. Yen, H. Blauvelt: Tech Digest, Top. Mtg. on Integ. Guided Wave Optics, (OSA, Washington DC, 1986) paper ThAA3
6.175 H.G. Bach, J. Krauser, H.P. Nolting, R.A. Logan, F.K. Reinhart: Appl. Phys. Lett. **42**, 692 (1983)
6.176 T.E. VanEck, L.M. Walpita, W.S.C. Chang, H.H. Wieder: Appl. Phys.Lett. **48**, 451 (1986)
6.177 F. Koyama, K. Iga: Tech. Digest, Top. Mtg. on Integ. and Fiber Optics (OSA, Washington DC, 1986) paper WBB5
6.178 R.H. Kingston: Proc. IEEE **72**, 954 (1984)
6.179 R.H. Kingston, F.J. Leonberger: IEEE J. **QE-19**, 1443 (1983)
6.180 D.A.B. Miller, J.S. Weiner, D.S. Chemla: IEEE J. **QE-22**, 1816 (1986)
6.181 D.A.B. Miller, D.S. Chemla, T.C. Damen, A.C. Gossard, W. Wiegmann, T.H. Wood, C.A. Burrus: Phys. Rev. Lett. **53**, 2173 (1984)
6.182 D.A.B. Miller, D.S. Chemla, T.C. Damen, A.C. Gossard, W. Wiegmann, T.H. Wood, C.A. Burrus: Phys. Rev. **B32**, 1043 (1985)
6.183 T.H. Wood, C.A. Burrus, R.S. Tucker, J.S. Weiner, D.A.B. Miller, D.S. Chemla, T.C. Damen, A.C. Gossard, W. Wiegmann: Electron. Lett. **21**, 693 (1985)
6.184 J.S. Weiner, D.A.B. Miller, D.S. Chemla, T.C. Damen, C.A. Burrus, T.H. Wood, A.C. Gossard, W. Wiegmann: Appl. Phys. Lett. **47**, 1148 (1985)
6.185 T.H. Wood: Appl. Phys. Lett. **48**, 1413 (1986)
6.186 M. Glick, F.K. Reinhart, G. Weimann: Helv. Phys. Acta **58**, 403 (1985)
6.187 M. Glick, F.K. Reinhart, G. Weimann, W. Schlopp: Appl. Phys. Lett. **48**, 989 (1986)
6.188 G.R. Olbright, N. Peyghambarian: Appl. Phys. Lett. **48**, 1184 (1986)
6.189 S.S. Yao, C. Karagaleff, A. Gabriel, R. Fortenberry, C.T. Seaton, G. Stegeman: Appl. Phys. Lett. **46**, 801 (1985)
6.190 See, for example: T. Findakly: Opt. Engrg. **24**, 244 (1985)
6.191 C.M. Ironside, J.F. Duffy, R.H. Hutchins, W.C. Bargai, C.T. Seaton, G. Stegeman: Proc. 5th Int'l. Conf. Integ. Optics and Optical Fiber Commun. Venice (Inst. Internl. d. Comm., Genoa, 1985) p. 237
6.192 D.S. Chemla, D.A.B. Miller, P.W. Smith, A.C. Gossard, W. Wiegmann: IEEE J. **QE-20**, 265 (1984)
6.193 P.L.K. Wa, J.E. Stitch, N.J. Matson, J.S. Roberts, P.N. Robson: Electron. Lett. **21**, 26 (1985)
6.194 Y. Silberberg, P.W. Smith, D.A.B. Miller, B. Tell, A.C. Gossard, W. Wiegmann: Appl. Phys. Lett. **46**, 701 (1985)
6.195 S.R. Forrest: J. Lightwave Tech. **LT-3**, 1248 (1985)
6.196 O. Wada, H. Hamaguchi, S. Miura, M. Makiuchi, K. Nakai, H. Horimatsu, T. Sakurai: Appl. Phys. Lett. **46**, 981 (1985)
6.197 M. Ito, O. Wada, K. Nakai, T. Sakurai: IEEE **EDL-5**, 531 (1984)
6.198 M. Makiuchi, H. Hamguchi, T. Kumai, M. Ito, O. Wada, T. Sakurai: IEEE **EDL-6**, 634 (1985)
6.199 I. Ury, S. Maraglit, M. Yust, A. Yariv: Appl. Phys. Lett. **34**, 430 (1979)
6.200 L.A. Coldren, B.I. Miller, K. Iga, J.A. Rentschler: Appl. Phys. Lett. **37**, 681 (1980)
6.201 H. Blauvelt, M. Bar-Chaim, D. FeKete, S. Margalit, A. Yariv: Appl. Phys. Lett. **40**, 289 (1982)
6.202 Z.L. Liau, J.N. Walpole: Appl. Phys. Lett. **46**, 115 (1985)
6.203 J.N. Walpole, Z.L. Liau: Appl. Phys. Lett. **48**, 1636 (1986)
6.204 T.H. Windhorn, W.D. Goodhue: Appl. Phys. Lett. **48**, 1675 (1986)
6.205 S. Uchiyama, K. Iga: Electron. Lett. **21**, 162 (1985)
6.206 K. Aiki, M. Nakamura, J. Umeda: IEEE J. **QE-13**, 220 (1977)
6.207 T.P. Tanaka, M. Hirao, M. Makamura: Tech. Digest, OFC 85 (OSA, Washington DC, 1985), paper TuC2, p. 30
6.208 J.K. Carney, M.J. Helix, R.H. Kolbas: Tech. Digest GaAs IC Symposium, Phoenix AZ (1983) p. 98

7. Recent Advances

T. Tamir

With 1 Figure

7.1 Introduction

The purpose of this chapter is to briefly review the more important advances that have occured since about 1987 in the areas covered by the preceding chapters. For this purpose, the sections below are titled and numbered so as to correspond with the preceding chapters. Most of the sections here were compiled primarily from material provided by the authors of the respective chapters. In this context, the authors of Chap. 6 wish to acknowledge the contribution of Dr. D.E. Bossi, United Technologies Research Center, who has played a major role in updating the material on semiconductor guided wave optics.

7.2 Theory of Optical Waveguides

Chapter 2 presents the fundamental theoretical material on optical waveguides. Because this material has been developed over the past 20 years or longer, the more recent efforts in this area have been very specialized. Within the framework of the present volume, the amount of updating that can be reported in this area is rather limited.

An interesting development is the use of nonlinear material for implementing optical guides having desirable functions, such as switching, second-harmonic generation, and others [7.1]. For conventional guides, the need for more accurate dispersion characteristics has stimulated increasingly sophisticated analytical or numerical techniques [7.2–6], as well as other related aspects, which are often published in special issues of the relevant journals [7.7,8].

7.3 Waveguide Transitions and Junctions

Since the original publication of this book, two issues that relate to Chap. 3 have attracted considerable interest in the integrated optics device community. They are: (1) the problem of crosstalk in directional-coupler-type devices, in particular due to mode conversion effects in the input and output transition structures, and (2) the description and analysis of the optical digital switch, recently introduced by *Silverberg* et al. [7.9], which relies on the adiabatic operation of waveguide branches and thus lends itself to a normal-mode analysis. We will provide a brief description and references for each of these topics.

Issues of crosstalk in directional couplers were raised in [7.10], due to an unequal excitation of normal modes at the input, and in [7.11], where a coupled mode theory approach was used to treat mode coupling in the input and output transition structures of a directional coupler. *Haus* and *Whitaker* [7.12] then showed that the use of suitable transition structures could eliminate the problem of unequal normal-mode excitation [7.10]. These papers led to several extensive analyses that treated the transition structures in waveguide devices in a general way [7.13–17]. The general conclusions are that transition structures do not cause crosstalk in ideal situations, but the addition of nonideal factors such as waveguide asymmetries, normal-mode coupling in nonadiabatic transition regions, etc., can lead to crosstalk. An exception to this statement is the effect of mode coupling in the transition region of a reversed $\Delta\beta$ directional coupler switch, which prohibits zero crosstalk in the bar state of the switch [7.11]. However, zero crosstalk in the bar state can still be achieved by operating the switch in a uniform mismatched (as opposed to a reversed phase mismatched) configuration. It follows then that the conventional switching symmetry associated with the reversed $\Delta\beta$ directional coupler switch is broken in that zero crosstalk switching cannot be obtained by reversing the polarity of an appropriate switching voltage [7.16]. Instead, the effect of the transition region is to result in switching voltages for the cross and bar state of a real switch which are slightly different, and these voltages must be tuned for ideally zero crosstalk.

The analysis of [7.15] expands the material of the chapter by using a scattering matrix analysis to treat a general waveguide transition with a four-port model. This treatment allows a normal-mode analysis for output and crosstalk to be applied to any type of waveguide transition that can be described by two normal modes. Interferometers and directional couplers are treated using this approach in [7.16].

The digital optical switch recently reported by *Silberberg* et al. [7.9] has reactivated interest in devices that achieve switching via the evolution of normal-mode shape in a waveguide branch, rather than the interference of normal modes in a directional coupler or a Mach-Zehnder interferometer. The digital switch may be considered as a combination of an active waveguide branch, demonstrated in [7.18] in a three-port configuration, and an asymmetric mode splitting branch, such as that demonstrated in [7.19] with a Mach-Zehnder interferometer. The passive version of the structure, shown in Fig. 3.13, was proposed in [7.20,21] as a 3-dB coupler. The active version shown here in Fig. 7.1, is of interest because its switching characteristic saturates in the on or off state, i.e., application of additional voltage does not change crosstalk. This is distinctly opposed to the operation of an interfer-

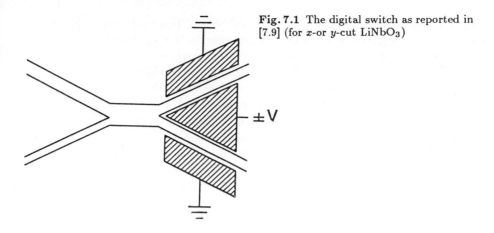

Fig. 7.1 The digital switch as reported in [7.9] (for x-or y-cut LiNbO₃)

ometric device, which requires a precise switching voltage. The device operates ideally as an adiabatic device with no coupling between normal modes. The active branch is made asymmetric electrically and switching is achieved by changing the sign of the applied voltage. This adiabatic mechanism is not strongly dependent on wavelength or polarization, except through the usual polarization dependence of the electro-optic coefficient. This feature allows the achievement of wavelength- and polarization-independent switching in a simpler manner than interferometric switching allows. 1 × 2 versions of these switches have been applied in a 4 × 4 switching array [7.22]. The price one pays for this operational simplicity is device length, because the adiabatic branches tend to be long. Voltage-length products can be defined for these structures and have been considered in [7.23].

7.4 Titanium-Diffused Lithium Niobate Waveguide Devices

Since the first printing of Chap. 4, the field of titanium-diffused lithium niobate waveguide devices has experienced sustained progress [7.24,25]. While a single high-volume commercial application has not yet emerged, prototype research devices, including modulators, small switch arrays, fiber gyro chips, and polarization controller devices, are presently being sold by several companies. In addition, the first commercial product, an optical test instrument, to include a lithium niobate waveguide device is being marketed [7.26], and a large number of multigigabit/s system experiments that untilize Ti:LiNibO₃ external modulators are being performed all over the world. The key issue for real commercialization of this technology, and indeed for all of integrated optics, remains the identification of high-volume applications to justify mass production. While efforts continue toward commercialization and applications, there have also been interesting and significant research advances as

399

well. These research advances are the subject of this update which, due to space limitations, must be brief and unfortunately incomplete.

Most of the recent advances represent improvements over earlier devices or implementation of previously proposed devices discussed in Chap. 4. However, one device not covered in the original chapter, the integrated acousto-optic (AO) tunable filter, has been demonstrated with very attractive performance characteristics [7.27]. The AO tunable filter is the acousto-optic analog of the electro-optic TE-TM mode converter of Sect. 4.5.1. Instead of coupling the orthogonal TE and TM modes using the off-diagonal element of the electro-optic tensor, coupling is achieved via the photoelastic effect, which is driven by a co-linear surface acoustic wave. The acoustic wavelength corresponds to the electrode period of (4.5.2) to provide phase-matched operation at λ_0. As a result, tuning of the center wavelength can be achieved simply by changing the acoustic wavelength by varying the drive frequency. Using the X-cut configuration and a 2.3 cm interaction length, 98% TE-TM coupling efficiency was achieved with a drive power of 340 mW. The drive frequency to phase-match at 1.5 μm was 175 MHz. The filter bandwidth for this interaction length was about 12 Å and the tuning range limited by the frequency response of the transducer was 1 400 Å [7.27]. One advantage of this filter is that a multiple band-pass response is achievable by driving the device with multiple frequencies [7.28]. By using the diversity geometry previously employed with electro-optic TE-TM filters, a polarization-independent acousto-optic filter has also been demonstrated [7.29]. By using an acoustic waveguide to concentrate the acoustic power, complete conversion has been achieved at $\lambda_0 = 0.63\,\mu$m with only 90 mW of electrical power over an interaction length of 0.9 cm [7.30]. Several systems experiments using acousto-optic tunable filters have been reported [7.31,32].

In the area of high-speed external modulators, important progress includes the applications in instrumentation, linear systems such as cable television, efficient band-pass modulators and switches and increased interest in very broadband (10 Gb/s) long haul digital systems as a result of progress in practical erbium-doped fiber amplifiers. For application to digital systems, broad bandwidth, low drive voltage and low loss coupling to fibers are important. Recent attempts to design optimized modulators have resulted in devices that provide excellent performance in a growing number of transmission system experiments at 10 Gbit/s and beyond. The basic result of this optimization is that, by increasing the silicon dioxide buffer layer thickness, the microwave index is increased, resulting in better velocity match and therefore greater bandwidth capability [7.33,34]. While the thicker buffer layer also increases the required modulation voltage, the increase is smaller than the increase in modulation bandwith. As a result, the electrode length can be increased to reduce the modulation voltage and still result in a modulation voltage smaller than for a shorter device with thin buffer layer. To reduce the elctrical loss of this longer device, a very thick gold electrode is employed, which also reduces the microwave index. As an example, a Y-branch interferometric modulator on Z-cut lithium niobate has been demonstrated with a 1 μm thick silicon dioxide buffer layer, a 10 μm thick gold electrode for the asymmetric coplanar strip (Fig. 4.10), a 9 μm wide hot line and a 15 μmwide electrode gap [7.34]. This structure has an impedance of 50 Ω and a microwave index of 2.55. For a device with a 20 mm long interaction length, the 3-dB bandwidth (electrical) is 12 GHz and the required modulation voltage is only 6.4 V for operation at 1.5 μm. Surface coating of an amorphous silicon film has improved thermal stability.

The importance of such modulators has increased because of the continued trend toward ever higher transmission data rates and the availability of fiber amplifiers. The latter, used as preamplifiers, increase the loss-limited span length, leaving chromatic dispersion as the chief limiter of the bit rate or span length. The effects of dispersion can be reduced by using a signal with information-limited bandwidth that can be readily achieved with an external modulator but not generally by direct intensity modulation of a single frequency semiconductor laser. Using the modulator above, researchers have transmitted information at a rate of 12 Gbit/s over 84 km error-free in dispersion-shifted fiber and over 161 km at 10 Gbit/s [7.35,36]. External modulators are even more important in systems where fiber amplifiers are used as in-line repeaters to provide several-hundred-kilometer spans without electrical regeneration. For example, a Ti:LiNbO₃ modulator has been used in a 516 km long span at 2.5 Gbit/s that included 12 amplifiers [7.37].

There is growing interest in external modulators for analog applications such as transmission of cable TV signals over fiber links. The third order nonlinearity of standard modulators is too large. However, by employing a low-noise, high-powered diode-pumped $Nd : Yag$ laser and an electrical circuit to provide some linearization, a system with Ti:LiNbO₃ external modulator has demonstrated acceptable carrier-to-noise ratio and distortion in a 50-channel cable TV transmission link [7.38].

Efficient low-frequency and high-frequency band-pass modulators have been demonstrated. The low-frequency devices are tuned lumped devices that have been used in analog links [7.39]. High-frequency devices using a microwave resonator electrode have been operated at 17 GHz with a band-pass bandwidth of 1.1 GHz [7.40,41]. Alternatively, using the periodic phase reversal electrode to achieve phase matching described in Sect. 4.2.2, low drive power (400 mW), high efficiency switching (100%) has been achieved at a center frequency of 36 GHz with roughly 6 GHz bandwidth [7.42]. With this device, optical pulses spaced in time by as little as 14 ps have been demultiplexed with low crosstalk ~ -20 dB) under sinusoidal electrical drive [7.42]. Active research continues on optical switching array circuits on lithium niobate. New architectures that grow in the number of 2 × 2 cross-point stages only as $\log_2 N$, where N is the number of input fibers, suggest the feasibility of space division switches that could handle several hundred lines each at a data rate of 1 Gbit/s [7.43, 44]. As an indication of improved fabrication technology, an 8 × 8 dilated Benes switch that includes 48 directional coupler devices exhibits a uniformity in its switching voltage of ±2% [7.45]. The dilated architecture substantially reduces the requirements on the crosspoint crosstalk [7.46]. This fact, together with the voltage uniformity, results in a switch whose individual crosspoints can be controlled by a common control voltage. This switch circuit consists of two lithium niobate substrates containing switch halves that have been epoxied together to form the complete circuit. New polarization-independent switch arrays have been fabricated using careful control over fabrication parameters [7.47] and by using a novel structure that offers coupling coefficient control [7.48]. A 4 × 4 array of asymmetric Y-branch (digital) switches that operates independently of polarization has recently been reported [7.22]. Research on switching systems based on Ti:LiNbO$_3$ switches has been very active. Several examples include a 32 line space division switch based on 8 × 8 and 4 × 8 arrays [7.49], a time-multiplexed center-stage switch to interconnect broadband local switches using the dilated Benes switch described above [7.50], progress toward a complete time-space-time photonic switch that includes optical time-slot interchange [7.51] and a high-speed hub switch that reconfigures periodically in less than 250 ps [7.52].

Considerable progress has been achieved on waveguide devices based upon proton exchange. In particular, by appropriately annealing after the exchange process, electro-optic devices that show no degradation in electro-optic coefficient have been reported [7.53]. These devices also exhibit propagation and fiber coupling losses comparable to titanium diffused waveguides and offer a higher immunity to the photorefractive effect. High-quality proton exchange waveguides in lithium tantalate have also been demonstrated with reduced photorefractive susceptibility [7.54]. Because proton exchange provides an index change only for the extraordinary index, these waveguides are single polarization. This is a strong advantage when making signal processing chips for fiber gyro applications [7.55].

A number of other interesting developments include a Nd-doped LiNbO$_3$ waveguide laser [7.56] and the proposal and demonstration of a truly reset-free, endless polarization controller [7.57]. As an indication of integration complexity, Ti:LiNbO$_3$ multifunction photonic circuits to provide signal processing for fiber sensor applications such as the gyroscope now include as many as 11 separate components [7.58, 59].

7.5 Mode-Controlled Semiconductor Lasers

Because it needed substantial revisions, Chap. 5 was updated by modifying and correcting the chapter itself rather than by adding a separate section here. Most of the modifications involve Sect. 5.7.5. Furthermore, Sect. 5.8

has been added, together with additional references, all of which have been inserted in the text of Chap. 5.

7.6 Semiconductor Integrated Optic Devices

The most fundamental achievement in the field of semiconductor integrated optics within the past five years has been the dramatic decrease in the propagation loss of optical waveguides produced by organometallic vapor phase epitaxy (OMVPE), also known as metal organic chemical vapor deposition (MOCVD), and by molecular beam epitaxy (MBE). Specifically, improvements in the quality and uniformity of substrates and growth techniques have allowed propagation losses in these structures to decrease from several dB per centimeter to tenths of a dB per centimeter [7.60–68]. Because of this extremely low propagation loss, a very accurate loss measurement technique, which involves measuring the finesse of the Fabry-Perot waveguide resonator, is commonly used to characterize these waveguide structures [7.69]. Dramatic reductions in the measured propagation loss of conventional and multiple-quantum-well (MQW) waveguides have been achieved in both the GaAs-based [7.60–63] and InP-based [7.64–68] material systems. In addition, low-loss heterostructure waveguide bends with curvature radii less than 1 mm have been demonstrated [7.70–72]. These low-loss bends have been formed by using a variety of fabrication techniques, such as reactive ion beam etching [7.70], wet chemical etching [7.71, 72] and impurity-induced disordering of multiple quantum wells as described below [7.73].

In addition to the demonstration of low-loss semiconductor integrated optical waveguides, a sizable effort has gone into the development of semiconductor optical waveguide modulators; numerous devices, including both intensity and phase modulators, have been demonstrated. The electro-absorption effect in semiconductor multiple-quantum-well structures, also known as the quantum-confined Stark effect (QCSE), is approximately 50 times greater than that in bulk semiconductors [7.74]. Therefore, significant interest has been generated in the use of MQW waveguides as optical intensity modulators [7.74–82]. However, electroabsorptive modulators have the drawback of heat dissipation and the inability to switch light between multiple output ports. In addition, high passive insertion losses and saturation effects can limit the usefulness of these devices.

Most present semiconductor modulator research appears to be concentrating on the development of waveguide devices that are based on phase modulation, both in conventional semiconductor waveguides and in multiple-quantum-well structures. Record levels of phase shift per volt per unit length have been achieved in conventional semiconductor waveguides using depletion-edge-translation (DET) modulators [7.83–86]. These devices combine a variety of electric field and carrier-related effects in the depletion region of a reverse biased $p - n$ junction to produce a maximum change in effective index per applied volt. Specifically, a phase shifting efficiency of 96°/Vmm has been reported in a four-layer GaAs/GaAlAs ridge-waveguide modulator operated at a wavelength of $1.06\,\mu$m [7.85]. Although the reported modulation efficiencies for these DET structures appear promising and these devices are well suited for on-chip coupling to integrated lasers, these structures have not to date been demonstrated in devices with low insertion loss or in integrated optic intensity modulators (such as Mach-Zehnder interferometers).

In addition to phase modulators in conventional semiconductor optical waveguides [7.83–91], a great deal of interest has recently been focused on the development of quantum-well electro-optic phase modulators [7.92–95]. These devices appear promising as a result of a strong electrorefractive effect that accompanies the quantum-confined Stark effect in these structures. For this electrorefractive effect, the induced index change exhibits a quadratic dependence on the applied electric field, in addition to the linear electro-optic index change found in zinc-blende crystals [7.93]. Electro-optic Mach-Zehnder interferometers with $\sim 10\,$dB extinction ratios have been demonstrated in InGaAsP/InP quantum-well waveguides with active lengths as short as $650\,\mu$m and with $\sim 17\,$V drive voltages [7.95]. Typically, most efficient phase modulation is achieved in a quantum-well device that is operated at a wavelength slightly beneath the bandgap of the semiconductor waveguide material. Unfortunately, under these operating conditions, the modulating electric field also creates an undesired intensity modulation of the optical signal; a tradeoff frequently exists between the phase-modulating ability of a device and the accompanying residual intensity modulation. In addition, the insertion loss in these devices is typically high even under zero bias conditions.

Semiconductor traveling-wave modulators hold promise for high-speed external modulation of optical signals without prohibitive drive power requirements, and significant advances in these devices have been reported within the past five years [7.96–101]. The velocity mismatch between the optical wave and the microwave becomes the limiting factor in determining the achievable bandwidth in these devices. Using a novel electrode configuration, a near match between the microwave and optical velocities has been achieved in III-V electro-optic modulators, resulting in 25 GHz bandwidths and drive voltages less than 5 V [7.100].

Semiconductor optical waveguide switches are critical components in the realization of integrated photonic circuits, and numerous achievements have recently been reported in these devices [7.102–117]. For example, an InGaAsP/InP optical waveguide switch, which relies upon a carrier-induced change in the refractive index of a conventional semiconductor waveguide, has been demonstrated [7.102,103] and a nonblocking 4 × 4 optical switch array has been produced [7.108–110]. Directional couplers and intersectional switches that utilize multiple-quantum-well electrorefraction have also been demonstrated in various MQW heterostructures [7.111–116]. Finally, a preliminary demonstration of a waveguide-crossing optical switch using the depletion edge translation concept has been reported [7.117], and a TE/TM polarization-independent directional coupler switch has been proposed for waveguides fabricated on (111)-oriented substrates [7.118].

The monolithic integration of various guided-wave photonic components has been subject of numerous reports throughout the past five years. Monolithic integration of semiconductor laser devices and transparent optical waveguides has been achieved by using a variety of novel device designs and material processing techniques [7.119–121]. These processing techniques, which include impurity-induced disordering of multiple quantum wells and epitaxial growth of quantum well layers on patterned nonplanar substrates, provide a means for lateral band-gap engineering of quantum-well heterostructures. Numerous examples of waveguide/photodetector integration have also been reported [7.122–125]. Most of these efforts have focused on evanescently coupled devices in which the absorbing photodetector layer is grown epitaxially on top of the transparent waveguide. Recently, an impedance-matching technique for enhanced waveguide/photodetector integration has been proposed and demonstrated [7.126, 127]. In terms of photonic integrated circuits, a monolithic 4-bit guided-wave analog-to-digital converter that uses a single tapped Mach-Zehnder interferometer has been fabricated in GaAs/GaAlAs [7.128]. In addition, a wavelength division multiplexing chip composed of three independently tunable lasers, a passive waveguide optical combiner, and an output optical amplifier monolithically integrated on a common InP substrate has been demonstrated [7.129].

The monlithic integration of photonic and electronic devices to form optoelectronic integrated circuits (OEICs) has remained an active area of investigation throughout the past five years. Several integrated transmitter and receiver chips have been developed [7.130–141], and a 4 × 4 GaAs OEIC switch module, consisting of a 4-channel OEIC receiver chip, a 4 × 4 GaAs electronic switch chip, and a 4-channel OEIC transmitter chip, has been demonstrated [7.142]. In addition, a GaAs transmitter chip and a companion OEIC receiver chip, containing 4 photodiodes and 8000 GaAs FETs, have been incorporated into a 950-Mbit/s fiber optic link for use in high performance data processing systems [7.143, 144].

All-optical switching has always been a topic of considerable interest, and several recent reports have begun to examine the possibilities for all-optical switching in GaAs and GaAlAs integrated optical waveguides. Pico-second all-optical switching has been demonstrated in GaAs/GaAlAs quantum-well nonlinear directional couplers [7.145]. In addition, femtosecond measurements of the nonresonant nonlinear index ($n_2 \sim 10$–$12\,\mathrm{cm}^2/\mathrm{W}$) in GaAlAs waveguides at room temperature have also been reported [7.146, 147]. However, the presence of two-photon absorption at wavelengths less than $\sim 30\,\text{nm}$ below the GaAlAs band edge may pose a serious limitation for high-speed all-optical switching in direct band-gap semiconductors. Finally, nondegenerate four-wave mixing experiments have been performed in GaAlAs ridge waveguides, yielding estimates for the various tensor components of the third-order nonlinear susceptibility in this zinc-blende material [7.148].

This review has largely focused on semiconductor integrated optical devices that have been developed for use in fiber-optic systems operating in the 0.8–1.5 μm wavelength range. However, because of their high power-handling capability and relatively broad transparency range (from the near to mid IR), semiconductor waveguide devices have also been developed for a variety of other laser-system applications. For example, several GaAs- and InP-based waveguide modulators, designed for CO_2 laser frequency shifting at 10.6 μm, have been reported [7.149–151]. These devices have been used either to up- and down-shift an input laser frequency by 8–18 GHz or to produce a sideband frequency that deviates from the input frequency by 0–5 GHz. In addition, for agile beam steering, the optical analog of a microwave phased-array radar, which is based upon the electro-optic effect in an epitaxially grown stack of semiconductor GaAlAs waveguides, has been demonstrated [7.152]. This device has been used as a wide-angle (26°) two-dimensional laser beam scanner and has also been shown to operate at frequencies as high as 1 Mhz.

The preceding progress in the design and performance of semiconductor integrated optic components has been accompanied by numerous recent developments in the fabrication techniques that are used to produce these photonic devices. For example, selective compositional disordering (or intermixing) of multiple-quantum-well materials has been an active area of investigation because of the important applications of this technology in forming integrated optical waveguide structures. Specifically, by using selective disordering, large changes in absorption and refractive index can be produced laterally across a wafer, thereby making the fabrication of a variety of quantum-well devices feasible. Several disordering techniques, including diffusion or implantation of impurities [7.153–156], annealing [7.157], ion bombardment [7.158–160], and laser melting [7.161], have been demonstrated and a variety of integrated optical devices has been produced. A few examples of devices produced using quantum-well disordering are: optical waveguides [7.162], TE/TM mode selective channel waveguides [7.163], and active devices, such as lasers, that are monolithically integrated with passive waveguides and modulators [7.119, 164].

In addition to selective compositional disordering, a number of other important developments have recently been reported in the fabrication of semi-conductor integrated optical devices. A majority of these techniques fall into the broad category of laser-assisted processing. Specifically, laser-induced photochemical etching has been employed to fabricate rib waveguides in the GaAs/GaAlAs heterostructure system [7.165]; a laser-assisted chemical vapor deposition technique has been used to produce a compositionally graded integrated GaAlAs lens [7.166]; laser-patterned desorption has enabled three-dimensional patterning of a GaAs quantum-well layer [7.167]; and laser-assisted selective chemical etching has enabled active trimming of GaAs waveguide devices [7.168]. Additionally, a MBE growth technique that employs graded substrate heating has been developed for producing compositionally and dimensionally tapered waveguide antennas [7.169]. These antennas serve as a means to increase the far-field directionality of beams emitted from GaAlAs waveguides and to increase the coupling efficiency between integrated optical waveguides and single-mode fibers.

Additional material systems have also been reported as a basis in which to form semiconductor integrated optical devices. For example, strained-layer InGaAs/AlGaAs multiple-quantum-well materials are being developed for laser diode applications and are likely to become useful for a variety of integrated optic devices [7.170]. In addition, several authors have reported schemes for making optical waveguides and waveguide devices in or on crystalline silicon [7.171–177]. A new type of optical waveguide, known as the antiresonant reflecting optical waveguide (ARROW), has been developed for use in the SiO$_2$-Si system [7.178–180]. This novel waveguide structure has been shown to produce low-loss, single polarization operation. Heteroepitaxy technology has been employed to produce GaAs-on-InP waveguides, which hold potential advantages for use in long wavelength OEICs and integrated optics [7.181]. However, the reported losses for waveguides obtained with this technique are quite high ($\sim 10\,\mathrm{db/cm}$). Slightly lower loss GaAlAs-on-InP waveguides ($\sim 7\,\mathrm{dB/cm}$) have recently been produced by using an epitaxial liftoff and transfer technique [7.182, 183]. In addition, GaAs/GaAlAs optical waveguides and phase modulators have been grown on Si substrates using MBE, and propagation losses as low as $\sim 1\,\mathrm{dB/cm}$ have been demonstrated in these ridge waveguide structures [7.184, 185].

The previous discussion of semiconductor integrated optics has focused primarily on the development of guided-wave photonic components. However, recent developments in closely related non-guided-wave III-V photonic technologies are likely to have a significant impact on future semiconductor integrated-optic devices. These areas of rapid development are briefly mentioned here, and the reader is referred to more detailed reviews in each of these important technological areas. A significant effort has recently been directed at the development of surface emitting semiconductor laser arrays with potentially widespread coherent and incoherent applications. Specifically, vertical-cavity arrays [7.186–191], grating-surface-emitting arrays [7.192–194], and surface emitting arrays that utilize deflecting mirrors [7.195–198] have all been demonstrated. Most vertical-cavity surface emitting lasers now utilize high-reflectivity dielectric stack reflectors to define the laser cavity. These dielectric stacks consist of monolithically grown alternating $\lambda/4$ semiconductor layers of different alloy composition. Dielectric mirrors of this type are also being investigated for resonant surface-normal quantum-well modulators [7.199–201] that can be used for two-dimensional optical signal processing and could prove useful for a variety of other devices. In addition, a category of optically controlled optoelectronic structures, known as self electro-optic effect devices (SEEDs), has been developed [7.202, 203]. These devices use an external bias and resistor (or illuminated silicon photodiode) in series with a multiple-quantum-well structure to obtain nonlinear bistable optical switching and could potentially be used to implement optical logic operations.

References

7.1   Special Feature on Guided-Wave Phenomena, J. Opt. Soc. Am. B **6**, 263 (1988)
7.2   P.K. Mishra, A. Sharma, S. Labroo, A.K. Ghatak: IEEE Trans. **Mtt-33**, 282 (1985)
7.3   M. Koshiba, M. Suzuki: J. Lightwave Technol. **LT-4**, 656 (1986)
7.4   E.A.J. Marcatili, A.A. Hardy: IEEE J. **Qe-24**, 766 (1988)
7.5   C.H. Henry, B.H. Verbeek: J. Lightwave Technol. **LT-7**, 308 (1989)
7.6   S.-T. Chu, S.K. Chaudhari: J. Lightwave Technol. **LT-7**, 2033 (1989)
7.7   Special Issue on Numerical Methods, IEEE Trans. **MTT-33**, 847 (1985)
7.8   Special Issue on Integrated Optics, J. Lightware Technol. **LT-6**, 984 (1988)
7.9   Y. Silberberg, P. Perlmutter, J.E. Baran: Appl. Phys. Lett. **51**, 1230 (1987)
7.10 K.L. Chen, S. Wang: Appl. Phys. Lett. **44**, 166 (1984)
7.11 T.K. Findakly, F.J. Leonberger: J. Lightwave Technol. **LT-6**, 36 (1988)
7.12 H.A. Haus, W.A. Whitaker, Jr.: Appl. Phys. Lett. **46** (1985)
7.13 K. Goel, W.S.C. Chang: IEEE J. **QE-23**, 2216 (1987)
7.14 J.P. Weber, L.Thylen, S. Wang: IEEE J. **QE-24**, 537 (1988)
7.15 W.K. Burns: J. Lightwave Technol. **LT-6**, 1051 (1988)
7.16 W.K. Burns: J. Lightwave Technol. **LT-6**, 1058 (1988); see also correction, J. Lightwave Technolg. **LT-7**, 1425 (1989)
7.17 S. Xu, S.T. Peng, F.K. Schwering: IEEE Trans. **MTT-37**, 686 (1989)
7.18 W.K. Burns, A.B. Lee, A.F. Milton: Appl. Phys. Lett. **29**, 790 (1976)
7.19 W.E. Martin: Appl. Phys. Lett. **26**, 562 (1975)
7.20 W.K. Burns, A.F. Milton, A.B. Lee, E.J. West: Appl. Opt. **15**, 1053 (1976)
7.21 M. Izutsu, A. Enokihara, T. Sueta: Opt. Lett. **7**, 549 (1982)
7.22 P. Granestrand, B. Lagerstrom, P. Svensson, L. Thylen, B. Stoltz, K. Bergvall, J.E. Falk, H. Olofsson: Electron. Lett. **26**, 4 (1990)
7.23 W.K. Burns: J. Lightwave Technol., to be published
7.24 E. Voges, A. Neyer: J. Lightwave Technol. **LT-5**, 1229 (1987)
7.25 L. Thylen: J. Lightwave Technol. **LT-6**, 847 (1988)
7.26 R.L. Jungerman, C.A. Johnsen, D.J. McQuatr, K. Salomaa, G. Conrad, D. Cropper, P. Hernday: Optical Fiber Communication Conf., San Francisco (1990), Paper FB2
7.27 B.L. Heffner, D.A. Smith, J.E. Baran, A.Yi-Yan, K.W. Cheung: Electron. Lett. **24**, 1563 (1988)
7.28 K.W. Cheung, D.A. Smith, J.E. Baran, B.L. Heffner: Electron. Lett. **25**, 375 (1989)
7.29 D.A. Smith, J.E. Baran, K.W. Cheung, J.J. Johnson: Appl. Phys. Lett **56**, 209 (1990)
7.30 J. Frangen, H. Hermann, R. Ricken, H. Seibert, W. Sohler, E. Strake: European Conf. on Integrated Optics, Paris (1989)
7.31 K.W. Cheung, S. C. Liew, C.N. Lo: Electron. Lett. **25**, 381 (1989)
7.32 K.W. Cheung, S.C. Liew, C.N. Lo: Lectron. Lett. **25**, 636 (1989)

7.33 S.K. Korotky: Topical Meeting on Numerical Simulation and Analysis in Guided-Wave Optics and Opto-Electronics, Houston (1989)

7.34 M. Seino, N. Mekada, T. Namiki, H. Nakajima: European Conf. on Optical Communications, San Francisco (1989) Paper ThB22-5

7.35 H. Nishimoto, I. Yokota, M. Suyama, T. Okiyama, M. Seino, T. Horimatsu, H. Kuwahara, T. Touge: Int. Conf. on Integrated Optics and Optical Communications (IOOC'89), Kobe (1989) Paper 20PDA-8

7.36 K. Hagimoto, Y. Miyagawa, Y. Miyamoto, M. Ohhashi, M. Ohhata, K. Aida, K. Nakagawa: IOOC'89, Kobe (1989) Paper 20PDA-6

7.37 K. Iwatsuki, S. Nishi, M. Saruwatari, M. Shimizu: IOOC'89, Kobe (1989) Paper 20PDA-1

7.38 R.B. Childs, U. O'Byrne: Optical Fiber Communications Conf. 90, San Francisco (1990) Paper PD23

7.39 G.E. Betts, L.M. Johnson, C.H. Cox, S.D. Lowney: IEEE Photon. Technol. Lett. 1, 404 (1989)

7.40 M. Izutsu, T. Sueta: IOOC'89, Kobe (1989) Paper 19D4-1

7.41 T. Mizuochi, T. Kitayama, S. Kawanaka, M. Izutsu, T. Sueta: Optical fiber Communications Conf.,'90, San Francisco (1990) Paper WM25

7.42 S.K. Korotky, J.J. Veselka: Optical Fiber Communications Conf.,'90, San Francisco (1990) Paper TUH2

7.43 M.J. Wale, P.J. Duthie, C.J. Groves-Kirkby, I. Bennion: Conf. on Lasers and Electro-Optics, Baltimore (1987) Paper WQ5

7.44 R.C. Alferness: IEEE J. SAC-6, 1117 (1988)

7.45 J.J. Veselka, T.O. Murphy, D.A. Herr, J.E. Watson, M.A. Milbrodt, K. Bahadori, M.F. Dautartas, C.T. Kemmerer, D.T. Moses, A.W. Schelling: Optical Fiber Communications Conf.,'89, Houston (1989) Paper ThB2

7.46 K. Padmanabhan, A.N. Netravali: IEEE Trans. Com-35, 1357 (1987)

7.47 II.S. Nishimoto, S. Suzuki, M. Kondo: Topical Meeting on Integrated and Guided-Wave Optics, Santa Fe (1988) Postdeadline Paper 6

7.48 P. Granestrand et al.. Topical Meeting on Integrated and Guided-Wave Optics, Santa Fe (1988) Postdeadline Paper 3

7.49 S. Suzuki, M. Kondo, K. Nagashima, M. Mitsuhasi, K. Komatsu, T. Miyakawa: Optical Fiber Communications Conf.,'87, Reno (1987) Paper WB4

7.50 R.A. Thompson, R.V. Anderson, J.V. Cambt, P.P. Giordano: Topical Meeting on Photonic Switching, Salt Lake City (1989) Paper ThF-3

7.51 N. Whitehead, N. Parsons, P. Vogel, G. Wicklund: Topical Meeting on Photonic Switching, Salt Lake City (1989) Paper WA2

7.52 S.K. Korotky, D.A. Herr, T.O. Murphy, J.J. Veselka, A. Azizi, R.W. Smith, B.L. Kasper: Optical Fiber Communicatios Conf.,'89, Houston (1989) Paper ThL3

7.53 P.G. Suchoski, T.K, Findakly, F.J. Leonberger: Opt. Lett. 13, 1058 (1988)

7.54 T. Findakly, P.G. Suchoski, F.J. Leonberger: Opt. Lett. 13, 797 (1988)

7.55 P.G. Suchoski, T.K. Findakly, F.J. Leonberger: Optical Fiber Communications Conf.,'90, San Francisco (1990) Paper FB3

7.56 E. Lallier, J.P. Pocholle, M. Papuchon, Grezes-Besset, E. Pelletier, M. De Micheli, M.J. Li, Q. He, D.B. Ostrowsky: IOOC'89, Kobe (1989) Paper 20PDB-1

7.57 F. Heismann, M.D. Divino, L.L. Buhl: Optical Fiber Communications Conf.,'90, San Francisco (1990) Paper TuH3

7.58 W. Minford et al.; Proc. SPIE 1169 (1989)

7.59 T. Findakly et al: Proc. SPIE 1169 (1989)

7.60 K. Hiruma, H. Inoue, K. Ishida, H. Matsumura: Appl. Phys. Lett. 47, 186 (1985)

7.61 H. Inoue, K. Hiruma, K. Ishida, T. Asai, H. Matsumura: J. Lightwave Technol. LT-3, 1270 (1985)

7.62 E. Kapon, R. Bhat: Appl. Phys. Lett. 50, 1628 (1987)

7.63 R.J. Deri, E. Kapon, L.M. Schiavone: Appl. Phys. Lett. 51, 789 (1987)

7.64 U. Koren, B.I. Miller, T.L. Koch, G.D. Boyd, R.J. Capik, C.E. Soccolich: Appl. Phys. Lett. 49, 1602 (1986)

7.65 C.H. Joyner, A.G. Dentai, R.C. Alferness, L.L. Buhl, M.D. Divino: Appl. Phys. Lett. 50 , 1509 (1987)

7.66 P.W.A. McIlroy, P.M. Rodgers, J.S. Singh, P.C. Spurdens, I.D. Henning: Electron. Lett. **23**, 701 (1987)
7.67 R.J. Deri, E. Kapon, R. Bhat, M. Seto, K. Kash: Appl. Phys. Lett. **54**, 1737 (1989)
7.68 R.J. Deri: Appl. Phys. Lett. **55**, 1495 (1989)
7.69 R.G. Walker: Electron. Lett. **21**, 581 (1985)
7.70 H. Takeuchi, K. Oe: Appl. Phys. Lett. **54**, 87 (1989)
7.71 R.J. Deri, M. Seto, A. Yi-Yan, E. Colas, R. Bhat: IEEE Photonics Technol. Lett. **1**, 46 (1989)
7.72 T.K. Tang, L.M. Miller, E. Andideh, T. Cockerill, P.D. Swanson, R. Bryan, T.A. DeTemple, I. Adesida, J.J. Coleman: IEEE Photonics Technol. Lett. **1**, 120 (1989)
7.73 P.D. Swanson, F. Julien, M.A. Emanuel, L. Sloan, T. Tang, T.A. DeTemple, J.J. Coleman: Opt. Lett. **13**, 245 (1988)
7.74 T.H. Wood: J. Lightwave Technol. **LT-6**, 743 (1988)
7.75 K. Wakita, Y. Kawamura, Y. Yoshikuni, H. Asahi: Electron. Lett. **22**, 907 (1986)
7.76 T.H. Wood, E.C. Carr, C.A. Burrus, R.S. Tucker, T.H. Chiu, W.T. Tsang: Electron. Lett. **23**, 540 (1987)
7.77 N.K. Dutta, N.A. Olsson: Electron. Lett. **23**, 853 (1987)
7.78 K. Wakita, S. Nojima, K. Nakashima, Y. Kawaguchi: Electron. Lett. **23**, 1067 (1987)
7.79 U. Koren, B.I. Miller, T.L. Koch, G. Eisenstein. R.S. Tucker, I. Bar-Joseph, D.S. Chemia: Appl. Phys. Lett. **51**, 1132 (1987)
7.80 T.H. Wood, E.C. Carr, C.A. Burrus, B.I. Miller, U. Koren: Electron. Lett. **24**, 840 (1988)
7.81 K. Wakita, I. Kotaka, H. Asai, S. Nojima, O. Mikami: Electron. Lett. **24**, 1324 (1988)
7.82 K. Wakita, I. Kotaka, O. Mitomi, Y. Kawamura, O. Mikami: In: Proc. 7th Int'l. Conf. Integrated Optics and Optical Fiber Communication, Kobe (Inst. of Elect., Info., and Commun. Engineers, Japan 1989) Vol. 2, p. 116
7.83 A. Alping, X.S. Wu, T.R. Hausken, L.A. Coldren: Appl. Phys. Lett. **48**, 1243 (1986)
7.84 L.A. Coldren, J.G. Mendoza-Alvarez, R.H. Yan: Appl. Phys. Lett. **51**, 792 (1987)
7.85 J.G. Mendoza-Alvarez, L.A. Coldren, A. Alping, R.H. Yan, T. Hausken, K. Lee, K. Pedrotti: J. Lightwave Technol. **LT-6**, 793 (1988)
7.86 X.S. Wu, X. Wu, Y. Li, Z. Lu, P. Zhou, L. Gong, C. Wu: Proc. 7th Int'l. Conf. Integrated Optics and Optical Fiber Communication, Kobe (Inst. of Elect., Info., and Commun. Engineers, Japan 1989) Vol. 3, p. 124
7.87 J. Faist, F.K. Reinhart, D. Martin, E. Tuncel: Appl. Phys. Lett. **50**, 68 (1987)
7.88 Y. Bourbin, A. Enard, R. Blondeau, M. Razeghi, D. Rondi, M. Papuchon, B. Decremoux: Electron. Lett. **24**, 221 (1988)
7.89 R.J. Deri, E. Kapon, J.P. Harbison, M. Seto, C.P. Yun, L.T. Florez: Appl. Phys. Lett. **53**, 1803 (1988)
7.90 S.S. Lee, R.V. Ramaswamy, V.S. Sundaram: Proc. 7th Int'l. Conf. Integrated Optics and Optical Fiber Communication, Kobe (Inst. of Elect., Info., and Commun. Engineers, Japan 1989) Vol. 1, p. 64
7.91 J. Pamulapati, P.K. Bhattacharya: Appl. Phys. Lett. **56**, 103 (1990)
7.92 U. Koren, T.L. Koch, H. Presting, B.I. Miller: Appl. Phys. Lett, **50** , 368 (1987)
7.93 J.E. Zucker, T.L. Hendrickson, C.A. Burrus: Appl. Phys. Lett. **52**, 945 (1988)
7.94 K. Wakita, O. Mitomi, I. Kotaka, S. Nojima, Y. Kawamura: IEEE Photon. Technol. Lett. **1**, 441 (1989)
7.95 J.E. Zucker, K.L. Jones, B.I. Miller, U. Koren: IEEE Photon. Technol. Lett. **2**, 32 (1990)
7.96 S.H. Lin, S.Y. Wang: Appl. Opt. **26**, 1696 (1987)
7.97 S.Y. Wang, S.H. Lin, Y.M. Houng: Appl. Phys. Lett. **51**, 83 (1987)
7.98 S.Y. Wang, S.H. Lin: J. Lightwave Technol. **LT-6**, 758 (1988)
7.99 R.G. Walker: Appl. Phys. Lett. **54**, 1613 (1989)
7.100 R.G. Walker, I. Bennion, A.C. Carter: Electron. Lett. **25** , 1549 (1989)
7.101 M.R.T. Tan, S.Y. Wang: In Technical Digest, Topical Meeting on Integrated Photonics Research (OSA, Washington, DC 1990) Paper ME4

7.102 S. Sakano, H. Inoue, H. Nakamura, T. Katsuyama, H. Matsumura: Electron. Lett. **22**, 594 (1986)

7.103 K. Ishida, H. Nakamura, H. Matsumura, T. Kadoi, H. Inoue: Appl. Phys. Lett. **50**, 141 (1987)

7.104 H. Inoue, K. Hiruma, K. Ishida, H. Sato, H. Matsumura: App. Opt. **25**, 1484 (1986)

7.105 F. Ito, T. Tanifuji: Appl. Phys. Lett. **54**, 134 (1989)

7.106 Y. Okada, R.H. Yan, L.A. Coldren, J.L. Merz, K. Tada: IEEE J. **QE-25**, 713 (1989)

7.107 F. Ito, M. Matsuura, T. Tanifuji: IEEE J. **QE-25**, 1677 (1989)

7.108 K. Inoue, H. Nakamura, K. Morosawa, Y. Sasaki, T. Katsujama, N. Chinone: IEEE J. **SAC-6**, 1262 (1988)

7.109 Y. Takahashi, H. Inoue, T. Kato, E. Amada: Electron. Lett. **25**, 964 (1989)

7.110 H. Inoue: Proc. 7th Int'l. Conf. Integrated Optics and Optical Fiber Communication, Kobe (Inst. of Elect., Info., and Commun. Engineers, Japan 1989) Vol. 4, p. 10

7.111 K.G. Ravikumar, K. Shimomura, T. Kikugawa, A. Izumi, S. Arai, Y. Suematsu, K. Matsubara: Electron. Lett. **24**, 415 (1988)

7.112 H. Yamamoto, M. Asada, Y. Suematsu: J. Lightwave Technol. **LT-6**, 1831 (1988)

7.113 T. Kikugawa, K.G. Ravikumar, K. Shimomura, A. Izumi, K. Matsubara, Y. Miyamoto, S. Arai, Y. Suematsu. IEEE Photon. Technol. Lett. **1**, 126 (1989)

7.114 M. Cada, B.P. Keyworth, J.M. Glinski, C. Rolland, A.J. SpringThorpe, K.O. Hill, R.A. Soref: Appl. Phys. Lett, **54**, 2509 (1989)

7.115 K.G. Ravikumar, T. Kikugawa, A. Izumi, K. Shimomura, K. Matsubara, Y. Miyamoto, S. ARai, Y. Suematsu: Proc. 7th Int'l. Conf. Integrated Optics and Optical Fiber Communication, Kobe (Inst. of Elect., Info., and Commun. Engineers, Japan 1989) Vol. 2, p. 132

7.116 J.E. Zucker, K.L. Jones, M.G. Young, B.I. Miller, U. Koren: Appl. Phys. Lett. **55**, 2280 (1989)

7.117 T.C. Huang, T. Hausken, K. Lee, N. Dagli, L.A. Coldren, D.R. Myers: IEEE Photon. Technol. Lett. **1**, 168 (1989)

7.118 K. Tada, H. Noguchi: Proc. 7th Int'l. Conf. Integrated Optics and Optical Fiber Communication, Kobe (Inst. of Elect., Info., and Commun. Engineers, Japan 1989) Vol. 1, p. 62

7.119 R.L. Thornton, J.E. Epler, T.L. Paoli: Appl. Phys. Lett **51**, 1983 (1987)

7.120 G.A. Vawter, J.L. Merz, L.A. Coldren: IEEE J. **QE-25**, 154 (1989)

7.121 C.J. Chang-Hasnain, E. Kapon, J.P. Harbison, L.T. Florez: Appl. Phys. Lett. **56**, 429 (1990)

7.122 C. Bornholdt, W. Doldissen, F. Fiedler, R. Kaiser, W. Kowalsky: Electron. Lett. **23**, 2 (1987)

7.123 S. Chandrasekhar, J.C. Campbell, A.G. Dentai, G.J. Qua: Electron. Lett. **23**, 501 (1987)

7.124 M. Erman, P. Jarry, R. Gamonal, J.L. Genter, P. Stephan, C. Guedon: J. Lightwave Technol. **LT-6**, 399 (1988)

7.125 K.Y. Liou, U. Koren, S. Chandrasekhar, T.L. Koch, A. Shahar, C.A. Burrus, R.P. Gnall: Appl. Phys. Lett. **54**, 114 (1989)

7.126 R.J. Deri, O. Wada: Appl. Phys. Lett. **55**, 2712 (1989)

7.127 R.J. Deri, N. Yasuoka, M. Makiuchi, O. Wada, A. Kuramata, H. Hamaguchi, R.J. Hawkins: Technical Digest, Topical Meeting on Integrated Photonics Research Optical Society of America, Washington, DC, 1990 Paper TUA6

7.128 R.G. Walker, A.C. Carter, I. Bennion: Proc. 7th Int'l. Conf. Integrated Optics and Optical Fiber Communication, Kobe (Inst. of Elect., Info., and Commun. Engineers, Japan 1989) Vol. 4, p. 78

7.129 U. Koren, T.L. Koch, B.I. Miller, G. Eisenstein, R.H. Bosworth: Appl. Phys. Lett. **54**, 2056 (1989)

7.130 M. Kuno, T. Sanada, H. Nobuhara, M. Makiuchi, T. Fujii, O. Wada, T. Sakurai: Appl. Phys. Lett. **49**, 1575 (1986)

7.131 N. Suzuki, H. Furuyama, Y. Hirayama, M. Morinaga, K. Eguchi, M. Kushibe, M. Funamizu, M. Nakamura: Electron. Lett. **24**, 467 (1988)

7.132 O. Wada, H. Nobuhara, T. Sanada, M. Kuno, M. Makiuchi, T. Fujii, T. Sakurai: J. Lightwave Technol. **LT-7**, 186 (1989)

7.133 H. Wang, D. Ankri: Electron. Lett. **22**, 391 (1986)

7.134 A. Suzuki, T. Itoh, T. Terakado, K. Kasahara, K. Asano, Y. Inomoto, H. Ishihara, T. Torikai, S. Fujita: Electron. Lett. **23**, 954 (1987)

7.135 A. Suzuki, K. Kasahara, M. Shikada: J. Lightwave Technol. **LT-5**, 1479 (1987)

7.136 O. Wada, S. Miura, T. Mikawa, O. Aoki, T. Kiyonaga: Electron. Lett **24**, 514 (1988)

7.137 H. Nobuhara, H. Hamaguchi, T. Fujii, O. Aoki, M. Makiuchi, O. Wada: Electron. Lett. **24**, 1246 (1988)

7.138 T. Suzaki, S. Fujita, Y. Inomoto, T. Terakado, K. Kasahara, K. Asano, T. Torikai, T. Itoh, M. Shikada, A. Suzuki: Electron. Lett, **24**, 1283 (1988)

7.139 S. Chandrasekhar, J.C. Campbell, A.G. Dentai, C.H. Joyner, G.J. Qua, A.H. Gnauck, M.D. Feuer: Electron. Lett. **24**, 1443 (1988)

7.140 N. Otsuka, K. Matsuda, J. Shibata: Proc. 7th Int'l. Conf. Integrated Optics and Optical Fiber Communication, Kobe (Inst. of Elect., Info., and Commun. Engineers, Japan 1989) Vol. 3, p. 166

7.141 J. Shimizu, T. Suzaki, T. Terakado, S. Fujita, K. Kasahara, Y. Inomoto, T. Itoh, A. Suzuki: Proc. 7th Int'l. Conf. Integrated Optics and Optical Fiber Communication, Kobe (Inst. of Elect., Info., and Commun. Engineers, Japan 1989) Vol. 3, p. 206

7.142 T. Iwama, T. Horimatsu, Y. Oikawa, K. Yamaguchi, M. Sasaki, T. Touge, M. Makiuchi, H. Hamaguchi, O. Wada: J. Lightwave Technol. **LT-6**, 772 (1988)

7.143 J.D. Crow et al.: IEEE Trans. **ED-36**, 236 (1989)

7.144 J.D. Crow: Proc. 7th Int'l. Conf. Integrated Optics and Optical Fiber Communication, Kobe (Inst. of Elect., Info., and Commun. Engineers, Japan 1989) Vol. 4, p. 86

7.145 R. Jin, C.L. Chuang, H.M. Gibbs, S.W. Koch, J.N. Polky, G.A. Pubanz: Appl. Phys. Lett. **53**, 1791 (1988)

7.146 M.J. LaGasse, K.K. Anderson, H.A. Haus, J.G. Fujimoto: Appl. Phys. Lett. **54** , 2068 (1989)

7.147 M.J. LaGasse, K.K. Anderson, C.A. Wang, H.A. Haus, J.G. Fujimoto: Appl. Phys. Lett. **56**, 417 (1990)

7.148 H.Q. Le, D.E. Bossi, K.B. Nichols, W.D. Goodhue: Appl. Phys. Lett. **56**, 1008 (1990)

7.149 P.K. Cheo, R.T. Brown: *Laser Radar III*. Proc. SPIE **999**, 27 (1988)

7.150 J.D. Farina, R.T. Brown, P.K. Cheo: In *Integrated and Guided-Wave Optics*, 1988 Technical Digest Series, Vol. 5 (Optical Society of America, Washington, DC 1988) p. 160

7.151 D. Delacourt, R. Blondeau, C. Brylinski, M.A. DiForte-Poisson, M. Papuchon: In *Integrated and Guided-Wave Optics*, 1988 Technical Digest Series, Vol.5 (Optical Society of America, Washington, DC 1988) p. 164

7.152 J.D. Farina, R. Grasso, R.H. Hobbs: Proc. IEEE Lasers and Electro-Optic Society Annual Meeting (IEEE, New York 1989) p. 95

7.153 W.D. Laidig, N. Holonyak, Jr., M.D. Camras, K. Hess, J.J. Coleman, P.D. Dapkus, J. Bardeen: Appl. Phys. Lett. **38**, 776 (1981)

7.154 K. Meehan, N. Holonyak, Jr., J.M. Brown, M.A. Nixon, P. Gavrilovic, R.D. Burnham: Appl. Phys. Lett. **45**, 549 (1984)

7.155 J.J. Coleman, P.D. Dapkus, C.G. Kirkpatrick, M.D. Camras, N. Holonyak, Jr.: Appl. Phys. Lett. **40**, 904 (1982)

7.156 T. Venkatesan, S.A. Schwarz, D.M. Hwang, R. Bhat, M. Koza, H.W. Yoon, P. Mei, Y. Arakawa, A. Yariv: Appl. Phys. Lett. **49**, 701 (1986)

7.157 L.J. Guido, N. Holonyak, Jr., K.C. Hsieh, R.W. Kalisi, W.E. Plano, R.D. Burnham, R.L. Thornton, J.E. Epler, T.L. Paoli: J. Appl. Phys. **61**, 1372 (1987)

7.158 J. Cibert, P.M. Petroff, D.J. Werder, S.J. Pearton, A.C. Gossard, J.H. English: Appl. Phys. Lett. **49**, 223 (1986)

7.159 P. Mei, T. Venkatesan, S.A. Schwarz, N.G. Stoffel, D.L. Hart, L.A. Florez: Appl. Phys. Lett. **52**, 1487 (1988)

7.160 K.K. Anderson, J.P. Donnelly, C.A. Wang, J.D. Woodhouse, H.A. Haus: Appl. Phys. Lett. **53**, 1632 (1988)

7.161 J.E. Epler, R.D. Burnham, R.L. Thornton, T.L. Paoli, M.C. Bashaw: Appl. Phys. Lett. **49**, 1447 (1986)

7.162 F. Julien, P.D. Swanson, M.A. Emanuel, D.G. Deppe, T.A. DeTemple, J.J. Coleman, N. Holonyak, Jr.: Appl. Phys. Lett. **50**, 866 (1987)

7.163 Y. Suzuki, H. Iwamura, O. Mikami: Appl. Phys. Lett. **56**, 19 (1990)

7.164 R.L. Thornton, W.J. Mosby, T.L.Paoli: J. Lightwave Technol. **LT-6**, 786 (1988)

7.165 A.E. Willner, M.N. Ruberto, D.J. Blumenthal, D.V. Podlesnik, R.M. Osgood, Jr.: Appl. Phys. Lett. **54**, 1839 (1989)

7.166 I. Yoshida, S.M. Bedair: Appl. Phys. Lett. **52**, 2208 (1988)

7.167 J.E. Epler, D.W. Treat, H.F. Chung, T. Tjoe, T.L. Paoli: Appl. Phys. Lett. **54**, 881 (1989)

7.168 R.T. Brown: Integrated and Guided Wave Optics, 1989 Technical Digest Series, Vol. 4 (Optical Society of America, Washington, DC 1989) p. 72

7.169 D.E. Bossi, W.D. Goodhue, M.C. Finn, K. Rauschenbach, J.W. Bales, R.H. Rediker: Appl. Phys. Lett. **56**, 420 (1990)

7.170 W. Dobbelaere, S. Kalem, D. Huang, M.S. Unlu, H. Morkoc: Electron. Lett. **24**, 295 (1988)

7.171 N. Takato, M. Yasu, M. Kawachi: Electron. Lett. **22**, 321 (1986)

7.172 R.A. Soref, J.P. Lorenzo: IEEE J. **QE-22**, 873 (1986)

7.173 S. Kaneda, Y. Fujisawa, K. Kikuiri: Electron. Lett. **22**, 922 (1986)

7.174 B.N. Kurdi, D.G. Hall: Opt. Lett. **13**, 175 (1988)

7.175 B.R. Hemenway, O. Solgaard, D.M. Bloom: Appl. Phys. Lett. **55**, 349 (1989)

7.176 Y. Shani, C.H. Henry, R.C. Kistler, K.J. Orlowsky, D.A. Ackerman: Appl. Phys. Lett. **55**, 2389 (1989)

7.177 Y. Shani, C.H. Henry, R.C. Kistler, R.F. Kazarinov, K.J. Orlowsky: Appl. Phys. Lett. **56**, 120 (1990)

7.178 M.A. Duguay, Y. Kokubun, T.L. Koch, L. Pfeiffer: Appl. Phys. Lett. **49**, 13 (1986)

7.179 Y. Kokubun, T. Baba, T. Sakaki, K. Iga: Electron. Lett. **22**, 892 (1986)

7.180 T. Baba, Y. Kokubun, T. Sakaki, K. Iga: J. Lightwave Technol. **LT-0**, 1440 (1988)

7.181 Y.H. Lo, R.J. Dai, J. Harbison, B.J. Skromme, M. Seto, D.M Hwang, T.P. Lee: Appl. Phys. Lett **53**, 1212 (1988)

7.182 A. Yi-Yan, W.K, Chan, T.J. Gmitter, M. Seto: Technical Digest, Topical Meeting on Integrated Photonics Research (OSA, Washington, DC 1990) Paper MI1

7.183 E. Yablonovitch, T. Gmitter, J.B. Harbison, R. Bhat: Appl. Phys. Lett. **51**, 2222 (1987)

7.184 Y.S. Kim, S.S. Lee, R.V. Ramaswamy, S. Sakai, Y.C. Kao, H. Shichijo: Proc. 7th Int'l. Conf. Integrated Optics and Optical Fiber Communication, Kobe (Inst. of Elect., Info., and Commun. Engineers, Japan 1989) Vol. 1, p. 68

7.185 Y.S. Kim, R.V. Ramaswamy, S. Sakai, R.J. Matyi, H. Shichijo: Appl. Phys. Lett. **53**, 1586 (1988)

7.186 K. Iga, F. Koyama, S. Kinoshita: IEEE J. **QE-24**, 1845 (1988)

7.187 M. Ogura, W. Hsin, M.C. Wu, S. Wang, J.R. Whinnery, S.C. Wang, J.J. Yang: Appl. Phys. Lett. **51**, 1655 (1987)

7.188 M.Y.A. Raja, S.R.J. Brueck, M. Osinski, C.F. Schaus, J.G. McInerney, T.M. Brennan, B.E. Hammons: IEEE J. **QE-25**, 1500 (1989)

7.189 S.W. Corzine, R.S. Geels, J.W. Scott, R.H. Yan, L.A. Coldren: IEEE J. **QE-25**, 1513 (1989)

7.190 D. Botez, L.M. Zinkiewicz, T.J. Roth, L.J. Mawst, G. Peterson: IEEE Photon. Technol. Lett. **1**, 205 (1989)

7.191 A. Scherer, J.L. Jewell, Y.H. Lee, J.P. Harbison, L.T. Florez: Appl. Phys. Lett. **55**, 2724 (1989)

7.192 G.A. Evans, N.W. Carlson, J.M. Hammer, M. Lurie, J.K. Butler, S.L. Palfrey, R. Amantea, L.A. Carr, F.Z. Hawrylo, E.A. James, C.J. Kaiser, J.B. Kirk, W.F. Reichert: IEEE J. **QE-25**, 1525 (1989)

7.193 J.S. Mott, S.H. Macomber: IEEE Photon. Technol. Lett. **1**, 202 (1989)

7.194 R. Parke, R. Waarts, D.F. Welch, A. Hardy, W. Streifer: Electron. Lett. **26**, 125 (1990)

7.195 Z.L. Liau, J.N. Walpole: Appl. Phys. Lett. **50**, 528 (1987)

7.196 J.P. Donnelly, W.D. Goodhue, T.H. Windhorn, R.J. Bailey, S.A. Lambert: Appl. Phys. Lett. **51**, 1138 (1987)
7.197 J.P. Donnelly, K. Rauschenbach, C.A. Wang, W.D. Goodhue, R.J. Bailey: Laser Diode Technology and Applications, ed. L. Figueroa. Proc. SPIE **1043**, 92 (1989)
7.198 J.J. Yang, M. Sergant, M. Jansen, S.S. Ou, L. Eaton W.W. Simmons: Appl. Phys. Lett. **49**, 1138 (1986)
7.199 R.H. Yan, R.J. Simes, L.A. Coldren: IEEE Photon. Technol. Lett. **1**, 273 (1989)
7.200 R.H. Yan, R.J. Simes, L.A. Coldren: IEEE J. **QE-25**, 2272 (1989)
7.201 R.H. Yan, R.J. Simes, L.A. Coldren: IEEE Photon. Technol. Lett. **2**, 118 (1990)
7.202 See, for example: D.A.B. Miller, D.S. Chemla, T.C. Damen, T.H. Wood, C.A. Burrus, Jr., A.C. Gossard, W. Wiegmann: IEEE J. **QE-21**, 1462 (1985)
7.203 I. Bar-Joseph, K.W. Goossen, J.M. Kuo, R.F. Kopf, D.A.B. Miller, D.S. Chemla: Appl. Phys. Lett. **55**, 340 (1989)

Subject Index

416